The Perspectives of Plant Natural Products for Mitigation of Obesity

The Perspectives of Plant Natural Products for Mitigation of Obesity

Editor

Fang Chen

MDPI • Basel • Beijing • Wuhan • Barcelona • Belgrade • Manchester • Tokyo • Cluj • Tianjin

MDPI

Editor
Fang Chen
China Agricultural University
China

Editorial Office
MDPI
St. Alban-Anlage 66
4052 Basel, Switzerland

This is a reprint of articles from the Special Issue published online in the open access journal *Nutrients* (ISSN 2072-6643) (available at: https://www.mdpi.com/journal/nutrients/special_issues/Perspectives_Plant_Natural_Products_Mitigation_Obesity).

For citation purposes, cite each article independently as indicated on the article page online and as indicated below:

LastName, A.A.; LastName, B.B.; LastName, C.C. Article Title. *Journal Name* **Year**, *Volume Number*, Page Range.

ISBN 978-3-0365-6970-3 (Hbk)
ISBN 978-3-0365-6971-0 (PDF)

Contents

About the Editor

Fang Chen

Fang Chen (Professor) currently serves as director of the Department of Food Nutrition and Safety, College of Food Science and Nutritional Engineering, China Agricultural University. She graduated from China Agricultural University with Bachelor's, Master's and Doctor's degrees. In 1998, she was appointed as Assistant Engineer at the Beijing Bee Product Institute. Since relocating to China Agricultural University in 2003, her research areas have included food nutrition and human health, with a focus on the molecular mechanisms of phytochemical compounds acting against chronic inflammation. She is currently editor of *eFood* and *Food Frontiers*, editorial board member of *Food Additives and Contaminants: Part A* and the *International Journal of Nutrition and Food Sciences*. She also has edited or co-edited 6 books, such as *Food Safety and Daily Diet*. Professor Chen has published more than 160 papers, including 92 papers in SCI, 2 papers in highly cited papers, and 5 papers at international conferences. She has presided over or participated in more than 20 projects such as the National Natural Science Foundation of China, National Science and Technology Support Program, and National Key R&D Project. She has obtained two second-class prizes of the National Science and Technology Progress award, and seven ministerial/provincial awards. She was presented "the Outstanding Youth Award of China Food Science and Technology Society" and "the 19th Mao Yisheng Beijing Youth Science and Technology Award", and received the first batch of the China Excellent Youth Science Fund of National Natural Science Foundation, the New Century Excellent Talents of the Ministry of Education, Beijing Science and Technology Star (B) Support Program, Ministry of Science and Technology "Young and Middle-aged Science and Technology Innovation Leaders", and the fourth batch of the "10,000 people" Plan "Scientific Leadership Talents".

nutrients

MDPI

Editorial

The Perspectives of Plant Natural Products for Mitigation of Obesity

Daotong Li and Fang Chen *

National Engineering Research Center for Fruit and Vegetable Processing, Key Laboratory of Fruits and Vegetables Processing, College of Food Science and Nutritional Engineering, Ministry of Agriculture, Engineering Research Centre for Fruits and Vegetables Processing, Ministry of Education, China Agricultural University, Beijing 100083, China
* Correspondence: chenfangch@sina.com; Tel.: +86-10-6273-7645

Obesity is a metabolic disease caused by an imbalance between energy intake and consumption, which leads to excessive fat accumulation in adipose tissues. The prevalence of obesity is increasing worldwide and its associated metabolic disorders have alarmingly become a global public health issue that severely affects the health and quality of life of people [1]. It is estimated that two billion people (nearly one-third of the world's population) have been classified as obese [2]. The obesity-associated hypertension, hyperglycemia and dyslipidemia are also risk factors that are closely linked to multiple diseases such as diabetes mellitus, cardiovascular disease, and cancer, which are major causes of death worldwide [3]. The rapid increase in the prevalence of obesity and the associated metabolic dysfunction highlights the urgent need for managing the detrimental health effects. Weight loss by surgical treatment is effective in reducing the risk of morbidity and mortality, while it may cause negative emotions in people. Some anti-obesity drugs have been available and approved for the treatment of obesity, whereas their use may trigger certain side effects [4]. Exploring and developing risk-free and efficient weight management treatments to prevent and treat obesity is still an unmet clinical need.

Recently, plant-derived natural compounds have attracted much attention due to their relatively high safety profile with fewer side effects. In our Special Issue entitled "The Perspectives of Plant Natural Products for Mitigation of Obesity", several articles have been published that focus on the crucial role of plant natural products in mitigating obesity. Flavonols are a group of flavonoids that include quercetin, kaempferol, myricetin, and isorhamnetin. Onions, tea, apples, kale, lettuce, tomatoes, broccoli, and grape skins are important foods of dietary flavonols. By investigating the relationship between dietary flavonols intake and obesity parameters (fat mass, waist circumference, and body mass index) in 40 obese and 40 healthy volunteers, the study by Joanna Popiolek-Kalisz [5] provided evidence that consumption of flavonol-enriched foods may play a protective role in obesity development. Kochujang is a Korean fermented, soybean-based red pepper paste that has been reported to have obesity-reducing effects in rats. In a randomized, double-blind clinical trial, the study of Han et al. [6] compared the anti-obesity effects of traditional Kochujang and commercially available Kochujang in overweight and obese patients. The results showed that the traditional Kochujang is superior to the commercial Kochujang in reducing body fat content and improving blood lipid profiles. These two human studies suggest that long-term consumption of functional foods enriched in natural dietary phytochemicals is effective to reduce the risk of obesity and metabolic syndrome.

The development of obesity is closely associated with adipogenesis, which is characterized by an increased number and size of adipocytes in adipose tissues. The differentiation of preadipocytes into mature adipocytes is a key step for cellular lipid metabolism and accumulation [7]. Inhibition of adipogenesis and lipid accumulation in adipocytes may be an effective strategy for regulating obesity. Warinhomhoun et al. [8] isolated the natural

Citation: Li, D.; Chen, F. The Perspectives of Plant Natural Products for Mitigation of Obesity. *Nutrients* **2023**, *15*, 1150. https://doi.org/10.3390/nu15051150

Received: 20 February 2023
Accepted: 23 February 2023
Published: 24 February 2023

metabolites from an orchid species in Thailand and evaluated their biological impact on the growth and differentiation of mouse embryonic 3T3-L1 pre-adipocytes. They found that a bibenzyl compound could suppress adipogenesis by inhibiting adipocyte differentiation and lipid accumulation. The underlying molecular mechanisms were linked to the down-regulation of adipogenic regulators such as peroxisome proliferators activated receptor γ (PPARγ) and CCAAT/enhancer-binding protein α (C/EBPα) and the suppression of mitogen-activated protein kinase (MAPK) pathway. Of note, the plant natural compounds can be used as a mixture to exert synergistic effects for the prevention and treatment of obesity. The study by Lee et al. [9] reported that a combination of *p*-synephrine, *p*-octopamine HCl, and hispidulin effectively reduced the expression of adipogenic marker proteins (PPARγ, C/EBPα, and C/EBPβ) in 3T3-L1 adipocytes and mitigated the body weight gain in high-fat diet (HFD)-fed mice. Xu et al. [10] explored the anti-obesity effect and mechanism of nuciferine, a main aporphine alkaloid component in lotus leaf, in HFD-fed mice. They found that treatment of nuciferine prevented body weight gain, improved glycolipid metabolism, and promoted energy expenditure in obese mice. Nuciferine suppressed lipogenesis and promoted fatty acid oxidation by activating the AMP-activated protein kinase (AMPK) pathway in hepatocytes and adipocytes. Liquiritigenin is a natural flavonoid isolated from the herb Glycyrrhiza uralensis Fisch. Qin et al. [11] studied the effects of liquiritigenin on lipogenesis in adipocytes and the results showed that liquiritigenin was able to inhibit lipid accumulation in 3T3-L1 cells via a mammalian target of rapamycin (mTOR)-mediated autophagy mechanism. Polymethoxyflavones are flavonoids found in citrus fruits with bioactive properties. Jin et al. [12] reported that the polymethoxyflavones isolated from citrus leaves effectively improved HFD-induced insulin resistance and dyslipidemia in obese mice. The polymethoxyflavones exert the anti-obesity effect via the modulation of fatty-acid oxidation and lipolysis. Fennel is a herbaceous and perennial plant cultivated in China and fennel essential oil contains volatile natural compounds including *p*-anisaldehyde, limonene, estragole, anethole, and trans-anethole. The study by Hong et al. [13] reported that inhalation of fennel essential oil exerted a protective effect against lipid and metabolic dysfunction in HFD-induced obese rats.

Overall, this Special Issue provides compelling evidence of the role of plant-derived natural compounds in the treatment and prevention of obesity. Studies are required to further explore the potential pharmacological and physiological activities of bioactive phytochemicals and understand their underlying mechanisms of action.

Author Contributions: Conceptualization, F.C.; writing-original draft preparation, D.L.; Writing—review and editing, F.C. All authors have read and agreed to the published version of the manuscript.

Funding: This research received no external funding.

Conflicts of Interest: The authors declare that they have no competing interests.

References

1. Seidell, J.C.; Halberstadt, J. The global burden of obesity and the challenges of prevention. *Ann. Nutr. Metab.* **2015**, *66* (Suppl. S2), 7–12. [CrossRef] [PubMed]
2. Chooi, Y.C.; Ding, C.; Magkos, F. The epidemiology of obesity. *Metabolism* **2019**, *92*, 6–10. [CrossRef] [PubMed]
3. Apovian, C.M. Obesity: Definition, comorbidities, causes, and burden. *Am. J. Manag. Care* **2016**, *22*, s176–s185. [PubMed]
4. Srivastava, G.; Apovian, C.M. Current pharmacotherapy for obesity. *Nat. Rev. Endocrinol.* **2018**, *14*, 12–24. [CrossRef] [PubMed]
5. Popiolek-Kalisz, J. The Impact of Dietary Flavonols on Central Obesity Parameters in Polish Adults. *Nutrients* **2022**, *14*, 5051. [CrossRef] [PubMed]
6. Han, A.L.; Jeong, S.J.; Ryu, M.S.; Yang, H.J.; Jeong, D.Y.; Park, D.S.; Lee, H.K. Anti-Obesity Effects of Traditional and Commercial Kochujang in Overweight and Obese Adults: A Randomized Controlled Trial. *Nutrients* **2022**, *14*, 2783. [CrossRef] [PubMed]
7. Hausman, D.B.; DiGirolamo, M.; Bartness, T.J.; Hausman, G.J.; Martin, R.J. The biology of white adipocyte proliferation. *Obes. Rev.* **2001**, *2*, 239–254. [CrossRef] [PubMed]
8. Warinhomhoun, S.; Khine, H.E.E.; Sritularak, B.; Likhitwitayawuid, K.; Miyamoto, T.; Tanaka, C.; Punsawad, C.; Punpreuk, Y.; Sungthong, R.; Chaotham, C. Secondary Metabolites in the *Dendrobium heterocarpum* Methanolic Extract and Their Impacts on Viability and Lipid Storage of 3T3-L1 Pre-Adipocytes. *Nutrients* **2022**, *14*, 2886. [CrossRef] [PubMed]

9. Lee, D.; Lee, J.H.; Kim, B.H.; Lee, S.; Kim, D.W.; Kang, K.S. Phytochemical Combination (*p*-Synephrine, *p*-Octopamine Hydrochloride, and Hispidulin) for Improving Obesity in Obese Mice Induced by High-Fat Diet. *Nutrients* **2022**, *14*, 2164. [CrossRef] [PubMed]
10. Xu, H.; Lyu, X.; Guo, X.; Yang, H.; Duan, L.; Zhu, H.; Pan, H.; Gong, F.; Wang, L. Distinct AMPK-Mediated FAS/HSL Pathway Is Implicated in the Alleviating Effect of Nuciferine on Obesity and Hepatic Steatosis in HFD-Fed Mice. *Nutrients* **2022**, *14*, 1898. [CrossRef] [PubMed]
11. Qin, H.; Song, Z.; Zhao, C.; Yang, J.; Xia, F.; Wang, L.; Ali, A.; Zheng, W. Liquiritigenin Inhibits Lipid Accumulation in 3T3-L1 Cells via mTOR-Mediated Regulation of the Autophagy Mechanism. *Nutrients* **2022**, *14*, 1287. [CrossRef] [PubMed]
12. Jin, Y.J.; Jang, M.G.; Kim, J.W.; Baek, S.; Ko, H.C.; Hur, S.P.; Kim, S.J. Anti-Obesity Effects of Polymethoxyflavone-Rich Fraction from Jinkyool (*Citrus sunki* Hort. ex Tanaka) Leaf on Obese Mice Induced by High-Fat Diet. *Nutrients* **2022**, *14*, 865. [CrossRef] [PubMed]
13. Hong, S.J.; Yoon, S.; Jo, S.M.; Jeong, H.; Youn, M.Y.; Kim, Y.J.; Kim, J.K.; Shin, E.C. Olfactory Stimulation by Fennel (*Foeniculum vulgare* Mill.) Essential Oil Improves Lipid Metabolism and Metabolic Disorders in High Fat-Induced Obese Rats. *Nutrients* **2022**, *14*, 741. [CrossRef] [PubMed]

nutrients MDPI

Article

The Impact of Dietary Flavonols on Central Obesity Parameters in Polish Adults

Joanna Popiolek-Kalisz [1,2]

[1] Clinical Dietetics Unit, Department of Bioanalytics, Medical University of Lublin, ul. Chodzki 7, 20-093 Lublin, Poland; joannapopiolekkalisz@umlub.pl
[2] Department of Cardiology, Cardinal Wyszynski Hospital in Lublin, al. Krasnicka 100, 20-718 Lublin, Poland

Abstract: Background: Central obesity is defined as the excessive fat tissue located in abdominal region accompanied by systemic inflammation, which drives to cardiovascular disease. Flavonols are antioxidative agents present in food. The aim of this study was investigating the relationship between dietary flavonols intake and central obesity. Methods and results: 80 participants (40 central obese and 40 healthy controls) were administered a food frequency questionnaire dedicated to flavonols intake assessment. Body composition was measured with bioelectrical impedance analysis. The analysis showed significant differences between central obese participants and healthy controls in total flavonol ($p = 0.005$), quercetin ($p = 0.003$), kaempferol ($p = 0.04$) and isorhamnetin ($p < 0.001$) habitual intake. Among central obese participants, there was a moderate inverse correlation between fat mass (FM) and total flavonol (R = −0.378; 95% CI: −0.620 to −0.071; $p = 0.02$), quercetin (R = −0.352; 95% CI: −0.601 to −0.041; $p = 0.03$), kaempferol (R = −0.425; 95% CI: −0.653 to −0.127; $p = 0.01$) and myricetin intake (R = −0.352; 95% CI: −0.601 to −0.041; $p = 0.03$). BMI was inversely correlated with total flavonol (R = −0.330; 95% CI: −0.584 to −0.016; $p = 0.04$) and quercetin intake (R = −0.336; 95% CI: −0.589 to −0.023; $p = 0.04$). Waist circumference was inversely correlated with total flavonol (R = −0.328; 95% CI: −0.586 to −0.009; $p = 0.04$), quercetin (R = −0.322; 95% CI: −0.582 to −0.002; $p = 0.048$) and myricetin intake (R = −0.367; 95% CI: −0.615 to −0.054; $p = 0.02$). Among flavonols' dietary sources, there was an inverse correlation between black tea consumption and FM (R: −0.511; 95% CI: −0.712 to −0.233; $p < 0.001$) and between coffee and waist circumference (R: −0.352; 95% CI: −0.604 to −0.036; $p = 0.03$) in central obese participants. Conclusions: The higher flavonol intake could play a protective role in abdominal obesity development. What is more, total and selected flavonol dietary intakes are inversely correlated with the parameters used for obesity assessment in central obese participants. The habitual consumption of products rich in flavonols, mainly tea and coffee, could possibly have a preventive role in abdominal obesity development.

Keywords: obesity; abdominal obesity; flavonols; quercetin; myricetin; isorhamnetin

Citation: Popiolek-Kalisz, J. The Impact of Dietary Flavonols on Central Obesity Parameters in Polish Adults. *Nutrients* **2022**, *14*, 5051. https://doi.org/10.3390/nu14235051

Academic Editor: Fang Chen

Received: 9 November 2022
Accepted: 24 November 2022
Published: 27 November 2022

1. Introduction

Civilizational diseases are the major cause of deaths worldwide, including ischemic heart disease responsible for 16% of world's total deaths, stroke responsible for 11% and diabetes, which followed a 70% increase from 2000 [1]. All these conditions are related to their risk factors, which involve elevated blood pressure, hyperglycemia and dyslipidemia, while central obesity is one of the causes that can lead to all of the mentioned conditions. What is more, it is proven that elevated body mass index (BMI) and waist circumference are strongly associated with, for example, cardiovascular disease (CVD) morbidity [2]. Higher values of BMI are also associated with higher mortality in CVD [3]. Throughout last years, there has been observed a clear trend toward obesity burden [4,5]. Thus, proper obesity controlling is one of the strategies not only for civilizational diseases prevention, but also for further management [6].

Obesity is the result of a misbalance between energy intake and increase in the triglycerides which are stored in adipocytes. In everyday practice, BMI is a low-cost parameter which enables easy obesity detection and classification. BMI is calculated by dividing body mass (in kilograms) by square of height (in meters). According to the World Health Organization, obesity is diagnosed when BMI is 30 kg/m^2 or higher. Then, obesity classes are also categorized on the basis of BMI value. Class 1 means BMI from 30 kg/m^2 to 34.99 kg/m^2, class 2 is BMI from 35 kg/m^2 to 39.99 kg/m^2 and class 3 is BMI 40 kg/m^2 or higher. On the other hand, the American Association of Clinical Endocrinologists and American College of Endocrinology guidelines propose different approach combining the application of BMI, waist circumference and adiposity measurements with available methods [7]. These guidelines underline that BMI alone cannot identify excess adiposity and serve for obesity diagnosis in all cases [7,8]. Fat distribution plays a role in metabolic consequences, as abdominal obesity is associated with higher risk of insulin resistance (IR), hypertension and dyslipidemia. Central obesity in Europe is clinically defined as a waist circumference $\geq$94 cm for men and $\geq$80 cm for women, while in other regions, the cut-off points may differ, e.g., >86 cm for women and >102 cm for men in the United States [7,9].

Bioelectrical impedance analysis (BIA) is a noninvasive method of body composition assessment. It is based on passing alternating electrical current through the body. Different body tissues present different electrical properties, thus the changes of impedance (Z), which is a combination of resistance (R) and reactance (Xc), can serve as direct bioelectrical parameter or can be transmuted into information about selected tissues content. This way BIA gives information about body fat mass (FM), total body water (TBW), extracellular water (ECW), intracellular water (ICW) and fat-free mass (FFM). The measurement can be taken at different electric current frequencies, usually at 5, 50, 100 or 200 kHz. BIA parameters can be used for general body composition analysis and for nutritional status assessment [10,11].

Excessive FM, particularly in the form of central obesity is associated with an inflammatory response which contributes to endothelium disfunction and atherosclerosis development. Adipose tissue inflammation, called *adipositis*, involves the production of cytokines and adipokines, such as tumor necrosis factor α, interleukin 1β and interleukin 6 [12–14]. These mechanisms are strongly presented in visceral fat tissue, as it contains more macrophages comparing to fat tissue in other regions [15]. It leads to numerous metabolic complications of obesity, including IR, which results in diabetes and dyslipidemia, whichthen results in atherosclerosis. Counteracting these conditions could be performed by targeting their primary mechanisms, including inflammation. Therefore, searching for agents with anti-inflammatory potential is one of the leading directions of dietary approach against obesity and its complications [16]. Flavonols are a group of flavonoids, and present antioxidative properties. The most abundant individual flavonol is quercetin, followed by kaempferol, myricetin, and isorhamnetin [17–19]. Nonetheless, this group includes also less prevalent ones, which are morin, galangin, fisetin, kaempferide, azaleatin, natsudaidain, pachypodol and rhamnazin [20]. The major products that contribute to dietary flavonols intake are onions [21], tea [22], apples [23], kale [24], lettuce [25], tomatoes [26], broccoli [27], grape skins [28] which are responsible for the flavonol contents in red wine [29], citrus fruits [30,31], and berries [19,20,32,33].

There are several papers that showed the positive impact of flavonols on obesity parameters; however there are not many of them which analyze the role of specific compounds [34,35]. This information would be helpful not only to indicate the direction of future studies regarding the dietary model, but also potential interventional studies and supplementation, which require investigating a single compound. What is more, there are not available any papers investigating the role of kaempferol, myricetin and isorhamnetin in obesity prevention in humans.

Therefore, the aim of this paper was to investigate the differences in dietary flavonol intake between participants with central obesity and healthy controls. Then the relationship between habitual intake of total and selected flavonols (quercetin, kaempferol, myricetin and isorhamnetin) and their sources and parameters used for obesity assessment and body composition in central obese patients was examined.

2. Materials and Methods

2.1. Study Group

In total, 80 participants (40 central obesity patients and 40 healthy controls) were enrolled in the study between March and November 2022. The inclusion criteria for the study group were as follows: (1) waist circumference $\geq$ 94 cm for men and $\geq$80 cm for women, (2) age 18–85 years, (3) written consent, (4) mental condition that enabled one-year retrospective dietary interview (i.e., answering the question: "How often during the last year did you consume the suggested portion of the following product?"), and (5) lack of conditions that interfere with BIA measurement (metal implants, e.g., orthopedical or pacemakers, abnormal body geometry, e.g., after amputations, pregnancy, and abnormal body temperature). The exclusion criteria were (1) lack of written consent, (2) abnormal mental condition, (3) presence of metal implants, (4) pregnancy, (5) amputations, (6) waist circumference < 94 cm for men and <80 cm for women, (7) age below 18 years or above 85 years, and (8) special diet due to health reasons.

Healthy controls were volunteers, who matched the following criteria: (1) waist circumference < 94 cm for men and <80 cm for women, (2) age 18–85 years, (3) written consent, (4) mental condition that enabled one-year retrospective dietary interview, and (5) lack of any chronic disease or chronic treatment. The exclusion criteria for the control group were (1) lack of written consent, (2) abnormal mental condition, (3) pregnancy, (4) special diet due to health reasons, (5) waist circumference $\geq$ 94 cm for men and $\geq$80 cm for women, (6) age below 18 years or above 85 years, and (7) chronic disease or treatment. All the participants were non-smokers.

2.2. Methods

2.2.1. Flavonol Intake Assessment

The food frequency questionnaire dedicated for specific flavonol long-term (one year) intake assessment was administered to the participants [36]. The questionnaire provided information upon the mean consumption of 140 products which are sources of flavonols during a preceding year. The suggested portions of the products were introduced on the basis of typical servings in everyday life (e.g., one piece, a glass) and described by both—a suggested serving (e.g., a piece, a glass) and a weight in grams. Then, on the basis of the declared frequency of selected products consumption (never or almost never, once a month, few times a month with a number of times per month given by the responder, once a week, few times a week with a number of times per week given by the responder, once a day, few times every day with a number of times per day given by the responder), the mean daily consumption of each product was calculated. The amounts of quercetin, kaempferol, isorhamnetin and myricetin in each product were based on the data available in the USDA database [37]. On the basis of it, the mean daily intake of quercetin, kaempferol, isorhamnetin, and myricetin was calculated for each participant. Total flavonol intake was calculated by adding the values for quercetin, kaempferol, isorhamnetin and myricetin. Dietary intakes were then calculated for body mass ratio by dividing the result of the mean daily intake of each compound by the measured body mass.

2.2.2. Anthropometrical and Body Composition Measurements

The anthropometrical measurements included body mass, height and waist circumference. The participants' body mass was measured with 0.05 kg accuracy with WTL-150A scale (Lubelskie Fabryki Wag, Lublin, Poland) by the trained professional. The participants were permitted to wear only underwear for this measurement. Waist circumference was measured with measuring tape with 0.5 cm accuracy in the mid-horizontal plane between the superior iliac crest and the lower margin of the last rib and height with stadiometer with 1.0 cm accuracy by the trained professional [38].

BIA measurement was performed with ImpediMed SFB7 (Impedimed Ltd., Brisbane, Australia) in accordance with the producer's instructions. The system was based on four self-adhesive surface electrodes connected to the analyzer. Electrodes were placed on the washed and dried skin of the left wrist and left ankle. The measurements were performed in the horizontal position with limbs resting loosely at 30–45 degrees to the body. Before the examination, participants had to lie in position for 5 min and were not allowed to drink, eat, or exert any physical effort in the preceding three hours. The measurements were taken in triplicate, and then the mean values were used for statistical analysis.

2.2.3. Statistical Analysis

Statistical analyses were performed with the RStudio software v. 4.2.0. The normality of the distribution of each parameter was checked by the Shapiro–Wilk test. The variables were presented as means (±SD). The Mann–Whitney test was used to compare the flavonol intakes between obese participants and healthy controls. p value below 0.05 was considered significant. The Pearson correlation was used to analyze the association between selected flavonol mean daily intake and BIA parameters, waist circumference or BMI. The cut-off points used for correlation coefficient were as follows: $\leq$0.29 as low, 0.30–0.49 as moderate and $\geq$0.50 as high correlation. Confidence interval (CI) level was 95%.

2.2.4. Ethics

The study was approved by the local Bioethics Committee of the Medical University of Lublin (consent no. KE-0254/9/01/2022). The study was conducted in line with the directives of the Declaration of Helsinki on Ethical Principles for Medical Research. All participants signed a written consent.

3. Results

3.1. General Characteristics of the Study Population

The final study group included 39 central obese participants (20 men and 19 women) and 40 healthy controls (11 men and 29 women). One obese participant was excluded from the analysis due to filling in the questionnaire incompletely. The mean body mass of central obese participants was 80.18 ± 10.35 kg, BMI was 29.18 ± 3.75 kg/m^2, and waist circumference was 102.72 ± 11.09 cm. Anthropometrical and body composition characteristics of the central obese group are presented in Table 1. The mean body mass of the healthy control group was 59.48 ± 10.43 kg, BMI was 20.94 ± 1.65 kg/m^2 and waist circumference was 65.6 ± 6.55 cm. The comparison of the anthropometrical characteristics of the central obesity and control group is presented in Table 2.

Table 1. Anthropometrical characteristics of the central obesity group.

Anthropometrical Parameter [Unit]	Mean Value	SD
Body mass [kg]	80.18	±10.35
BMI [kg/m^2]	29.18	±3.75
Waist circumference [cm]	102.72	±11.09
TBW [kg]	41.67	±6.00
TBW% [%]	52.26	±5.70
ECF [kg]	18.48	±2.95
ECF% [%]	44.39	±3.26
ICF [kg]	23.19	±3.71
ICF% [%]	55.63	±3.23
FFM [kg]	56.94	±8.17
FFM% [%]	71.36	±7.80
FM [kg]	23.12	±7.46
FM% [%]	28.61	±7.79

BMI—body mass index; TBW—total body water mass; TBW%—total body water percentage; ECF—extracellular fluid mass; ECF%—extracellular fluid percentage; ICF—intracellular fluid mass; ICF%—intracellular fluid percentage; FFM—fat-free mass; FFM%—fat-free mass percentage; FM—fat mass; FM%—fat mass percentage.

Table 2. Comparison of the anthropometrical characteristics of the central obese and control group.

Anthropometrical Parameter [Unit]	Central Obesity (n = 39)	SD	Healthy Control (n = 40)	SD	*p*
Body mass [kg]	80.18	±10.35	59.48	±10.43	<0.001
BMI [kg/m^2]	29.18	±3.75	20.94	±1.65	<0.001
Waist circumference [cm]	102.72	±11.09	65.6	±6.55	<0.001

BMI—body mass index.

3.2. Central Obesity and Healthy Participants Comparison

The analysis revealed significant differences in most of the flavonol intake between central obesity and healthy participants. The central obese participants presented significantly lower mean daily total flavonols intake (0.82 ± 0.44 mg/kg vs. 1.18 ± 0.66 mg/kg; $p = 0.005$), quercetin (0.52 ± 0.29 mg/kg vs. 0.77 ± 0.45 mg/kg; $p = 0.003$), kaempferol (0.19 ± 0.11 mg/kg vs. 0.25 ± 0.15 mg/kg; $p = 0.04$) and isorhamnetin (0.03 ± 0.02 mg/kg vs. 0.08 ± 0.07) mg/kg; $p < 0.001$. The detailed results are presented in Table 3. The boxplots showing these results are presented in Figure 1.

Table 3. The comparison of flavonol intake between central obesity participants and healthy control.

Mean Daily Intake [mg/kg]	Central Obesity (n = 39)	SD	Healthy Control (n = 40)	SD	*p*
Total flavonols	0.82	±0.44	1.18	±0.66	0.005
Quercetin	0.52	±0.29	0.77	±0.45	0.003
Kaempferol	0.19	±0.11	0.25	±0.15	0.04
Isorhamnetin	0.03	±0.02	0.08	±0.07	<0.001
Myricetin	0.07	±0.05	0.09	±0.06	0.19

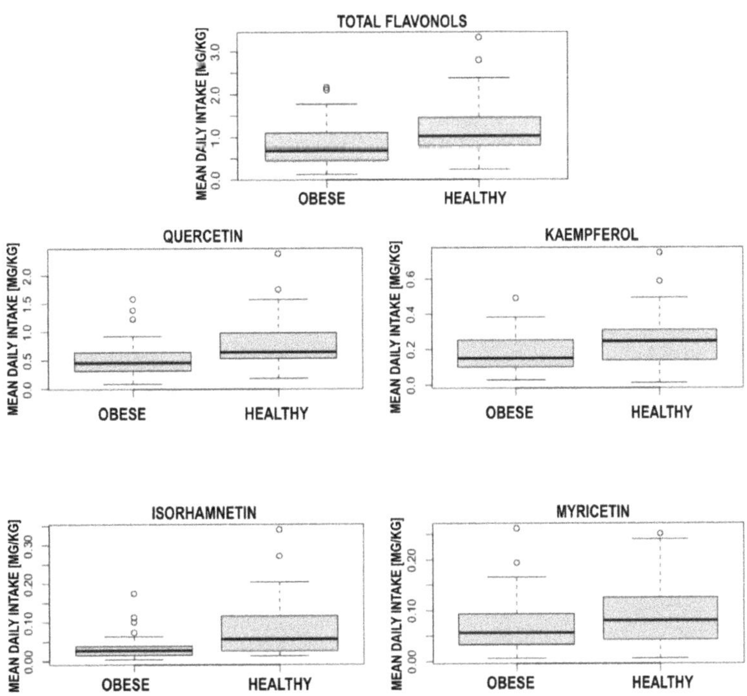

Figure 1. The boxplots presenting differences in flavonol intake between central obese participants and healthy control.

3.3. *Body Composition in Central Obese Participants*

The analysis showed significant moderate inverse correlation between quercetin intake and FM (R = −0.352; 95% CI: −0.601 to −0.041; p = 0.03). Kaempferol intake negatively correlated with TBW percentage (R = 0.330; 95% CI: 0.016 to 0.585; p = 0.04), FFM percentage (R = 0.321; 95% CI: 0.006 to 0.578; p = 0.046), FM (R = −0.425; 95% CI: −0.653 to −0.127; p = 0.01) and FM percentage (R = −0.330; 95% CI: −0.585 to −0.016; p = 0.04). Myricetin intake inversely correlated with FM (R = −0.352; 95% CI: −0.601 to −0.041; p = 0.03). Total flavonol intake correlated with FM (R = −0.378; 95% CI: −0.620 to −0.071; p = 0.02). Detailed results are presented in Table 4.

Table 4. Correlation between flavonol intake and BIA parameters, BMI and waist circumference.

	Quercetin		
	R	**95% CI**	***p***
TBW	−0.127	−0.425; 0.197	0.44
TBW%	0.225	−0.097; 0.505	0.17
ECF	−0.131	−0.428; 0.193	0.43
ECF%	−0.011	−0.326; 0.305	0.95
ICF	−0.099	−0.402; 0.223	0.55
ICF%	0.004	−0.312; 0.319	0.98
FFM	−0.126	−0.424; 0.198	0.45
FFM%	0.215	−0.108; 0.497	0.19
FM	−0.352	−0.601; −0.041	0.03
FM%	−0.225	−0.505; 0.097	0.17
BMI	−0.336	−0.589; −0.023	0.04
Waist circumference	−0.322	−0.582; −0.002	0.05
	Kaempferol		
	R	95% CI	*p*
TBW	−0.062	−0.371; 0.258	0.71
TBW%	0.330	0.016; 0.585	0.04
ECF	−0.083	−0.388; 0.239	0.62
ECF%	−0.024	−0.337; 0.294	0.89
ICF	−0.034	−0.345; 0.285	0.84
ICF%	0.021	−0.297; 0.334	0.90
FFM	−0.062	−0.370; 0.259	0.71
FFM%	0.321	0.006; 0.578	0.05
FM	−0.425	−0.653; −0.127	0.01
FM%	−0.330	−0.585; −0.016	0.04
BMI	−0.285	−0.551; 0.03	0.08
Waist circumference	−0.293	−0.560; 0.03	0.07
	Isorhamnetin		
	R	95% CI	*p*
TBW	−0.034	−0.345; 0.285	0.84
TBW%	−0.029	−0.342; 0.289	0.86
ECF	0.013	−0.304; 0.327	0.94
ECF%	0.094	−0.229; 0.397	0.57
ICF	−0.064	−0.372; 0.257	0.70
ICF%	−0.102	−0.405; 0.220	0.54
FFM	−0.034	−0.346; 0.284	0.84
FFM%	−0.039	−0.350; 0.280	0.81
FM	0.003	−0.313; 0.318	0.99
FM%	0.029	−0.289; 0.342	0.86
BMI	−0.110	−0.411; 0.213	0.51
Waist circumference	−0.033	−0.349; 0.290	0.84

Table 4. *Cont.*

	Myricetin		
	R	95% CI	p
TBW	−0.079	−0.385; 0.243	0.63
TBW%	0.254	−0.067; 0.527	0.12
ECF	−0.079	−0.385; 0.243	0.63
ECF%	0.011	−0.305; 0.325	0.95
ICF	−0.063	−0.371; 0.258	0.70
ICF%	−0.016	−0.330; 0.301	0.92
FFM	−0.078	−0.384; 0.244	0.64
FFM%	0.246	−0.076; 0.521	0.13
FM	−0.352	−0.601; −0.041	0.03
FM%	−0.254	−0.527; 0.067	0.12
BMI	−0.279	−0.546; 0.040	0.09
Waist circumference	−0.367	−0.615; −0.054	0.02
	Total flavonols		
	R	95% CI	p
TBW	−0.109	−0.410; 0.214	0.51
TBW%	0.259	−0.062; 0.531	0.11
ECF	−0.114	−0.415; 0.209	0.49
ECF%	−0.007	−0.322; 0.309	0.97
ICF	−0.084	−0.389; 0.238	0.61
ICF%	0.0003	−0.315; 0.316	0.99
FFM	−0.108	−0.410; 0.215	0.51
FFM%	0.249	−0.073; 0.523	0.13
FM	−0.378	−0.620; −0.071	0.02
FM%	−0.259	−0.531; 0.061	0.11
BMI	−0.330	−0.584; −0.016	0.04
Waist circumference	−0.328	−0.586; −0.009	0.04

BMI—body mass index; TBW—total body water mass; TBW%—total body water percentage; ECF—extracellular fluid mass; ECF%—extracellular fluid percentage; ICF—intracellular fluid mass; ICF%—intracellular fluid percentage; FFM—fat-free mass; FFM%—fat-free mass percentage; FM—fat mass; FM%—fat mass percentage.

3.4. Anthropometrical Parameters in Central Obese Participants

The analysis showed a significant moderate inverse correlation between quercetin intake and BMI (R = −0.336; 95% CI: −0.589 to −0.023; $p = 0.04$) and waist circumference (R = −0.322; 95% CI: −0.582 to −0.002; $p = 0.048$). Myricetin intake inversely correlated with waist circumference (R = −0.367; 95% CI: −0.615 to −0.054; $p = 0.02$). Total flavonol intake correlated negatively with BMI (R = −0.330; 95% CI: −0.584 to −0.016; $p = 0.04$) and waist circumference (R = −0.328; 95% CI: −0.586 to −0.009; $p = 0.04$). Detailed results are presented in Table 4.

3.5. Flavonols' Sources in Central Obese Participants

The major contributors to flavonol intake were onions (white and red), tomatoes, blueberries, apples, tea (black and green), coffee and wine. The analysis on the relationship between their consumption and mentioned above parameters used for obesity assessment (FM, FM percentage, waist circumference and BMI) revealed significant strong correlation between black tea consumption and FM (R: −0.511 95% CI: −0.712 to −0.233; $p < 0.001$) or FM% (R: −0.522; 95% CI: −0.719 to −0.247; $p < 0.001$). Waist circumference was significantly correlated with tomatoes (R: 0.370; 95% CI: 0.057 to 0.617; $p = 0.02$) and coffee consumption (R: −0.352; 95% CI: −0.604 to −0.036; $p = 0.03$). Detailed results are presented in Table 5.

Table 5. Correlation between flavonols' sources mean consumption and obesity assessment parameters.

	Fat Mass		
Product	**R**	**95% CI**	***p***
White onion	0.178	−0.146; 0.467	0.28
Red onion	−0.074	−0.381; 0.247	0.65
Tomatoes	0.283	−0.036; 0.549	0.08
Blueberry	0.043	−0.276; 0.354	0.79
Apple	−0.088	−0.393; 0.234	0.59
Black tea	−0.511	−0.712; −0.233	<0.001
Green tea	0.007	−0.310; 0.321	0.97
Coffee	−0.003	−0.318; 0.313	0.98
Wine	−0.100	−0.403; 0.223	0.55
	Fat mass %		
Product	R	95% CI	*p*
White onion	0.145	−0.178; 0.441	0.38
Red onion	−0.120	−0.419; 0.204	0.47
Tomatoes	0.166	−0.158; 0.457	0.31
Blueberry	0.235	−0.087; 0.512	0.15
Apple	0.012	−0.305; 0.326	0.94
Black tea	−0.522	−0.719; −0.247	<0.001
Green tea	0.057	−0.263; 0.366	0.73
Coffee	0.183	−0.141; 0.471	0.26
Wine	−0.092	−0.396; 0.230	0.58
	Waist circumference		
Product	R	95% CI	*p*
White onion	0.169	−0.159; 0.464	0.31
Red onion	0.031	−0.291; 0.347	0.86
Tomatoes	0.370	0.057; 0.617	0.02
Blueberry	−0.277	−0.548; 0.047	0.09
Apple	−0.018	−0.336; 0.303	0.91
Black tea	−0.201	−0.489; 0.127	0.23
Green tea	−0.141	−0.441; 0.187	0.40
Coffee	−0.352	−0.604; −0.036	0.03
Wine	−0.025	−0.342; 0.297	0.88
	BMI		
Product	R	95% CI	*p*
White onion	−0.069	−0.377; 0.252	0.67
Red onion	−0.050	−0.360; 0.270	0.76
Tomatoes	0.283	−0.035; 0.550	0.08
Blueberry	−0.062	−0.370; 0.259	0.71
Apple	−0.052	−0.361; 0.268	0.76
Black tea	−0.311	−0.570; 0.005	0.05
Green tea	−0.010	−0.325 0.306	0.95
Coffee	−0.137	−0.434; 0.187	0.41
Wine	0.045	−0.275; 0.355	0.79

4. Discussion

Obesity, particularly abdominal obesity, is one of the conditions leading to the development of civilizational diseases, which is why obesity prevention could potentially modify their progress [39]. Obesity is most often the result of excessive caloric intake; however, qualitative differences could also play a potential role in obesity development. Depending on the used parameter, obesity is defined as BMI $\geq$ 30 kg/m^2, while abdominal obesity is diagnosed for waist circumference $\geq$ 94 cm for men and $\geq$80 cm for women in Europe [9]. Even though obesity is described as the presence of excessive fat tissue, there is not any consensus for cut-off points for FM or FM percentage to diagnose obesity [40]. Obesity is one of the most common health problems. In Poland, the central obesity prevalence is 45.7% for women and 32.2% for men in the total population [41].

Systemic inflammation, which accompanies obesity, leads to endothelium dysfunction and promotes atherosclerosis progression. Central obesity, which is the excessive fat tissue in the abdominal area, is a particular risk factor of diseases such as CVD or diabetes [42]. As obesity is a condition accompanied by the chronic low-grade inflammation of adipose tissue, the potential role of antioxidative agents has been emerging in recent years [43,44]. That is why flavonols' impact on civilizational diseases and their risk factors was suggested in numerous studies [45–47]; however, its relationship with central obesity has not been established yet. It is an important point of view, as central obesity is one of the CVD and diabetes risk factors. What is more, there are no available studies conducted in humans that have examined the relationship between flavonols other than quercetin and central obesity. This study analyzed the correlation between myricetin, isorhamnetin and kaempferol intake and central obesity. The presented study also showed that there is an association between selected flavonol intake and the anthropometrical and body composition parameters is central obese patients.

This study showed the significant differences in flavonol habitual intakes between participants with central obesity and healthy controls. This relationship was observed for total flavonols and for selected compounds, such as quercetin, kaempferol and isorhamnetin. What is more, among the central obese participants, the general flavonol intake was also moderately inversely correlated with FM, BMI and waist circumference. Thus, we can provide a suggestion that the habitual consumption of products rich in flavonols could possibly have a preventive role in central obesity development. This observation is consistent with the results from the Korean population, in which dietary flavonol intake was inversely correlated with abdominal obesity prevalence in women [34]. However, in the same study, flavonol intake was also positively correlated with obesity based on BMI in men [34]. This difference might be caused by the fact that the authors of the mentioned study used the absolute flavonol intake instead of mean intake related to, for example, body mass. What is more, there are methodological differences between the study of Kim et al. and the presented research, as the study in the Korean group was based on 24 h recall and dual X-ray absorptiometry, while in the presented study, the intake of flavonols was investigated in one-year perspective, and body composition was assessed with BIA, which is more available in clinical background [34]. Moreover, these both studies are observational types, and they need interventional continuation to support their results.

Flavonols are a group of flavonoids that share a 3-hydroxyflavone backbone; nonetheless, the single compounds differ, e.g., by the presence and position of hydroxyl groups. These structural differences impact the bioactivity of these compounds [48]. Quercetin is more reactive than kaempferol due to the presence of an additional hydroxyl group at the R1 position [48]. This could be one of the potential reasons for the diverse impact strength on obesity parameters of different flavonols in central obese participants. What is more, the heat treatments, storage systems of the foods or cultivation practices could also impact the flavonols' bioavailability and bioactivity. It was shown that cooking spinach leads to quercetin and kaempferol residues' reduction [49]. A similar observation was made in berries, as cooking lead to quercetin loss—minor in strawberries and significant in bilberries [50]. On the other hand, the heat treatment of blueberries [51] or white beans [52] did not change their quercetin and kaempferol content, while in wine pomace, it even increased the flavonol content [53]. Storage could also impact flavonol content in some cases, as after 9 months of storage, quercetin loss was observed in bilberries and lingonberries, but not in black currants or red raspberries [50]. What is more, kaempferol and myricetin were even more susceptible to losses during storage [50]. Cultivation conditions could also impact phenolic content of foods, as higher UV exposition leads to higher quercetin content in red lettuce [54]. The differences in flavonol content could be also affected by the cultivar. These variabilities were described, for example, in sea buckthorn [55], plums [56] or grapes [29].

Quercetin is the most widespread individual flavonol in the everyday diet, as it is responsible for most of the observed relationships. It is present mainly in tea [22], onions [21] and berries (with whortleberry, lingonberry and cranberry being the most rich ones) [32]. Quercetin intake was inversely correlated with FM, BMI and waist circumference in central obese participants. Although the study by Nishimura et al. showed that quercetin supplementation (60 mg/day) did not change the amount of abdominal fat in general healthy group, the authors observed a reduction in the visceral fat amount in participants with lower levels of high density cholesterol (HDL-C) [35]. When high doses of quercetin were supplemented (500 or 1000 mg), they did not have impact on body mass or composition in healthy individuals [57]. On the other hand, quercetin supplementation was reported as being beneficial in term of hormonal balance in overweight and obese women [58]. It also reduced blood pressure and improved the lipid profile in overweight patients at risk of CVD [45,59]; however, the strength of its impact could also depend on genotype [60]. This may suggest that the quercetin impact on selected parameters could be present mainly in patients with excessive body mass or in combination with other flavonols, as they also presented potential impact on obesity parameters.

Kaempferol, the second prevalent flavonol, which is present mainly in kale [61] and spinach [49], intake was moderately inversely correlated with FM and FM percentage in central obese participants. Even though its role does not seem as critical as quercetin in terms of central obesity prevention (p = 0.003 for quercetin vs. p = 0.04 for kaempferol), they could possibly co-operate in their clinical effects. This corresponds with the results from animal model studies [62,63]. This study is the first one to link kaempferol to central obesity prevention in humans.

Isorhamnetin intake, whose main dietary contributor is onion [64], differed between central obese participants and healthy control, so we might assume that it could present protective properties against central obesity development; however, this is the first observation upon this topic in humans ever. On the other hand, once the central obesity condition is present, isorhamnetin intake was not correlated with the measured parameters. This observation is in opposition to the results presented by Rodriguez-Rodriguez et al. who observed metabolic benefits of isorhamnetin supplementation in obese mice [65]. These differences might be caused by the fact the in the mentioned study, the authors analyzed the metabolic changes only at the cellular level, such as glucose transporter 2 and peroxisome proliferator-activated receptor gamma mRNA content. The other reason could be the animal model used for the mentioned study. There are no available studies which focus on isorhamnetin and central obesity occurrence or parameters in humans, so more of them are definitely needed to support these observations.

Myricetin, which is less present in the everyday diet, but present mainly due to coffee consumption [64], intake was moderately inversely correlated with FM and waist circumference in central obese participants. Nonetheless, it was the only individual flavonol whose consumption did not differ significantly between central obese participants and the healthy control; thus, we can assume that its effects refer more to progression inhibition than to primary protection against central obesity. This observation is also in alignment with the results from animal model studies [66]. Nonetheless, this is also the first study that proved the potential relationship between obesity parameters and myricetin intake in humans.

The dietary sources of flavonols usually contain a set of them instead of one separated compound. As mentioned above, the co-operation between different individual flavonols could possibly also interfere with their biological effects and affect the strength of the impact on obesity parameters. That is why the relationship between the main dietary sources of flavonols and obesity parameters was also investigated. It showed that black tea consumption was strongly inversely correlated with FM and FM percentage. As tea is one of the main sources of quercetin, this observation is in line with the results referring to individual compounds and with animal model studies [67]. Numerous studies proved the antioxidative properties of tea polyphenols [68–70]. Nonetheless, the mechanism of tea impact on obesity development is complex, and aside from antioxidative potential, it included, for example, microbiome and nutrient intake interactions or protein kinase activation [67,71]. Black and green tea polyphenols also impact lipid metabolism from the level of intestinal digestion, e.g., emulsification to lipid accumulation, as black tea polyphenols can activate AMP-kinase involved in lipid metabolism or act on the nuclear receptors [72–74]. The majority of studies which suggest the beneficial role of tea for obesity focused on green tea [75]. Green tea is non-fermented in opposition to fermented black tea, which is why green tea presents higher antioxidative potential [76]. It is interesting that green tea consumption, which is even more rich in flavonols, did not present such a relationship with the obesity parameters in this study. It could be possibly the result of the much lower local consumption of green tea compared to black tea (0.48 portions/day vs. 1.81 portions/day). On the other hand, black tea polyphenols present some properties crucial in obesity development at a higher level compared to green tea, e.g., the inhibition of lipid emulsification during digestion [73].

As mentioned above, it is worth noting that mean tea consumption and quercetin intake did present different impact strength on obesity parameters, as quercetin intake was moderately inversely correlated with BMI, waist circumference and FM, while black tea consumption was strongly inversely correlated with FM and FM percentage. It could be the results of the fact that, even though black tea is one of the major quercetin sources, it contains also other bioactive compounds, such as catechins or gallocatechins, which also present these properties [77–79]. That is why the strength of this impact on FM could be potentially a result of the co-operation between quercetin and other compounds.

The preventive role of coffee consumption on central obesity is also suggested on the basis of the presented results. Mean coffee intake was moderately inversely associated with waist circumference, which is consistent with the results of the meta-analysis by Lee et al. [80]. Coffee is an important source of flavonols; however, it contains also other bioactive compounds, such as chlorogenic acid, caffeine, trigonelline, and magnesium, which are also associated with anti-obesity benefits [81].

Limitations of the Study

This study has its limitations. The presented results are based on the questionnaire responses, so it shares all the limitations of this type of study. That is why the results should be taken with great caution. It was aimed to analyze the long-term dietary habits, so support by multiple biochemical tests regarding flavonol blood levels throughout a one-year period would be helpful. What is more, this study is also a retrospective model. A prospective and ideally interventional human study with biochemical support should confirm the presented results. The healthy controls were volunteers, so the majority of female participants was noted in the healthy control group; thus, the gender proportions in the healthy and central obese groups do not match the populational trends. The BIA parameters were also measured only in obese participants, so confirming the detailed relationships in healthy participants would be valuable. Nonetheless, this is the first study which analyzed the relationship between dietary flavonols as the single compounds and central obesity. This approach is highly valuable, as it provides a direction for future interventional studies regarding the preventive role of selected flavonols in obesity. This study was aimed to analyze general trends; however, a detailed investigation of the mechanisms responsible for these observations is needed, as it might be beyond the antioxidative properties.

5. Conclusions

Dietary flavonol intake could potentially present anti-obesity benefits. This study showed that participants with central obesity habitually consumed fewer flavonols (total flavonols, quercetin, kaempferol and isorhamnetin) than healthy participants. What is more, among the central obese participants, quercetin, kaempferol and myricetin intakes were inversely associated with parameters used for obesity assessment and classification, such as FM, waist circumference and BMI. Among major flavonol sources, tea presented strong correlation and coffee, moderate correlation, with obesity parameters in central obese participants. These results could suggest that a diet rich in flavonols, such as quercetin, kaempferol and isorhamnetin, with their main sources (tea and coffee), could be possibly protective against central obesity development.

Funding: This study was supported by Ministry of Education and Science in Poland within statutory activity of Medical University of Lublin (DS 472/2022).

Institutional Review Board Statement: The study was conducted in accordance with the Declaration of Helsinki and approved by the local Ethics Committee of the Medical University of Lublin (consent no. KE-0254/9/01/2022).

Informed Consent Statement: Informed consent was obtained from all patients involved in the study.

Data Availability Statement: The data that support the findings of this study are available from the corresponding author upon reasonable request.

Acknowledgments: The author would like to thank Emilia Fornal for funding acquisition.

Conflicts of Interest: The author declares no conflict of interest.

References

1. World Health Organization. The Top 10 Causes of Death—Factsheet. Available online: https://www.who.int/news-room/fact-sheets/detail/the-top-10-causes-of-death (accessed on 14 October 2022).
2. Wormser, D.; Kaptoge, S.; Di Angelantonio, E.; Wood, A.M.; Pennells, L.; Thompson, A.; Sarwar, N.; Kizer, J.R.; Lawlor, D.A.; Nordestgaard, B.G.; et al. Separate and combined associations of body-mass index and abdominal adiposity with cardiovascular disease: Collaborative analysis of 58 prospective studies. *Lancet* **2011**, *377*, 1085–1095. [CrossRef] [PubMed]
3. Sun, Y.-Q.; Burgess, S.; Staley, J.R.; Wood, A.M.; Bell, S.; Kaptoge, S.K.; Guo, Q.; Bolton, T.R.; Mason, A.M.; Butterworth, A.S.; et al. Body mass index and all cause mortality in HUNT and UK Biobank studies: Linear and non-linear mendelian randomisation analyses. *BMJ* **2019**, *364*, l1042. [CrossRef] [PubMed]
4. Bentham, J.; Di Cesare, M.; Bilano, V.; Bixby, H.; Zhou, B.; Stevens, G.A.; Riley, L.M.; Taddei, C.; Hajifathalian, K.; Lu, Y.; et al. Worldwide Trends in Body-Mass Index, Underweight, Overweight, and Obesity from 1975 to 2016: A Pooled Analysis of 2416 Population-Based Measurement Studies in 128.9 Million Children, Adolescents, and Adults. *Lancet* **2017**, *390*, 2627–2642. [CrossRef]
5. Liu, B.; Du, Y.; Wu, Y.; Snetselaar, L.G.; Wallace, R.B.; Bao, W. Trends in obesity and adiposity measures by race or ethnicity among adults in the United States 2011-18: Population based study. *BMJ* **2021**, *372*, n365. [CrossRef]
6. Visseren, F.L.J.; Mach, F.; Smulders, Y.M.; Carballo, D.; Koskinas, K.C.; Bäck, M.; Benetos, A.; Biffi, A.; Boavida, J.-M.; Capodanno, D.; et al. 2021 ESC Guidelines on cardiovascular disease prevention in clinical practice: Developed by the Task Force for cardio-vascular disease prevention in clinical practice with representatives of the European Society of Cardiology and 12 medical societies with the special contribution of the European Association of Preventive Cardiology (EAPC). *Eur. Heart J.* **2021**, *42*, 3227–3337. [CrossRef]
7. Garvey, W.T.; Mechanick, J.I.; Brett, E.M.; Garber, A.J.; Hurley, D.L.; Jastreboff, A.M.; Nadolsky, K.; Pessah-Pollack, R.; Plodkowski, R.; Reviewers of the AACE/ACE Obesity Clinical Practice Guidelines. American association of clinical endocrinologists and american college of endocrinology comprehensive clinical practice guidelines for medical care of patients with obesity. *Endocr. Pract.* **2016**, *22*, 1–203. [CrossRef] [PubMed]
8. Romero-Corral, A.; Somers, V.K.; Sierra-Johnson, J.; Thomas, R.; Collazo-Clavell, M.L.; Korinek, J.; Allison, T.G.; A Batsis, J.; Kuniyoshi, F.S.; Lopez-Jimenez, F. Accuracy of body mass index in diagnosing obesity in the adult general population. *Int. J. Obes.* **2008**, *32*, 959–966. [CrossRef] [PubMed]
9. Yumuk, V.; Tsigos, C.; Fried, M.; Schindler, K.; Busetto, L.; Micic, D.; Toplak, H. European Guidelines for Obesity Management in Adults. *Obes. Facts* **2015**, *8*, 402–424. [CrossRef] [PubMed]
10. Popiołek, J.; Teter, M.; Kozak, G.; Powrózek, T.; Mlak, R.; Karakuła-Juchnowicz, H.; Małecka-Massalska, T. Anthropometrical and Bioelectrical Impedance Analysis Parameters in Anorexia Nervosa Patients' Nutritional Status Assessment. *Medicina* **2019**, *55*, 671. [CrossRef]
11. Wannamethee, S.G.; Atkins, J.L. Muscle loss and obesity: The health implications of sarcopenia and sarcopenic obesity. *Proc. Nutr. Soc.* **2015**, *74*, 405–412. [CrossRef]
12. Kawai, T.; Autieri, M.V.; Scalia, R. Adipose tissue inflammation and metabolic dysfunction in obesity. *Am. J. Physiol. Cell Physiol.* **2021**, *320*, C375–C391. [CrossRef]
13. Jiang, N.; Li, Y.; Shu, T.; Wang, J. Cytokines and inflammation in adipogenesis: An updated review. *Front. Med.* **2019**, *13*, 314–329. [CrossRef] [PubMed]
14. Coppack, S.W. Pro-inflammatory cytokines and adipose tissue. *Proc. Nutr. Soc.* **2001**, *60*, 349–356. [CrossRef] [PubMed]

15. Okamoto, Y.; Higashiyama, H.; Rong, J.X.; McVey, M.J.; Kinoshita, M.; Asano, S.; Hansen, M.K. Comparison of mitochondrial and macrophage content between subcutaneous and visceral fat in db/db mice. *Exp. Mol. Pathol.* **2007**, *83*, 73–83. [CrossRef] [PubMed]
16. Pérez-Torres, I.; Castrejón-Téllez, V.; Soto, M.E.; Rubio-Ruiz, M.E.; Manzano-Pech, L.; Guarner-Lans, V. Oxidative Stress, Plant Natural Antioxidants, and Obesity. *Int. J. Mol. Sci.* **2021**, *22*, 1786. [CrossRef]
17. Zamora-Ros, R.; Andres-Lacueva, C.; Lamuela-Raventós, R.M.; Berenguer, T.; Jakszyn, P.; Barricarte, A.; Ardanaz, E.; Amiano, P.; Dorronsoro, M.; Larrañaga, N.; et al. Estimation of Dietary Sources and Flavonoid Intake in a Spanish Adult Population (EPIC-Spain). *J. Am. Diet. Assoc.* **2010**, *110*, 390–398. [CrossRef]
18. Sampson, L.; Rimm, E.; Hollman, P.C.; de Vries, J.H.; Katan, M.B. Flavonol and Flavone Intakes in US Health Professionals. *J. Am. Diet. Assoc.* **2002**, *102*, 1414–1420. [CrossRef]
19. Zamora-Ros, R.; Knaze, V.; Luján-Barroso, L.; Slimani, N.; Romieu, I.; Fedirko, V.; de Magistris, M.S.; Ericson, U.; Amiano, P.; Trichopoulou, A.; et al. Estimated dietary intakes of flavonols, flavanones and flavones in the European Prospective Investigation into Cancer and Nutrition (EPIC) 24 hour dietary recall cohort. *Br. J. Nutr.* **2011**, *106*, 1915–1925. [CrossRef]
20. Panche, A.N.; Diwan, A.D.; Chandra, S.R. Flavonoids: An overview. *J. Nutr. Sci.* **2016**, *5*, e47 [CrossRef]
21. Sagar, N.A.; Pareek, S.; Benkeblia, N.; Xiao, J. Onion (*Allium cepa* L.) bioactives: Chemistry, pharmacotherapeutic functions, and industrial applications. *Food Front.* **2022**, *3*, 380–412. [CrossRef]
22. Dwyer, J.T.; Peterson, J. Tea and flavonoids: Where we are, where to go next. *Am. J. Clin. Nutr.* **2013**, *98*, 1611S–1618S. [CrossRef] [PubMed]
23. Awad, M.A.; de Jager, A.; van Westing, L.M. Flavonoid and chlorogenic acid levels in apple fruit: Characterisation of variation. *Sci. Hortic.* **2000**, *83*, 249–263. [CrossRef]
24. Olsen, H.; Aaby, K.; Borge, G.I.A. Characterization and Quantification of Flavonoids and Hydroxycinnamic Acids in Curly Kale (*Brassica oleracea* L. Convar. *acephala* Var. *sabellica*) by HPLC-DAD-ESI-MS^n. *J. Agric. Food Chem.* **2009**, *57*, 2816–2825. [CrossRef] [PubMed]
25. Zivcak, M.; Brückova, K.; Sytar, O.; Brestic, M.; Olsovska, K.; Allakhverdiev, S.I. Lettuce flavonoids screening and phenotyping by chlorophyll fluorescence excitation ratio. *Planta* **2017**, *245*, 1215–1229. [CrossRef] [PubMed]
26. Slimestad, R.; Fossen, T.; Verheul, M.J. The Flavonoids of Tomatoes. *J. Agric. Food Chem.* **2008**, *56*, 2436–2441. [CrossRef]
27. Wu, X.; Zhao, Y.; Haytowitz, D.B.; Chen, P.; Pehrsson, P.R. Effects of domestic cooking on flavonoids in broccoli and calculation of retention factors. *Heliyon* **2019**, *5*, e01310. [CrossRef] [PubMed]
28. Giuffrè, A.M. HPLC-DAD detection of changes in phenol content of red berry skins during grape ripening. *Eur. Food Res. Technol.* **2013**, *237*, 555–564. [CrossRef]
29. Zhu, L.; Zhang, Y.; Lu, J. Phenolic Contents and Compositions in Skins of Red Wine Grape Cultivars among Various Genetic Backgrounds and Originations. *Int. J. Mol. Sci.* **2012**, *13*, 3492–3510. [CrossRef]
30. Maria, G.A.; Riccardo, N. Citrus bergamia, Risso: The peel, the juice and the seed oil of the bergamot fruit of Reggio Calabria (South Italy). *Emir. J. Food Agric.* **2020**, *32*, 522–523. [CrossRef]
31. Costanzo, G.; Vitale, E.; Iesce, M.R.; Naviglio, D.; Amoresano, A.; Fontanarosa, C.; Spinelli, M.; Ciaravolo, M.; Arena, C. Antioxidant Properties of Pulp, Peel and Seeds of Phlegrean Mandarin (*Citrus reticulata* Blanco) at Different Stages of Fruit Ripening. *Antioxidants* **2022**, *11*, 187. [CrossRef]
32. Häkkinen, S.H.; Kärenlampi, S.O.; Heinonen, I.M.; Mykkänen, H.M.; Törrönen, A.R. Content of the Flavonols Quercetin, Myricetin, and Kaempferol in 25 Edible Berries. *J. Agric. Food Chem.* **1999**, *47*, 2274–2279. [CrossRef] [PubMed]
33. Cassidy, A.; O'Reilly, É.J.; Kay, C.; Sampson, L.; Franz, M.; Forman, J.P.; Curhan, G.; Rimm, E.B. Habitual intake of flavonoid subclasses and incident hypertension in adults. *Am. J. Clin. Nutr.* **2011**, *93*, 338–347. [CrossRef] [PubMed]
34. Kim, S.-A.; Kim, J.; Jun, S.; Wie, G.-A.; Shin, S.; Joung, H. Association between dietary flavonoid intake and obesity among adults in Korea. *Appl. Physiol. Nutr. Metab.* **2020**, *45*, 203–212. [CrossRef] [PubMed]
35. Nishimura, M.; Muro, T.; Kobori, M.; Nishihira, J. Effect of Daily Ingestion of Quercetin-Rich Onion Powder for 12 Weeks on Visceral Fat: A Randomised, Double-Blind, Placebo-Controlled, Parallel-Group Study. *Nutrients* **2020**, *12*, 91. [CrossRef] [PubMed]
36. Popiolek-Kalisz, J.; Fornal, E. Dietary Isorhamnetin Intake Is Inversely Associated with Coronary Artery Disease Occurrence in Polish Adults. *Int. J. Environ. Res. Public Health* **2022**, *19*, 12546. [CrossRef]
37. Bhagwat, S.; Haytowitz, D.B.; Holden, J.M. *USDA Database for the Flavonoid Content of Selected Foods Release 3*; U.S. Department of Agriculture: Beltsville, MD, USA, 2011. Available online: https://www.ars.usda.gov/arsuserfiles/80400525/data/flav/flav_r03.pdf (accessed on 14 October 2022).
38. Després, J.-P.; Lemieux, I. Abdominal obesity and metabolic syndrome. *Nature* **2006**, *444*, 881–887. [CrossRef]
39. World Health Organization (WHO). *Obesity: Preventing and Managing the Global Epidemic. Report of a WHO Consultation*; WHO Technical Report Series 894; World Health Organization: Geneva, Switzerland, 2000.
40. Cederholm, T.; Barazzoni, R.; Austin, P.; Ballmer, P.; Biolo, G.; Bischoff, S.C.; Compher, C.; Correia, I.; Higashiguchi, T.; Holst, M.; et al. ESPEN guidelines on definitions and terminology of clinical nutrition. *Clin. Nutr.* **2017**, *36*, 49–64. [CrossRef]
41. Stepaniak, U.; Micek, A.; Waśkiewicz, A.; Bielecki, W.; Drygas, W.; Janion, M.; Kozakiewicz, K.; Niklas, A.; Puch-Walczak, A.; Pająk, A. Prevalence of general and abdominal obesity and overweight among adults in Poland. Results of the WOBASZ II study (2013–2014) and comparison with the WOBASZ study (2003–2005). *Pol. Arch. Med. Wewn.* **2016**, *126*, 662–671. [CrossRef]

42. Zhang, C.; Rexrode, K.; van Dam, R.; Li, T.Y.; Hu, F.B. Abdominal Obesity and the Risk of All-Cause, Cardiovascular, and Cancer Mortality: Sixteen Years of Follow-up in US Women. *Circulation* **2008**, *117*, 1658–1667. [CrossRef]
43. Furukawa, S.; Fujita, T.; Shimabukuro, M.; Iwaki, M.; Yamada, Y.; Nakajima, Y.; Nakayama, O.; Makishima, M.; Matsuda, M.; Shimomura, I. Increased oxidative stress in obesity and its impact on metabolic syndrome. *J. Clin. Investig.* **2004**, *114*, 1752–1761. [CrossRef]
44. Keaney, J.F.; Larson, M.; Vasan, R.S.; Wilson, P.W.; Lipinska, I.; Corey, D.; Massaro, J.; Sutherland, P.; Vita, J.; Benjamin, E. Obesity and Systemic Oxidative Stress: Clinical Correlates of Oxidative Stress in the Framingham Study. *Arter. Thromb. Vasc. Biol.* **2003**, *23*, 434–439. [CrossRef]
45. Popiolek-Kalisz, J.; Fornal, E. The Effects of Quercetin Supplementation on Blood Pressure—Meta-Analysis. *Curr. Probl. Cardiol.* **2022**, *47*, 101350. [CrossRef] [PubMed]
46. Popiolek-Kalisz, J.; Fornal, E. The Impact of Flavonols on Cardiovascular Risk. *Nutrients* **2022**, *14*, 1973. [CrossRef] [PubMed]
47. Popiolek-Kalisz, J.; Blaszczak, P.; Fornal, E. Dietary Isorhamnetin Intake Is Associated with Lower Blood Pressure in Coronary Artery Disease Patients. *Nutrients* **2022**, *14*, 4586. [CrossRef]
48. Sharma, A.; Sharma, P.; Singh Tuli, H.; Sharma, A.K. Phytochemical and Pharmacological Properties of Flavonols. In *eLS*; John Wiley & Sons, Ltd.: Hoboken, NJ, USA, 2018.
49. Kuti, J.O.; Konuru, H.B. Antioxidant Capacity and Phenolic Content in Leaf Extracts of Tree Spinach (*Cnidoscolus* spp.). *J. Agric. Food Chem.* **2004**, *52*, 117–121. [CrossRef] [PubMed]
50. Häkkinen, S.H.; Kärenlampi, S.O.; Mykkänen, H.M.; Törrönen, A.R. Influence of Domestic Processing and Storage on Flavonol Contents in Berries. *J. Agric. Food Chem.* **2000**, *48*, 2960–2965. [CrossRef]
51. Rodriguez-Mateos, A.; Cifuentes-Gomez, T.; George, T.W.; Spencer, J.P.E. Impact of Cooking, Proving, and Baking on the (Poly)phenol Content of Wild Blueberry. *J. Agric. Food Chem.* **2014**, *62*, 3979–3986. [CrossRef] [PubMed]
52. Huber, K.; Brigide, P.; Bretas, E.B.; Canniatti-Brazaca, S.G. Effect of Thermal Processing and Maceration on the Antioxidant Activity of White Beans. *PLoS ONE* **2014**, *9*, e99325. [CrossRef]
53. Del Pino-García, R.; González-SanJosé, M.L.; Rivero-Pérez, M.D.; García-Lomillo, J.; Muñiz, P. The effects of heat treatment on the phenolic composition and antioxidant capacity of red wine pomace seasonings. *Food Chem.* **2017**, *221*, 1723–1732. [CrossRef]
54. García-Macías, P.; Ordidge, M.; Vysini, E.; Waroonphan, S.; Battey, N.H.; Gordon, M.H.; Hadley, P.; John, P.; Lovegrove, J.A.; Wagstaffe, A. Changes in the Flavonoid and Phenolic Acid Contents and Antioxidant Activity of Red Leaf Lettuce (Lollo Rosso) Due to Cultivation under Plastic Films Varying in Ultraviolet Transparency. *J. Agric. Food Chem.* **2007**, *55*, 10168–10172. [CrossRef]
55. Ma, X.; Laaksonen, O.; Zheng, J.; Yang, W.; Trépanier, M.; Kallio, H.; Yang, B. Flavonol glycosides in berries of two major subspecies of sea buckthorn (*Hippophaë rhamnoides* L.) and influence of growth sites. *Food Chem.* **2016**, *200*, 189–198. [CrossRef] [PubMed]
56. Liaudanskas, M.; Okulevičiūtė, R.; Lanauskas, J.; Kviklys, D.; Zymonė, K.; Rendyuk, T.; Žvikas, V.; Uselis, N.; Janulis, V. Variability in the Content of Phenolic Compounds in Plum Fruit. *Plants* **2020**, *9*, 1611. [CrossRef] [PubMed]
57. Knab, A.M.; Shanely, R.A.; Jin, F.; Austin, M.D.; Sha, W.; Nieman, D.C. Quercetin with vitamin C and niacin does not affect body mass or composition. *Appl. Physiol. Nutr. Metab.* **2011**, *36*, 331–338. [CrossRef] [PubMed]
58. Khorshidi, M.; Moini, A.; Alipoor, E.; Rezvan, N.; Gorgani-Firuzjaee, S.; Yaseri, M.; Hosseinzadeh-Attar, M.J. The effects of quercetin supplementation on metabolic and hormonal parameters as well as plasma concentration and gene expression of resistin in overweight or obese women with polycystic ovary syndrome. *Phytother. Res.* **2018**, *32*, 2282–2289. [CrossRef] [PubMed]
59. Egert, S.; Bosy-Westphal, A.; Seiberl, J.; Kürbitz, C.; Settler, U.; Plachta-Danielzik, S.; Wagner, A.E.; Frank, J.; Schrezenmeir, J.; Rimbach, G.; et al. Quercetin reduces systolic blood pressure and plasma oxidised low-density lipoprotein concentrations in overweight subjects with a high-cardiovascular disease risk phenotype: A double-blinded, placebo-controlled cross-over study. *Br. J. Nutr.* **2009**, *102*, 1065–1074. [CrossRef]
60. Egert, S.; Boesch-Saadatmandi, C.; Wolffram, S.; Rimbach, G.; Muller, M.J. Serum Lipid and Blood Pressure Responses to Quercetin Vary in Overweight Patients by Apolipoprotein E Genotype. *J. Nutr.* **2010**, *140*, 278–284. [CrossRef]
61. Kaur, D.; Shri, R.; Kamboj, A. Bioactivity-directed isolation, characterization, and quantification of an anxiolytic flavonoid from *Brassica oleracea* L. *J. Food Biochem.* **2021**, *45*, e13608. [CrossRef]
62. Wang, T.; Wu, Q.; Zhao, T. Preventive Effects of Kaempferol on High-Fat Diet-Induced Obesity Complications in C57BL/6 Mice. *BioMed Res. Int.* **2020**, *2020*, 4532482. [CrossRef]
63. Bian, Y.; Lei, J.; Zhong, J.; Wang, B.; Wan, Y.; Li, J.; Liao, C.; He, Y.; Liu, Z.; Ito, K.; et al. Kaempferol reduces obesity, prevents intestinal inflammation, and modulates gut microbiota in high-fat diet mice. *J. Nutr. Biochem.* **2022**, *99*, 108840. [CrossRef]
64. Somerset, S.M.; Johannot, L. Dietary Flavonoid Sources in Australian Adults. *Nutr. Cancer* **2008**, *60*, 442–449. [CrossRef]
65. Rodríguez-Rodríguez, C.; Torres, N.; Gutiérrez-Uribe, J.A.; Noriega, L.G.; Torre-Villalvazo, I.; Leal-Díaz, A.M.; Antunes-Ricardo, M.; Márquez-Mota, C.; Ordaz, G.; Chavez-Santoscoy, R.A.; et al. The effect of isorhamnetin glycosides extracted from Opuntia ficus-indica in a mouse model of diet induced obesity. *Food Funct.* **2015**, *6*, 805–815. [CrossRef] [PubMed]
66. Akindehin, S.; Jung, Y.-S.; Kim, S.-N.; Son, Y.-H.; Lee, I.; Seong, J.K.; Jeong, H.W.; Lee, Y.-H. Myricetin Exerts Anti-Obesity Effects through Upregulation of SIRT3 in Adipose Tissue. *Nutrients* **2018**, *10*, 1962. [CrossRef] [PubMed]
67. Xu, J.; Li, M.; Zhang, Y.; Chu, S.; Huo, Y.; Zhao, J.; Wan, C. Huangjinya Black Tea Alleviates Obesity and Insulin Resistance via Modulating Fecal Metabolome in High-Fat Diet-Fed Mice. *Mol. Nutr. Food Res.* **2020**, *64*, e2000353. [CrossRef] [PubMed]

68. Ben Lagha, A.; Grenier, D. Black tea theaflavins attenuate *Porphyromonas gingivalis* virulence properties, modulate gingival keratinocyte tight junction integrity and exert anti-inflammatory activity. *J. Periodontal Res.* **2017**, *52*, 458–470. [CrossRef]
69. Ramadan, G.; El-Beih, N.M.; Talaat, R.M.; El-Ghffar, E.A.A. Anti-inflammatory activity of green versus black tea aqueous extract in a rat model of human rheumatoid arthritis. *Int. J. Rheum. Dis.* **2017**, *20*, 203–213. [CrossRef]
70. Liu, L.; Wu, X.; Zhang, B.; Yang, W.; Li, D.; Dong, Y.; Yin, Y.; Chen, Q. Protective effects of tea polyphenols on exhaustive exercise-induced fatigue, inflammation and tissue damage. *Food Nutr. Res.* **2017**, *61*, 1333390. [CrossRef]
71. Pan, S.; Deng, X.; Sun, S.; Lai, X.; Sun, L.; Li, Q.; Xiang, L.; Zhang, L.; Huang, Y. Black tea affects obesity by reducing nutrient intake and activating AMP-activated protein kinase in mice. *Mol. Biol. Rep.* **2018**, *45*, 689–697. [CrossRef]
72. Pan, M.-H.; Gao, Y.; Tu, Y. Mechanisms of Body Weight Reduction by Black Tea Polyphenols. *Molecules* **2016**, *21*, 1659. [CrossRef]
73. Shishikura, Y.; Khokhar, A.S.; Murray, B.S. Effects of Tea Polyphenols on Emulsification of Olive Oil in a Small Intestine Model System. *J. Agric. Food Chem.* **2006**, *54*, 1906–1913. [CrossRef]
74. Kudo, N.; Arai, Y.; Suhara, Y.; Ishii, T.; Nakayama, T.; Osakabe, N. A Single Oral Administration of Theaflavins Increases Energy Expenditure and the Expression of Metabolic Genes. *PLoS ONE* **2015**, *10*, e0137809. [CrossRef]
75. Ohishi, T.; Fukutomi, R.; Shoji, Y.; Goto, S.; Isemura, M. The Beneficial Effects of Principal Polyphenols from Green Tea, Coffee, Wine, and Curry on Obesity. *Molecules* **2021**, *26*, 453. [CrossRef] [PubMed]
76. Tang, G.-Y.; Meng, X.; Gan, R.-Y.; Zhao, C.-N.; Liu, Q.; Feng, Y.-B.; Li, S.; Wei, X.-L.; Atanasov, A.G.; Corke, H.; et al. Health Functions and Related Molecular Mechanisms of Tea Components: An Update Review. *Int. J. Mol. Sci.* **2019**, *20*, 6196. [CrossRef] [PubMed]
77. Khan, N.; Mukhtar, H. Tea Polyphenols in Promotion of Human Health. *Nutrients* **2019**, *11*, 39. [CrossRef] [PubMed]
78. Tanaka, T.; Matsuo, Y. Production Mechanisms of Black Tea Polyphenols. *Chem. Pharm. Bull.* **2020**, *68*, 1131–1142. [CrossRef] [PubMed]
79. Hibi, M.; Takase, H.; Iwasaki, M.; Osaki, N.; Katsuragi, Y. Efficacy of tea catechin-rich beverages to reduce abdominal adiposity and metabolic syndrome risks in obese and overweight subjects: A pooled analysis of 6 human trials. *Nutr. Res.* **2018**, *55*, 1–10. [CrossRef]
80. Lee, A.; Lim, W.; Kim, S.; Khil, H.; Cheon, E.; An, S.; Hong, S.; Lee, D.H.; Kang, S.-S.; Oh, H.; et al. Coffee Intake and Obesity: A Meta-Analysis. *Nutrients* **2019**, *11*, 1274. [CrossRef]
81. Higdon, J.V.; Frei, B. Coffee and Health: A Review of Recent Human Research. *Crit. Rev. Food Sci. Nutr.* **2006**, *46*, 101–123. [CrossRef]

nutrients

MDPI

Article

Barley Leaf Ameliorates *Citrobacter rodentium*-Induced Colitis through Preventive Effects

Yu Feng, Daotong Li, Chen Ma, Meiling Tian, Xiaosong Hu and Fang Chen *

Key Laboratory of Fruits and Vegetables Processing, Ministry of Agriculture, Engineering Research Centre for Fruits and Vegetables Processing, National Engineering Research Center for Fruit and Vegetable Processing, College of Food Science and Nutritional Engineering, China Agricultural University, Beijing 100083, China
* Correspondence: chenfangch@sina.com; Tel.: +86-10-62737645 (ext. 18)

Abstract: The incidence and prevalence of inflammatory bowel disease (IBD) have been increasing globally and progressively in recent decades. Barley leaf (BL) is a nutritional supplement that is shown to have health-promoting effects on intestinal homeostasis. Our previous study demonstrated that BL could significantly attenuate *Citrobacter rodentium* (CR)-induced colitis, but whether it exerts a prophylactic or therapeutic effect remains elusive. In this study, we supplemented BL before or during CR infestation to investigate which way BL acts. The results showed that BL supplementation prior to infection significantly reduced the disease activity index (DAI) score, weight loss, colon shortening, colonic wall swelling, and transmissible murine colonic hyperplasia. It significantly reduced the amount of CR in the feces and also markedly inhibited the extraintestinal transmission of CR. Meanwhile, it significantly reduced the levels and expression of tumor necrosis factor-alpha (TNF-α), interferon-gamma (IFNγ), and interleukin-1β (IL1β). In addition, pretreatment with BL improved CR-induced gut microbiota dysbiosis by reducing the content of Proteobacteria, while increasing the content of *Lactobacillus*. In contrast, the effect of BL supplementation during infestation on the improvement of CR-induced colitis was not as good as that of pretreatment with BL. In conclusion, BL protects against CR-caused colitis in a preventive manner.

Keywords: barley leaf; colitis; *Citrobacter rodentium*; gut microbiota; prophylactic effect

Citation: Feng, Y.; Li, D.; Ma, C.; Tian, M.; Hu, X.; Chen, F. Barley Leaf Ameliorates *Citrobacter rodentium* Induced Colitis through Preventive Effects. *Nutrients* **2022**, *14*, 3833. https://doi.org/10.3390/nu14183833

Academic Editor: Jon A. Vanderhoof

Received: 28 August 2022
Accepted: 14 September 2022
Published: 16 September 2022

1. Introduction

Inflammatory bowel disease (IBD), a non-specific chronic recurrent inflammatory disease of the intestine, is characterized by abdominal pain and diarrhea, bloody stool, weight loss, and potential risk of cancer [1]. Epidemiological data indicated that the prevalence and incidence of IBD are increasing every year, and IBD has become a worldwide public health threat [2,3]. However, most of the current pharmacological treatments for IBD lack of specificity and may produce adverse effects such as exorbitant costs, immunosuppression, and drug resistance [4]. Therefore, developing a novel protective or therapeutic approach to IBD is of urgent need.

Although the precise pathogenesis of IBD is unclear, many studies have revealed that gut microbiota and eating habits perform crucial roles in the initiation and progression of IBD [5,6]. The mammalian gastrointestinal tract resides with trillions of microorganisms, and the makeup and diversity of the gut microbiota are pivotal for defending against pathogenic bacteria, regulating host immunity, and promoting nutrients' absorption [7]. Previous studies indicate that the type, quality, and origin of food is the predominant factor influencing the structure of the gut microbiota, which, in turn, could affect host health and intestinal homeostasis [8,9]. Multiple studies have described variations in the intestinal flora structure between IBD patients and healthy individuals [10,11]. For example, the Enterobacteriaceae family, which contains many common Gram-negative pathogens, such as *Salmonella*, *Escherichia coli*, and *Shigella*, is elevated in IBD patients [10]. Meanwhile, there is a decrease in the amount of butyrate-producing intestinal bacteria in IBD patients [11].

Intriguingly, clinical trials have shown that donor fecal microbiota transplantation can effectively improve the symptoms of IBD patients and restore the intestinal flora structure of IBD to that of healthy individuals [12]. Thus, the regulation of gut microbiota might be a promising option for preventing and curing IBD.

Citrobacter rodentium (CR) is a natural mouse Gram-negative pathogenic bacterium that shares sixty-seven percent of its genes with human enteropathogenic *Escherichia coli* and enterohaemorrhagic *E. coli* [13]. Therefore, CR-induced bacterial colitis has long been utilized as a model to study the pathogenesis of enteric-infection-induced inflammation [14]. The infection of mice with CR triggers the formation of attaching and effacing (A/E) lesions, and this is distinguished by shedding of the brush border microvilli and intensive bacterial attachment [13]. Additionally, CR causes epithelial cell proliferation, colonic crypt hyperplasia, and mucosal thickening in mice. Following oral inoculation, CR causes a profound gut microbiota dysbiosis that is characterized by an overgrowth of Proteobacteria and a decline in the richness and overall diversity of the resident microbiota [14]. Initially, it colonizes the cecum patch, where it is considered to adapt to the intestinal environment in vivo. Then the pathogen migrates to the distal colon and undergoes rapid expansion by day 4 post-infection (p.i.). The CR burden peaks on day 7 p.i. and remains high up to day 12 p.i., and it is usually cleared by day 21 p.i. [14,15]. The capacity of the gut microbiota to colonize and outcompete CR is determined on nutritional availability [16], suggesting that dietary nutrients may serve as crucial factors to protect against CR-induced enteric infection.

Barley leaf (BL), the young leaf of *Hordeum vulgare* L., is a traditional Chinese herb that has been historically recorded to have potential beneficial health effects. The primary components of BL include dietary fiber, protein, polyphenol, vitamins, and minerals [17]. Multiple studies have revealed that BL exhibited antioxidant, lipid-lowering, and other physiological functions, which are effective in the treatment of several chronic disorders, such as diabetes and bone loss [18,19]. Yamaura et al. found that BL could partially inhibit the increase in NGF levels in the hippocampus and be antidepressant [20]. In addition, supplementation with BL should have health benefits by preventing diseases caused by oxidative damage, such as colorectal cancer and cardiovascular disease [21,22]. Our previous studies have demonstrated that dietary BL protects against CR-induced colitis (unpublished); however, whether it exerts a prophylactic or therapeutic effect remains elusive. In this study, we supplemented BL before (beCR) or during CR (duCR) infestation to evaluate the preventive or therapeutic impacts of BL on CR-induced colonic inflammation and gut microbiota dysbiosis.

2. Materials and Methods

2.1. Preparation of BL Powder

The powder of BL was prepared according to our previous studies [23,24]. Briefly, the fresh leaves of *Hordeum vulgare* L. (cultivated in Hangzhou, China) were cleaned, sliced, and dried. The dried BLs were then pounded into powder and filtered through a 300-mesh sieve. The nutritional composition of barley leaves used in this experiment is shown in Supplementary Table S1.

2.2. Animals

The mice utilized in this study were 4–6-week-old male C3H/HeN mice purchased from Beijing Vital River Laboratory Animal Technology Co., Ltd. (Beijing, China). After one-week of acclimatization, the mice were randomly divided into four groups: CD+CR, BL+CR, duCR, and beCR (n = 6 per group). CD+CR and BL+CR groups were fed a standard chow diet (CD) and isocaloric diet, in which 2.5% BL was added, respectively; the beCR group was fed BL chow before CR infection and switched to CD after infection, while the duCR group was fed CD before infection and BL chow at the time of infection. The macronutrient composition of the two diets is shown in Supplementary Table S2. Animal experiments were conducted following the National Institutes of Health guide for the care and use of Laboratory Animals (NIH Publications No. 8023, revised 1978), and the protocols were reviewed and approved by the Animal Care and Ethics Committee of China Agricultural University (Ethics reference number: AW32602202-4-2).

2.3. *CR Infection*

To induce bacterial colitis, mice were inoculated with approximately 1×10^9 colony-forming units (CFUs)/mouse of CR strain DBS 100 (ATCC 51459). Briefly, a sterile Luria-Bertani medium was injected into an individual colony of CR grown on a fresh MacConkey agar plate (Solarbio, Beijing, China) at 37 °C and shaken overnight. After mice with CD and BL diet for 3 weeks, mice were infected.

2.4. *Quantification of CR in Mouse Feces and Tissues*

On days 1, 4, 7, and 10 post-infection, the fresh stool pellets of mice were harvested, weighed, and homogenized by using a BeadMill 24 benchtop bead-based homogenizer (Servicebio, Wuhan, China). Then sample homogenates were plated on MacConkey agar in a serial ten-fold dilution and CR colonies were counted during the following day. The CR colony has unique characteristics around a red center and a white edge, which could be clearly identified. Spleens and livers were collected and processed for organ bacterial-burden analysis.

2.5. *Disease Activity Index (DAI)*

The severity of colitis was evaluated by using the DAI score. Briefly, mouse DAI was calculated by weight loss (0 = none; 1 = 1–5%; 2 = 6–10%; 3 = 11–15%; 4 = >16%), stool consistency (0 = normal; 2 = loose stool; 4 = diarrhea), and general health status (0 = general well; 1 = slightly under average; 2 = poor; 3 = very poor; 4 = terrible), as previously described [25]. The maximum score was 12.

2.6. *Histological Staining*

The colonic and cecal tissues were embedded in paraffin after being fixed in 4% formalin solution. Following that, samples were sectioned and stained with hematoxylin and eosin (H&E). Tissue sections were estimated as previously described [25].

For goblet cell analysis, samples were stained with Alcian blue, using commercial kits (Servicebio Technology CO., Ltd., Wuhan, China). Image analysis was performed by utilizing Image J software.

2.7. *Immunofluorescence Staining*

The paraffin-embedded slices were deparaffinized and rehydrated before antigen retrieval with citric acid antigen retrieval solution (pH 6.0). Next, slides were blocked in goat serum. The primary antibody used was rabbit antisera generated against Ki67 (1:200; Abcam); after that, the CY3-conjugated goat anti-rabbit IgG (1:400; Jackson) as a secondary antibody was added to specially label the marker. Then sections were stained with 4′,6-diamidino-2-phenylindole (DAPI; Solarbio) and sealed with an anti-fluorescence quenching agent. The images were observed and collected under the Nikon fluorescence microscope.

2.8. *Inflammatory Cytokine Analysis*

The enzyme-linked immunosorbent assay (ELISA) kit (Nanjing Jiancheng Institute of Biological Engineering, Nanjing, China) was applied to evaluate cytokine levels in the colon. First, ~100 mg colonic tissue was ground in liquid nitrogen and dissolved in tissue protein lysate (Solarbio, China). The supernatant was then centrifuged and measured for protein concentration, using a BCA protein assay kit (Solarbio, China). Afterward, IL-4, IL-1β, TNF-α, and IFN-γ were tested.

2.9. 16S rRNA Gene Sequencing

Extracted total DNA from mouse feces was utilized as a template for PCR amplification of the V3/V4 region of the bacterial 16S rRNA gene. PCR amplification products obtained by amplification on an ABI GeneAmp® 9700 PCR system (AppliedBiosystems, Foster City, CA, USA) were sequenced on the Illumina Miseq platform in Majorbio BioPharm Technology Co., Ltd., (Shanghai, China) and further classified into operational taxonomic units (OTUs) within 0.03 (equivalent to 97% similarity) difference. The taxonomic community structure and phylogeny were assessed by visualizing datasets of microbial diversity and abundance of different samples. All data analyses were performed online on the Majorbio Cloud Platform. All raw sequence data were archived in the NCBI Short Read Archive database under the Bioproject accession number PRJNA870411.

2.10. Statistics

Data are presented as MEAN $\pm$ SEM. Statistical significance was determined by one-way or two-way analysis of variance (ANOVA), followed by the Duncan test, and the Bonferroni statistical test was applied post hoc. A p-value < 0.05 was considered statistically significant.

3. Results

3.1. Effect of beCR and duCR on CR-Induced Colitis

To investigate the preventive or therapeutic effects of BL on ameliorating CR-induced colitis, we designed the experiment in which mice were given BL before and during CR infection (Figure 1A). Compared with the CD+CR group, the mice in the BL+CR group exhibited significant improvement in body weight reduction, elevated DAI index, shortened colon length and swollen spleen due to CR infestation (Figure 1B–F). Notably, in comparison to mice supplemented with BL during CR infection, those supplemented with BL prior to infestation showed improvements in body weight loss and DAI severity, producing similar effects to the BL+CR group (Figure 1B,C). Despite the fact that there was no significant change in colon length between the duCR and beCR groups, the intestinal condition of mice in the beCR group was better than that in the duCR group, as shown by the presence of formed fecal pellets in the colon, improved intestinal wall swelling, and relieved cecum wrinkling (Figure 1D,E). Additionally, both beCR and duCR interventions could significantly improve splenomegaly due to CR, but no significant difference was found between them. Furthermore, there was no significant variation in diet and water consumption between groups, indicating that BL had no deleterious effects on the mice's nutritional and water habits (Supplementary Figure S1A,B). Taken together, BL improvement of CR-caused enterocolitis might be largely in a preventive manner.

3.2. Effect of beCR and duCR on CR-Induced Intestinal Pathology

Transmissible murine colonic hyperplasia (TMCH) is the hallmark feature in CR-infected mice, exhibiting epithelial cell proliferation, colonic crypt hyperplasia, and mucosal thickening [14]. To investigate the effect of beCR and duCR on CR-induced intestinal pathology, we performed H&E staining, Alcian blue staining, and Ki67^{+} immunofluorescence staining on mouse colonic tissues. As compared with the CD+CR group, inflammatory cell infiltration, colonic crypt hyperplasia, and goblet cell deficiency were significantly alleviated in the colonic tissue of the BL+CR group mice (Figure 2A–E). The H&E results showed that beCR intervention could significantly improve the increase in mucosal thickness, reduce histopathological scores, and produce comparable intervention effects as BL+CR, whereas duCR intervention failed to alleviate the colonic crypt hyperplasia due to CR (Figure 2A–C). Ki67 is an associated antigen of immature proliferating cells and is essential in cell proliferation. Consistent with the H&E results, beCR intervention significantly ameliorated CR-induced goblet cell deficiency and Ki67^{+} cell overexpression, similar to the effect of BL+CR intervention; although duCR treatment reduced Ki67^{+} cell overproliferation, it was dramatically different from BL+CR and duCR intervention

(Figure 2A,D,E). Furthermore, we performed H&E staining of the mouse cecum and observed that BL+CR and beCR interventions also significantly improved inflammatory cell infiltration, ulceration and loss of epithelial integrity in the cecum caused by CR, whereas duCR treatment failed to ameliorate these conditions (Supplementary Figure S2A,B). In conclusion, the beCR intervention was superior to the duCR intervention in improving mouse histopathology.

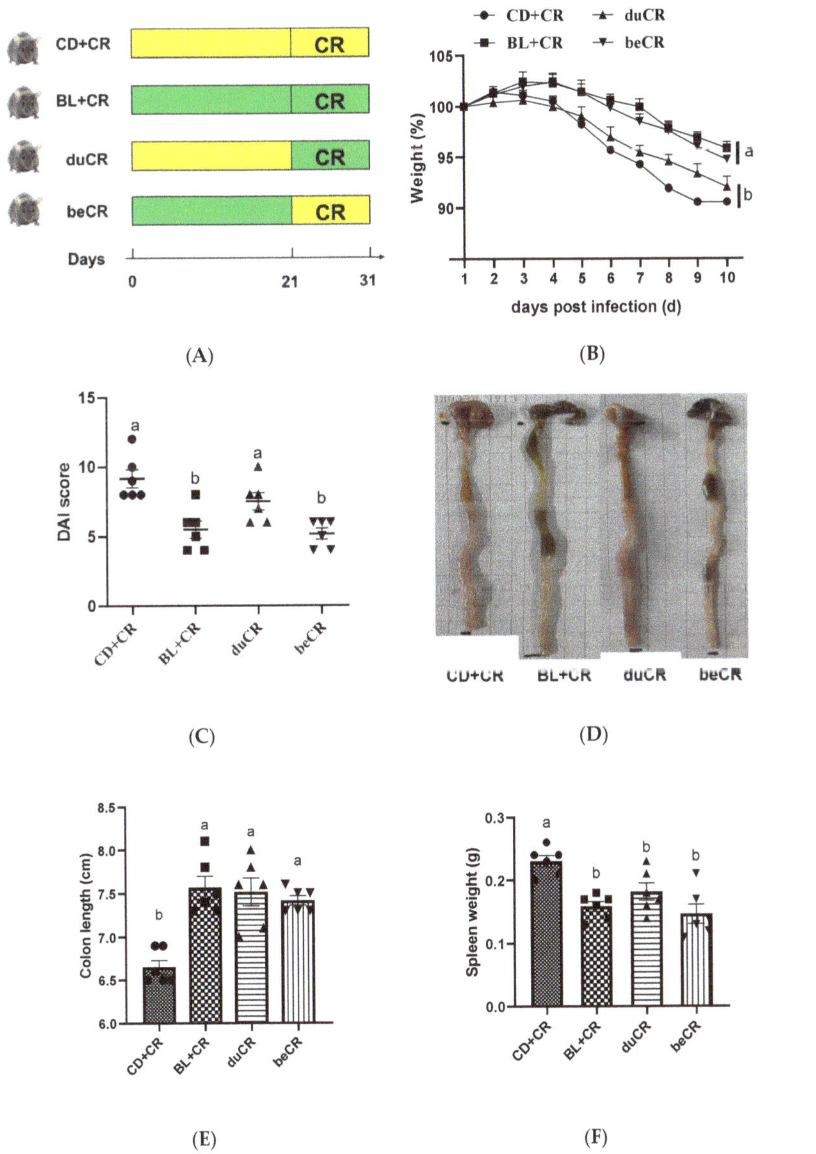

Figure 1. Effect of BL supplementation on mice before and during CR infestation. (**A**) Study design of mice grouping and treatment ($n = 6$ per group). (**B**) Weight loss. (**C**) Disease activity index (DAI) scores. (**D**) Representative images showing the gross appearance of the intestinal tissues; (**E**) Colon length. (**F**) Spleen weight. The a, b means in the same bar without a common letter differ at $p < 0.05$.

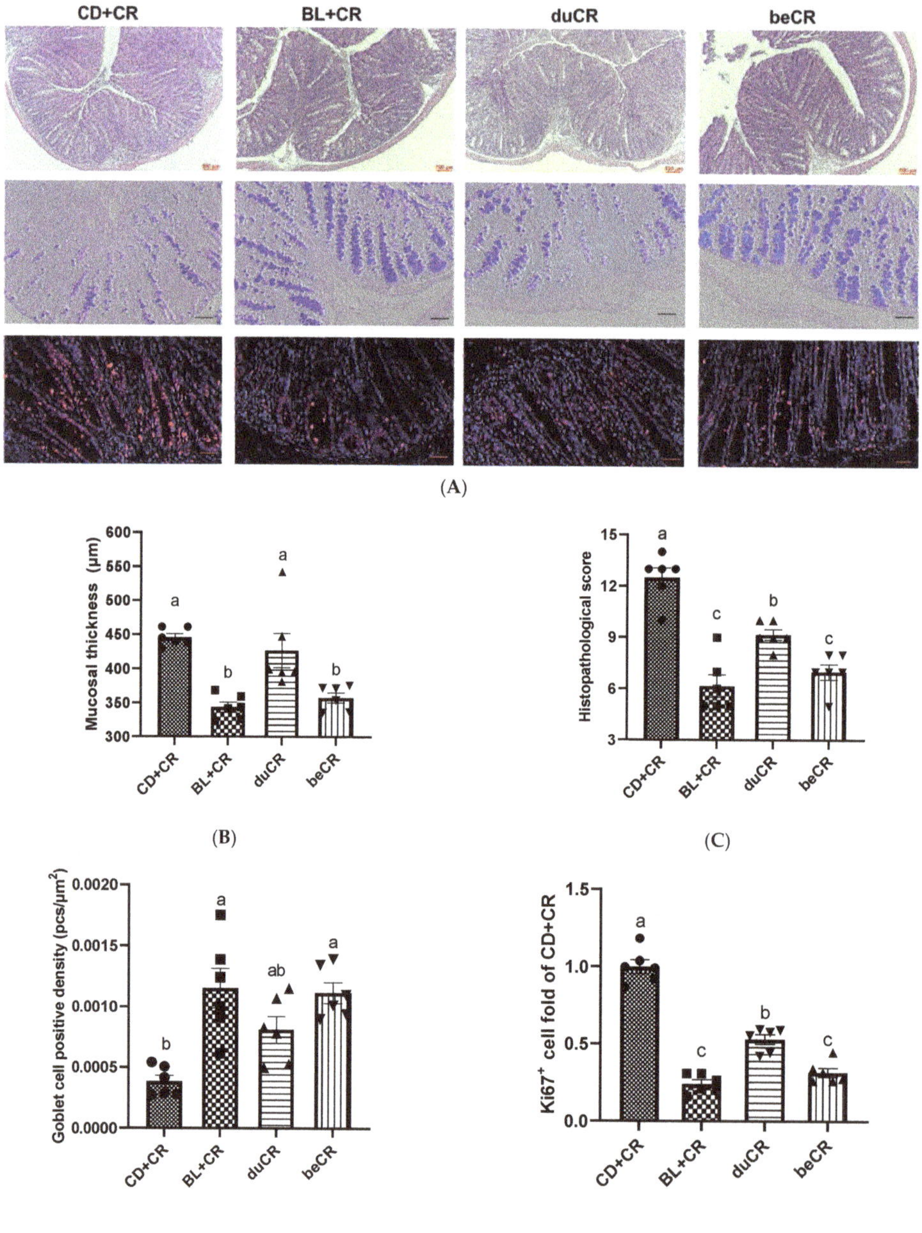

Figure 2. Effect of the beCR and duCR intervention on improving colonic histopathology. (**A**) Representative H&E staining images of colon tissues. Scale bar, 100 μm. Representative Alcian blue staining images of colonic sections. Scale bar, 100 μm. Representative Ki67-stained immunofluorescence images of colon tissues. Ki67 immunofluorescence is indicated in red, and DAPI nuclear staining is blue. Scale bar, 50 μm. (**B**) Mucosal thickness. (**C**) Histopathology score of colonic section. (**D**) Goblet cell positive density. (**E**) Quantification of Ki67 positive cells. The a, b, c means in the same bar without a common letter differ at $p < 0.05$.

3.3. Effect of beCR and duCR on Inflammatory Cytokines in Colon

Cytokines are inflammatory regulators that are essential intermediary phenotypes in inflammatory diseases. Therefore, we detected the level of inflammatory cytokines in the colon by ELISA. As shown in Figure 3, dietary supplementation with BL did not have any impact on the content of the anti-inflammatory cytokine IL-4. However, BL+CR and beCR interventions remarkably decreased the concentrations of TNF-α, IL-1β, and IFN-γ, and the effects of both interventions were equivalent; meanwhile, the duCR intervention, although it reduced the elevated levels of pro-inflammatory cytokines due to CR, did not differ significantly from the CD+CR group.

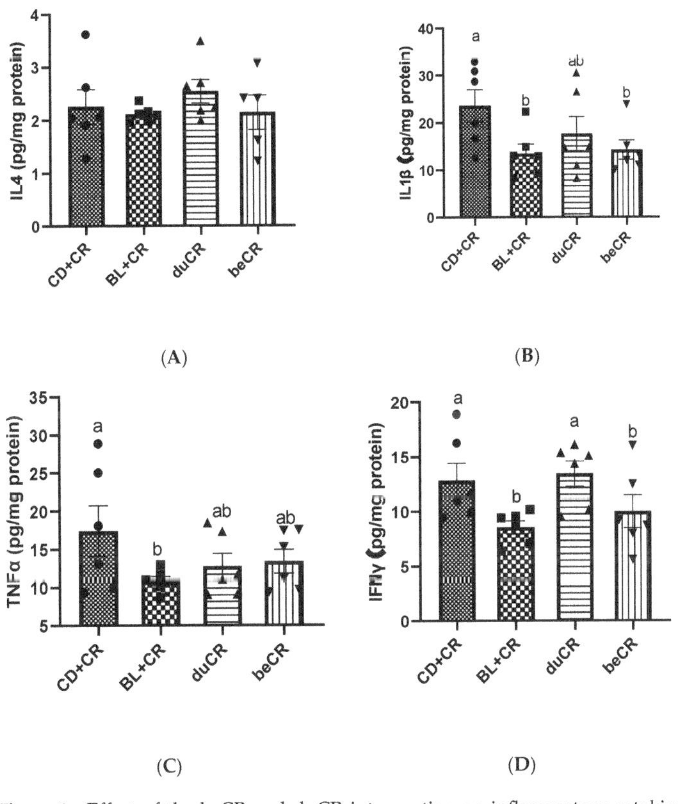

Figure 3. Effect of the beCR and duCR intervention on inflammatory cytokines in colon. The concentrations of anti-inflammatory cytokines (IL-4) (**A**) and pro-inflammatory cytokines IL-1β (**B**), TNF-α (**C**), and IFNγ (**D**) in colonic tissue of mice. The a, b means in the same bar without a common letter differ at $p < 0.05$.

3.4. Effect of beCR and duCR on CR Colonization

To investigate the effect of beCR and duCR interventions on CR colonization in the intestine of mice, we collected mouse feces on days 1, 4, 7, and 10 after CR infection, and bacterial CFUs were enumerated by dilution and plate counts. On day 1 p.i., there was no significant difference in CR burden between groups of mice; however, on days 4, 7, and 10 p.i., beCR intervention significantly inhibited CR colonization in the intestine, producing an intervention effect similar to that of BL+CR; meanwhile, the CR burden in duCR-intervened mice was not significantly distinct from that in the CD+CR group (Figure 4A). Furthermore, we also examined the extraintestinal transmission of CR. As shown in Figure 4B,C, the beCR intervention significantly suppressed the amount of CR in the spleen and liver, exhibiting comparable effects to the BL+CR intervention, whereas the duCR intervention only significantly depressed the amount of CR in the spleen and had no significant effect on the amount of CR in the liver.

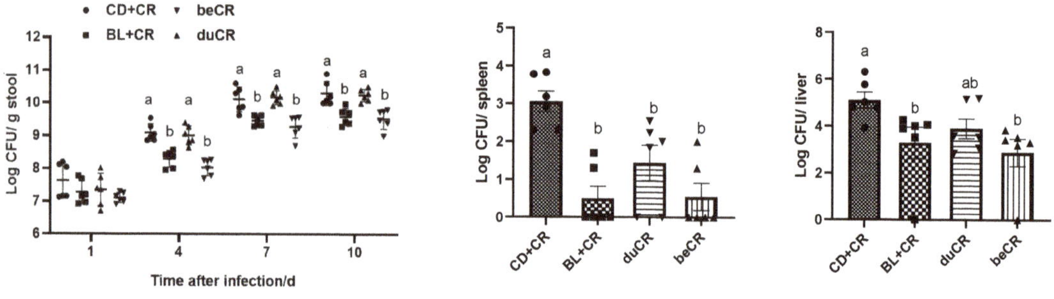

(A) (B) (C)

Figure 4. Effect of the beCR and duCR intervention on CR colonization. (**A**) Number of CR in feces on day 1, 4, 7, and 10 p.i. (**B**) Number of CR in spleen on day 10 p.i. (**C**) Quantification of CR in liver on day 10 p.i. The a, b means in the same bar without a common letter differ at $p < 0.05$.

3.5. *Effect of beCR and duCR on Modulating Alpha and Beta Diversity of Gut Microbiota*

Numerous studies have demonstrated that the composition and structure of the gut microbiota perform an essential role in the progression and development of IBD [5,6]. Therefore, in this experiment, 16S rRNA sequencing was employed to investigate the effects of beCR and ducR on microbial community changes. Alpha diversity phenotypes the richness (Ace and Chao indexes) and diversity (Shannon and Simpson indexes) of the intestinal flora. In comparison to the CD+CR group, both BL+CR and beCR interventions significantly increased the ACE index and Chao index, indicating that both interventions could increase the richness of the intestinal flora; meanwhile, the duCR intervention had no effect on either index (Figure 5A,B). For improving gut microbiota diversity, the effect of beCR treatment was equivalent to that of BL+CR, whereas there was no significant difference between duCR and CD+CR groups (Figure 5C,D). The non-metric multidimensional scaling (NMDS) analysis results showed that mice in the BL+CR and beCR groups had a similar gut microbiota composition and structure; meanwhile, those in the duCR and CD+CR groups showed a comparable bacterial community in composition and structure, and this was consistent with the results of Partial-Least-Squares Discriminant Analysis (PLS-DA) analysis (Figure 5E,F).

3.6. *Effect of beCR and duCR on Regulating Taxonomic Microbial Community Profiles*

At the phylum level, dietary supplementation with BL could dramatically increase the abundance of Firmicutes, while suppressing the abundance of Proteobacteria (Figure 6A,B). At the genus level, in comparison with the CD+CR group, the beCR intervention could dramatically inhibit *Citrobacter* content, and the intervention effect was similar to that of BL+CR; meanwhile, in regard to the duCR intervention, although it could reduce *Citrobacter* content, it was not significantly distinguishable from the CD+CR group (Figure 6C–E and Supplementary Figure S3). Notably, both the beCR and duCR intervention increased the abundance of *Lactobacillus* and did not differ significantly from the effect of BL+CR intervention, suggesting that BL might be a prebiotic enriched with *Lactobacillus* (Figure 6C–E and Supplementary Figure S3). Moreover, the beCR treatment also increased the level of *unclassified_f_Lachnospiraceae*, *norank_f_ Lachnospiraceaeand*, and *Enterorhabdus*, and it decreased the level of *Erysipelatoclostridium*; however, there were no significant differences with the CD+CR and duCR groups (Figure 6C). Then we correlated the above-mentioned genera with indexes related to colitis in all groups. Spearman's correlation analysis showed that

Citrobacter was significantly and positively correlated with mucosal thickness, pathological histological score, DAI index, spleen weight, pro-inflammatory cytokine IL1β, and quantification of Ki67 positive cells, while *Lactobacillus* was substantially and negatively correlated with these indices. In addition, *unclassified_f_Lachnospiraceae, norank_f_ Lachnospiraceaeand,* and *Enterorhabdus* were also negatively correlated with these colitis-related indicators; in contrast, *Erysipelatoclostridium* was in positive correlation with these indicators, although there was no significant relationship between these bacteria and these indicators (Figure 6F).

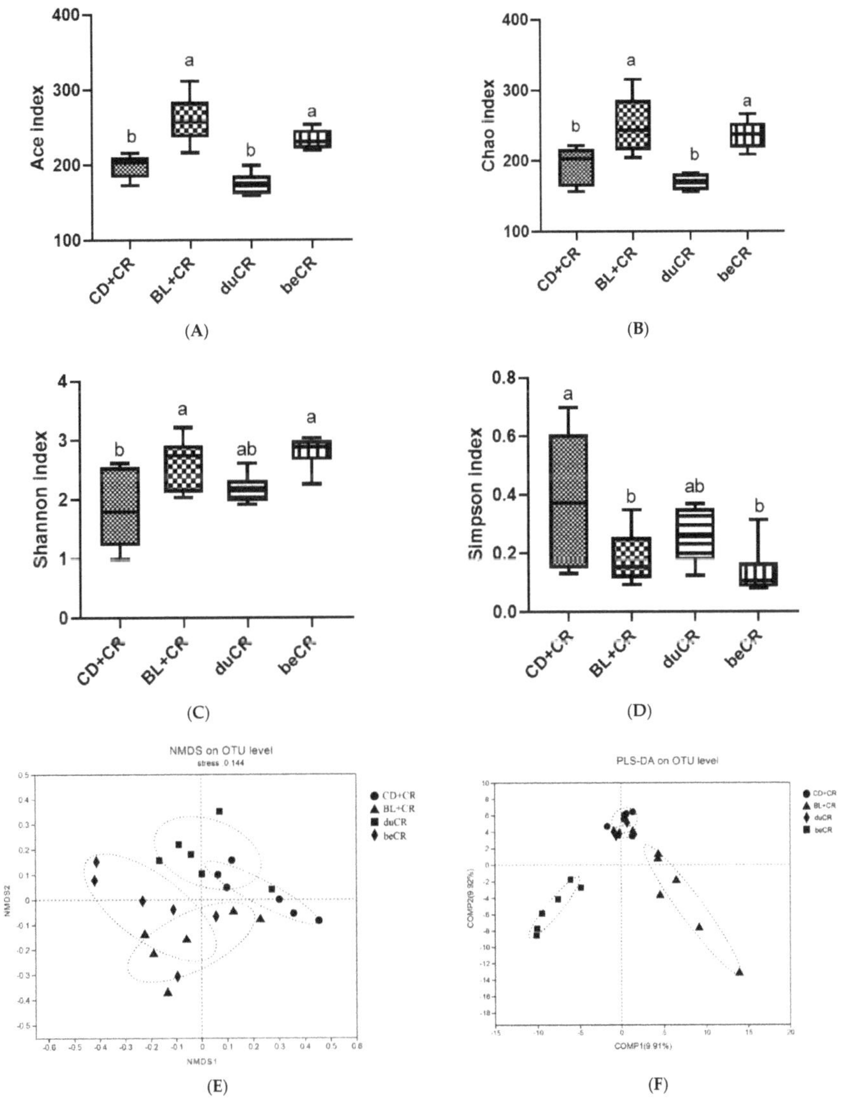

Figure 5. Effect of the beCR and duCR intervention on modulating alpha and beta diversity of intestinal flora. The richness of gut microbiota evaluated by (**A**) ACE index and (**B**) Chao index. The diversity of gut microbiota evaluated by (**C**) Shannon index and (**D**) Simpson index. (**E**) NMDS analysis on OTU level. (**F**) PLS-DA analysis on OTU level. The a, b means in the same bar without a common letter differ at $p < 0.05$.

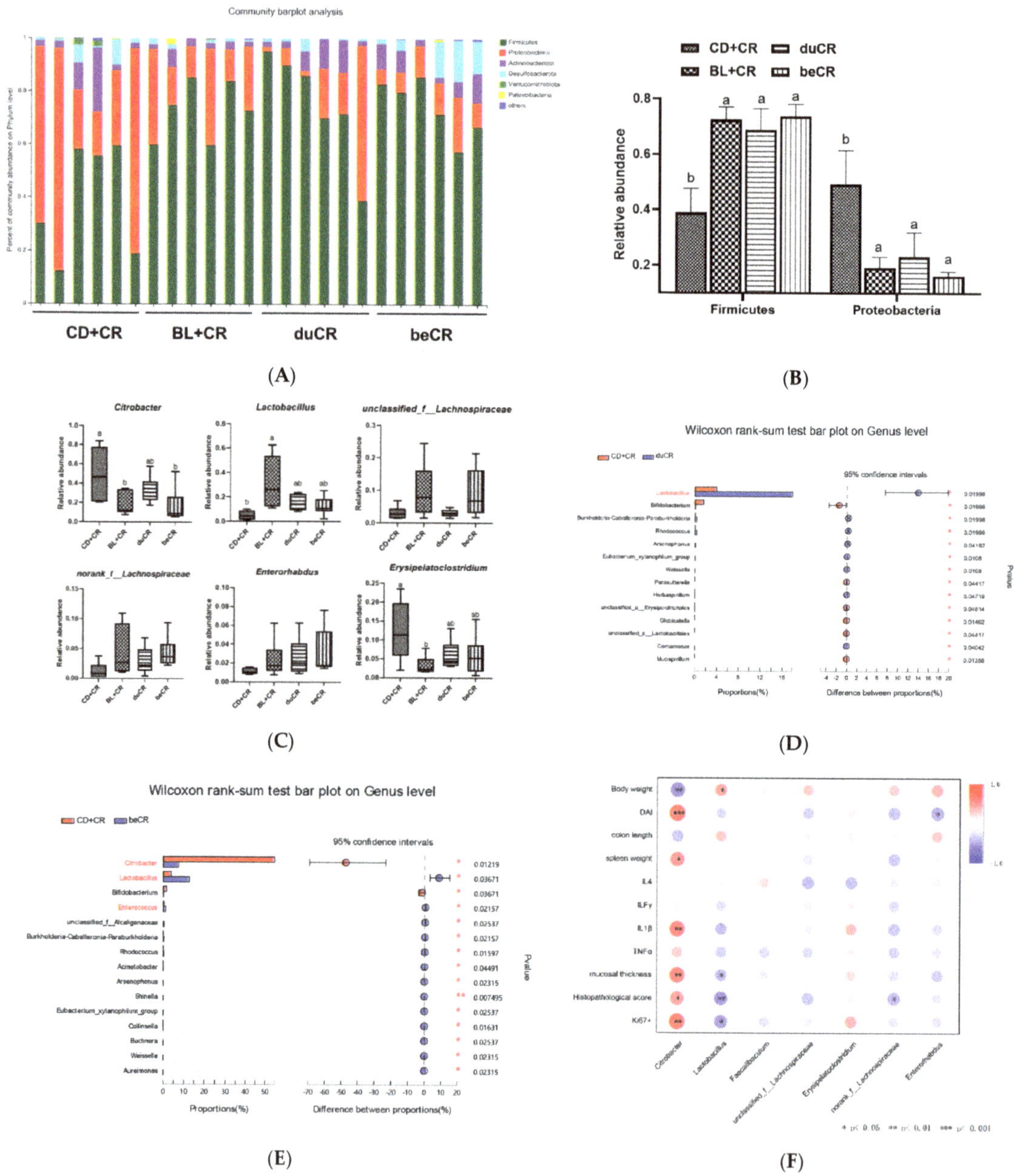

Figure 6. Effect of the beCR and duCR intervention on regulating taxonomic microbial community profiles. (**A**) Taxonomic distributions of gut bacterial composition at the phylum level. (**B**) Relative abundance of Firmicutes and Proteobacteria. (**C**) Relative abundance of *Citrobacter, Lactobacillus, unclassified_f_Lachnospiraceae, norank_f_Lachnospiraceae, Enterorhabdus,* and *Erysipelatoclostridium*. (**D**) Wilcoxon rank-sum test of group CD+CR and duCR at the genus level. (**E**) Wilcoxon rank-sum test of group CD+CR and beCR at the genus level. (**F**) Spearman correlations analysis between the microbiota and colitis-related index. The a, b means in the same bar without a common letter differ at $p < 0.05$.

3.7. Effect of beCR and duCR on Key Microbial Phylotypes

To further identify microbial taxa that supplement the BL community biomarkers before and during CR infestation, we performed liner discriminate analysis (LDA) in conjunction with effect size measurements (LEfSe) with an LDA score > 2. Cladogram analysis revealed that the levels of specific bacterial taxa varied from phylum to genus within each category (Figure 7A). The LefSe analysis results demonstrated that the CD+CR group mice displayed a higher abundance of *Citrobacter*, and *Lactobacillus* was more abundant in the duCR group. *Enterorhabdus* and *Mucispirillum* were found in greater abundance in the beCR group (Figure 7B).

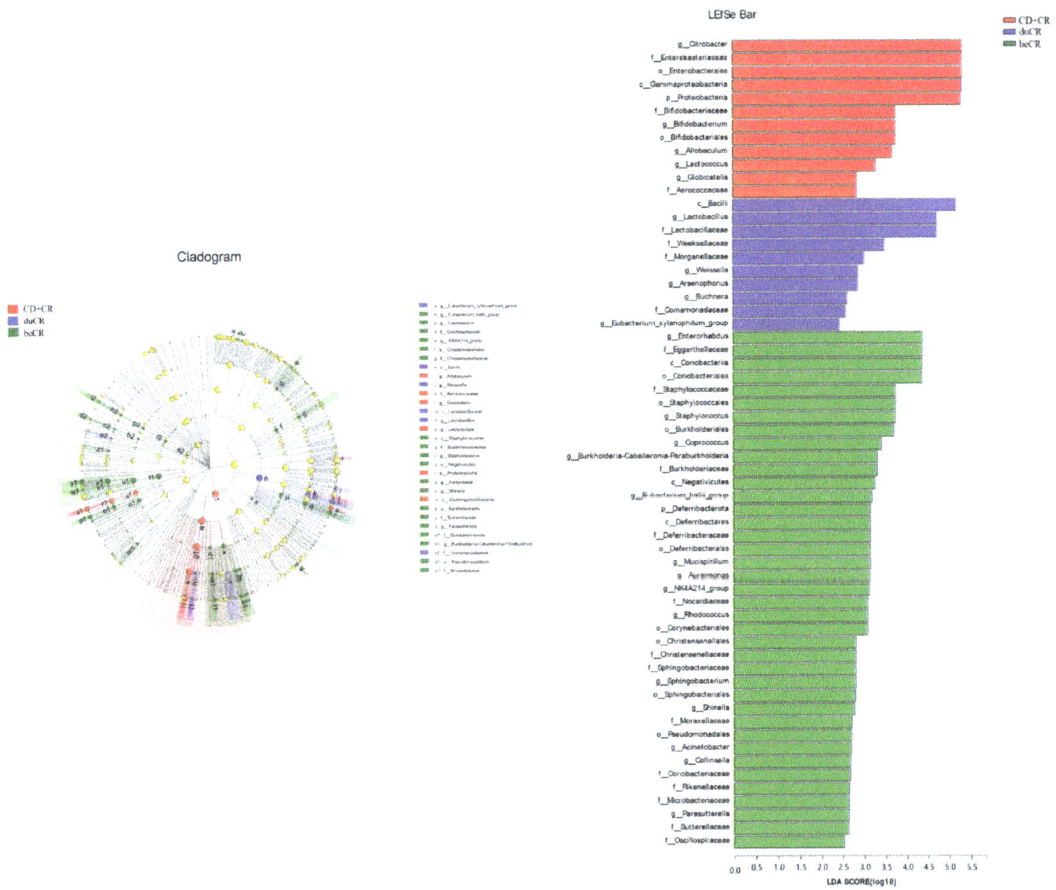

(**A**) (**B**)

Figure 7. Effect of the beCR and duCR intervention on key microbial phylotypes. (**A**) Cladogram and (**B**) LDA scores derived from LEfSe analysis.

4. Discussion

The implications of rising global IBD prevalence and incidence for human public health are fundamental and challenging [2,3]. Thus, it is desperate for a concerted effort in disease prevention and health-service innovation to reduce the global burden of IBD. Accumulating evidence has demonstrated that dietary intervention, a primary element in regulating gut microbiota, is effective in the clinical remission and therapy of IBD [6].

However, the mechanism and role of dietary intervention remain elusive. Li et al. found that BL as a dietary supplement significantly improved chemical-injury-induced colitis and colorectal cancer [22,24]; and Tian et al. found that barley leaf insoluble dietary fiber improved dextran sulfate sodium (DSS)-induced colitis by modulating intestinal flora [23]. Our previous study suggested that BL could effectively ameliorate CR-induced colitis (unpublished), and to further elaborate on how BL exerts its ameliorative effect, in this experiment, we supplemented BL before and during CR infection in mice, respectively. Our experimental results indicated that the effect of BL supplementation before infection on the CR-induced colitis was superior to that of BL supplementation upon infection, and this improvement was comparable to that of BL supplementation throughout, suggesting that BL mitigates CR-induced colitis to a large extent in a preventive manner. The dosage of BL utilized in this study is a human equivalent dose that has been reported without observed adverse effects [26]. Our study highlights the potential use of BL as a nutritional supplement that is suitable for the clinical management of IBD in humans.

CR is a mouse natural A/E bacterium that causes severe colitis when infected with C3H/HeN mice [13]. Consistent with previously reported studies [14], CR resulted in reduced body weight, increased DAI index, diarrhea, and a shortened colon in mice. Our results indicated that BL supplementation during infection was not effective in improving CR-induced colitis symptoms, whereas BL supplementation before infection could significantly ameliorate CR-induced colitis and produced similar ameliorative effects to those of BL supplementation throughout (Figure 1B–E). CR infestation achieves extraintestinal transmission through the lymphatic circulation, resulting in splenomegaly. Our results suggested that supplementation with BL before CR infestation significantly inhibited the extraintestinal spread of CR, which might be related to its ability to reduce the CR burden in mice (Figure 4A–C). Therefore, dietary supplementation with BL prior to CR infestation significantly alleviated the CR-caused splenomegaly (Figure 1F). Pathogenic bacteria or exogenous agents would induce a host immune response, so we detected the content of inflammatory cytokines in mouse colonic tissue. The results showed that pre-infestation dietary supplementation with BL significantly suppressed the elevation of pro-inflammatory cytokines TNF-α, IL-1β, and IFN-γ caused by CR, suggesting that BL supplementation before CR infestation improved splenomegaly associated with a decrease in circulating pro-inflammatory cytokines (Figure 3). In contrast, supplementation with BL during CR infestation was less effective in improving colitis caused by CR than supplementation with BL before infection.

CR colonization of the intestine could trigger a series of intestinal injuries. TMCH, the hallmark feature of CR-infected mice, is a consequence in which the CR utilizes its type III secretion system to inject virulence factors into enterocytes, resulting in an over-proliferation of immaturely differentiated cells, and is characterized by crypt hyperplasia, apical enterocyte surface heterogeneity, and mucosal thickening [13]. In addition, CR can destroy host mitochondria, thus increasing the oxygen content in the intestinal lumen and accelerating its proliferation to compete with the commensal intestinal flora [27]. CR initially colonizes the cecum patch to adapt to the in vivo environment and then migrates to the distal colon. Our results indicated that pretreatment with BL significantly ameliorated CR-induced colonic crypt hyperplasia, which was associated with a reduction in the excessive proliferation of Ki67 cells (Figure 2A,E). In addition, dietary supplementation of BL prior to infestation also ameliorated CR-induced colonic and cecal inflammatory cell infiltration, ulceration, and loss of epithelial integrity (Figure 2A–C and Supplementary Figure S2). The goblet cell, a mucus-producing cell, is an important component of the intestinal mucosal barrier. Previous studies have found that Muc2 synthesis is essential for host protection during A/E bacterial infection because it reduces the overall number of pathogens and symbionts associated with the colonic mucosal surface [28]. Our data revealed that dietary supplementation of BL prior to infestation was effective in ameliorating the reduction of goblet cell numbers caused by CR (Figure 2A,D), and this partially explained why it suppressed CR load in feces and other tissues of mice (Figure 3). However, the improvement of intestinal histopathology by BL supplementation at the time of CR infestation was far inferior to that of BL supplementation before infestation (Figure 2).

It has been observed that CR causes a profound gut microbiota dysbiosis which is defined by an overgrowth of CR and a decline in the richness and general diversity of the resident microbiota [14]. An alpha and beta diversity analysis of the bacterial community showed that pre-infection dietary supplementation of BL significantly ameliorated the reduced richness and diversity of the intestinal flora and improved gut microbiota dysbiosis with similar flora composition and structure to that produced by feeding BL throughout the experiment (Figure 5). Multiple studies have demonstrated an elevated proportion of Proteobacteria in the gut microbiota of IBD patients [29]. Proteobacteria exploits the host's immune system to facilitate the recruitment of pro-inflammatory cells, leads to dysbiosis, and ultimately promotes the progression of IBD [29]. Our data suggested that the dietary supplementation of BL prior to infestation could significantly suppress the relative abundance of Proteobacteria, while increasing the content of Firmicutes (Figure 6A,B). At the genus level, pretreatment with BL supplementation significantly suppressed CR levels, and this result is consistent with the results of reducing the CR burden in feces (Figures 4A and 6C,E). Interestingly, dietary supplementation with BL, either before or during infestation, could enhance the relative abundance of *Lactobacillus* (Figure 6C–E), indicating that BL might act as a prebiotic to enrich *Lactobacillus*. Numerous studies have demonstrated that *Lactobacillus* could improve CR-induced colitis through competitive inhibition, replacement rejection, and moderate modulation of the Wnt/β-catenin signaling pathway to avoid over-activation [30]. Using in vivo fluorescence imaging experiments, Waki et al. found that single feeding of *Lactobacillus brevis KB290* could reduce the susceptibility of mice to CR infection [31]. In another study, pretreatment of C57BL/6 mice with *Lactobacillus rhamnosus GG* (LGG) improved CR-induced colitis in a TLR2-dependent manner [32]. *Lactobacillus acidophilus* could reduce the diarrheal phenotype of CR infection in mice [33], and early pre-inoculation with *L. acidophilus* could enhance host defense against CR infection and attenuate colitis [34]. Apart from *Lactobacillus*, dietary supplementation with BL prior to infestation also elevated the relative abundance of *Lachnospiraceae* and *Enterorhabdus* (Figure 6C). *Lachnospiraceae* is known to perform an essential function in maintaining intestinal health and protecting its hosts. Desen et al. found that angiogenin could mitigate the severity of colitis by modulating *Lachnospiraceae* and α-Proteobacteria [35]. A recent clinical trial has shown that Crohn's disease patients exhibit a strong adaptive immune response to human-derived *Lachnospiraceae* flagellin, suggesting that *Lachnospiraceae* might be a target for prognosis and future personalized therapy [36]. Another study showed that mice with resected cecum tissue, resulting in a reduced level of *Lachnospiraceae*, exacerbated CR-induced colitis [37]. Noticeably, dietary supplementation with BL prior to infestation suppressed the relative abundance of *Erysipelatoclostridium* besides CR content (Figure 6C). *Erysipelatoclostridium* is a Gram-positive bacterium that causes invasive infections in various tissues, especially in immunocompromised individuals [38]. Manabu et al. found that *Erysipelatoclostridium* enhanced Th1 response after colonization of germ-free mice, leading to colonic mucosal damage in mice [39]. However, the effect of BL supplementation during infestation on the regulation of CR-induced intestinal flora dysbiosis was not as good as that of BL supplementation before infestation. BL powder, rich in dietary fiber, could increase the content of short-chain fatty acids and aromatic metabolites ferulic acid (FA) in the intestinal tract [40]. Hwang et al. found that FA could be used as an intestinal protective agent to improve the intestinal epithelial barrier environment [41]. Moreover, it was shown that the intraperitoneal administration of butyric acid significantly improved CR-caused colitis [42], and Wang et al. found that butyric acid improved intestinal immune homeostasis by enhancing IL-22 production [43]. Niacin, a fermentation product of intestinal flora, ameliorates colitis and colorectal cancer in a Gpr109a-dependent manner [44]. Moreover, our previous studies have demonstrated that the intestinal flora metabolites of BL, namely inosine and secondary bile acids, also could ameliorate DSS-induced colitis [23,24]. Thus, more research is needed to identify the individual *Lactobacillus* species and metabolite that contribute to the protective benefits of BL against CR-induced enteric infection and to examine the underlying mechanisms.

5. Conclusions

In this experiment, our results revealed that pretreatment with BL could dramatically improve CR-induced body weight loss and attenuate CR-induced intestinal damage; it also significantly reduced host CR burden and improved CR-induced gut microbiota disorders (Figure 8). However, the effect of BL supplementation during infestation on the improvement of CR-induced colitis was not as good as that of pretreatment with BL. In conclusion, BL improvement of CR-caused colitis might be largely in a preventive manner. Our works provide a reference for the therapy and prevention of IBD.

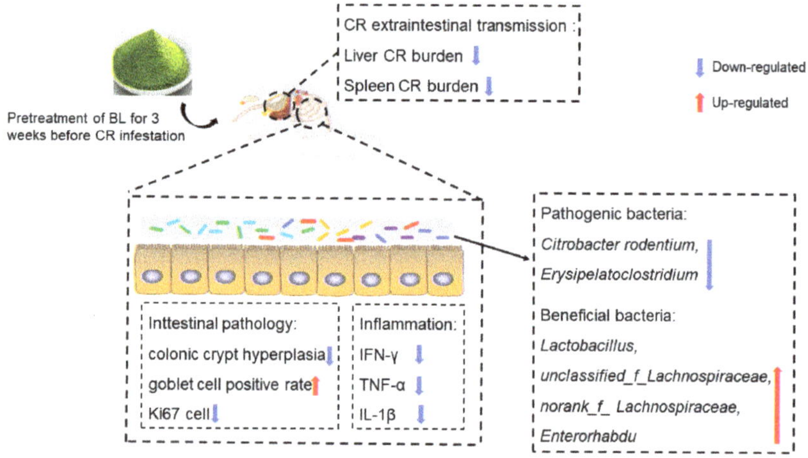

Figure 8. Schematic diagram of BLs' role in improving CR-causing colitis through a preventive approach.

Supplementary Materials: The following supporting information can be downloaded at https://www.mdpi.com/article/10.3390/nu14183833/s1. Figure S1: Daily food consumption (A) and water intake (B) of each group. Figure S2: (A) Representative H&E staining images of cecum tissues. Scale bar, 100 μm. (B) Histopathology score of cecal section. Lowercase a, b, and c mean in the same bar without a common letter differ at $p < 0.05$. Figure S3: Taxonomic distributions of intestinal flora at the genus level. Table S1: The nutritional composition of barley leaf. Table S2: The macronutrient composition of chow diet (CD) and an isocaloric barley leaf (BL)-supplemented diet.

Author Contributions: Conceptualization, Y.F., C.M. and F.C.; methodology, Y.F., D.L. and F.C.; software, Y.F., C.M. and F.C.; validation, Y.F. and F.C.; formal analysis, Y.F. and F.C.; investigation, Y.F. and M.T.; resources, F.C.; data curation, F.C.; writing—original draft preparation, Y.F.; writing—review and editing, Y.F. and F.C.; visualization, Y.F. and F.C.; project administration, X.H. and F.C. All authors have read and agreed to the published version of the manuscript.

Funding: This work was supported by the National Natural Science Foundation of China (grant number 32001677), and the China Postdoctoral Science Foundation (grant number 2020M680256).

Institutional Review Board Statement: Animal experiments were conducted following the National Institutes of Health guide for the care and use of Laboratory Animals (NIH Publications No. 8023, revised 1978), and the protocols were reviewed and approved by the Animal Care and Ethics Committee of China Agricultural University (Ethics reference number: AW32602202-4-2).

Informed Consent Statement: Not applicable.

Acknowledgments: We appreciate the technical assistance provided by Majorbio Bio-Pharm Technology Co., Ltd.

Conflicts of Interest: The authors declare no conflict of interest.

References

1. Seyed Tabib, N.S.; Madgwick, M.; Sudhakar, P.; Verstockt, B.; Korcsmaros, T.; Vermeire, S. Big data in IBD: Big progress for clinical practice. *Gut* **2020**, *69*, 1520–1532. [CrossRef]
2. Kotze, P.G.; Underwood, F.E.; Damião, A.O.M.C.; Ferraz, J.G.P.; Saad-Hossne, R.; Toro, M.; Iade, B.; Bosques-Padilla, F.; Teixeira, F.V.; Juliao-Banos, F.; et al. Progression of Inflammatory Bowel Diseases Throughout Latin America and the Caribbean: A Systematic Review. *Clin. Gastroenterol. Hepatol.* **2020**, *18*, 304–312. [CrossRef]
3. Ye, Y.; Manne, S.; Treem, W.R.; Bennett, D. Prevalence of Inflammatory Bowel Disease in Pediatric and Adult Populations: Recent Estimates from Large National Databases in the United States, 2007–2016. *Inflamm. Bowel Dis.* **2020**, *26*, 619–625. [CrossRef]
4. Bruscoli, S.; Febo, M.; Riccardi, C.; Migliorati, G. Glucocorticoid Therapy in Inflammatory Bowel Disease: Mechanisms and Clinical Practice. *Front. Immunol.* **2021**, *12*, 691480. [CrossRef]
5. Zuo, T.; Ng, S.C. The Gut Microbiota in the Pathogenesis and Therapeutics of Inflammatory bowel disease. *Front. Microbiol.* **2018**, *9*, 2247. [CrossRef]
6. Mentella, M.C.; Scaldaferri, F.; Pizzoferrato, M.; Gasbarrini, A.; Miggiano, G.A.D. Nutrition, IBD and Gut Microbiota: A Review. *Nutrients* **2020**, *12*, 944. [CrossRef]
7. Pickard, J.M.; Zeng, M.Y.; Caruso, R.; Núñez, G. Gut microbiota: Role in pathogen colonization, immune responses, and inflammatory disease. *Immunol. Rev.* **2017**, *279*, 70–89. [CrossRef]
8. Montrose, D.C.; Nishiguchi, R.; Basu, S.; Staab, H.A.; Zhou, X.K.; Wang, H.; Meng, L.; Johncilla, M.; Cubillos-Ruiz, J.R.; Morales, D.K.; et al. Dietary Fructose Alters the Composition, Localization, and Metabolism of Gut Microbiota in Association With Worsening Colitis. *Cell. Mol. Gastroenterol. Hepatol.* **2021**, *11*, 525–550. [CrossRef]
9. Makki, K.; Deehan, E.C.; Walter, J.; Bäckhed, F. The Impact of Dietary Fiber on Gut Microbiota in Host Health and Disease. *Cell Host Microbe* **2018**, *23*, 705–715. [CrossRef]
10. Ni, J.; Wu, G.D.; Albenberg, L.; Tomov, V.T. Gut microbiota and IBD: Causation or correlation? *Nat. Rev. Gastroenterol. Hepatol.* **2017**, *14*, 573–584. [CrossRef]
11. Ferrer-Picón, E.; Dotti, I.; Corraliza, A.M.; Mayorgas, A.; Esteller, M.; Perales, J.C.; Ricart, E.; Masamunt, M.C.; Carrasco, A.; Tristán, E.; et al. Intestinal Inflammation Modulates the Epithelial Response to Butyrate in Patients with Inflammatory Bowel Disease. *Inflamm. Bowel Dis.* **2020**, *26*, 43–55. [CrossRef] [PubMed]
12. Kelly, C.R.; Khoruts, A.; Staley, C.; Sadowsky, M.J.; Abd, M.; Alani, M.; Bakow, B.; Curran, P.; McKenney, J.; Tisch, A.; et al. Effect of fecal microbiota transplantation on recurrence in multiply recurrent clostridium difficile infection a randomized trial. *Ann. Intern. Med.* **2016**, *165*, 609–616. [CrossRef] [PubMed]
13. Crepin, V.F.; Collins, J.W.; Habibzay, M.; Frankel, G. Citrobacter rodentium mouse model of bacterial infection. *Nat. Protoc.* **2016**, *11*, 1851–1876. [CrossRef]
14. Collins, J.W.; Keeney, K.M.; Crepin, V.F.; Rathinam, V.A.K.; Fitzgerald, K.A.; Finlay, B.B.; Frankel, G. Citrobacter rodentium: Infection, inflammation and the microbiota. *Nat. Rev. Microbiol.* **2014**, *12*, 612–623. [CrossRef]
15. Wiles, S.; Clare, S.; Harker, J.; Huett, A.; Young, D.; Dougan, G.; Frankel, G. Organ specificity, colonization and clearance dynamics in vivo following oral challenges with the murine pathogen Citrobacter rodentium. *Cell. Microbiol.* **2004**, *6*, 963–972. [CrossRef]
16. Kamada, N.; Kim, Y.G.; Sham, H.P.; Vallance, B.A.; Puente, J.L.; Martens, E.C.; Núñez, G. Regulated virulence controls the ability of a pathogen to compete with the gut microbiota. *Science* **2012**, *336*, 1325–1329. [CrossRef]
17. Ferreres, F.; Kršková, Z.; Gonçalves, R.F.; Valentão, P.; Pereira, J.A.; Dušek, J.; Martin, J.; Andrade, P.B. Free water-soluble phenolics profiling in barley (*Hordeum vulgare* L.). *J. Agric. Food Chem.* **2009**, *57*, 2405–2409. [CrossRef] [PubMed]
18. Hwang, Y.-H.; Ha, H.; Kim, R.; Cho, C.-W.; Song, Y.-R.; Hong, H.-D.; Kim, T. Protective effects of a polysaccharide BLE0 isolated from barley leaf on bone loss in ovariectomized mice. *Int. J. Biol. Macromol.* **2019**, *123*, 314–321. [CrossRef]
19. Yu, Y.M.; Chang, W.C.; Chang, C.T.; Hsieh, C.L.; Tsai, C.E. Effects of young barley leaf extract and antioxidative vitamins on LDL oxidation and free radical scavenging activities in type 2 diabetes. *Diabetes Metab.* **2002**, *28*, 107–114.
20. Yamaura, K.; Shimada, M.; Fukata, H.; Nakayama, N.; Bi, Y.; Ueno, K. Antidepressant-like effects of young green barley leaf (*Hordeum vulgare* L.) in the mouse forced swimming test. *Pharmacogn. Res.* **2012**, *4*, 22–26. [CrossRef]
21. Kamiyama, M.; Shibamoto, T. Flavonoids with potent antioxidant activity found in young green barley leaves. *J. Agric. Food Chem.* **2012**, *60*, 6260–6267. [CrossRef] [PubMed]
22. Li, D.; Feng, Y.; Tian, M.; Hu, X.; Zheng, R.; Chen, F. Dietary barley leaf mitigates tumorigenesis in experimental colitis-associated colorectal cancer. *Nutrients* **2021**, *13*, 3487. [CrossRef] [PubMed]
23. Tian, M.; Li, D.; Ma, C.; Feng, Y.; Hu, X.; Chen, F. Barley leaf insoluble dietary fiber alleviated dextran sulfate sodium-induced mice colitis by modulating gut microbiota. *Nutrients* **2021**, *13*, 846. [CrossRef]
24. Li, D.; Feng, Y.; Tian, M.; Ji, J.; Hu, X.; Chen, F. Gut microbiota-derived inosine from dietary barley leaf supplementation attenuates colitis through PPARγ signaling activation. *Microbiome* **2021**, *9*, 83. [CrossRef]

25. Zhang, Y.; Jiang, D.; Jin, Y.; Jia, H.; Yang, Y.; Kim, I.H.; Dai, Z.; Zhang, J.; Ren, F.; Wu, Z. Glycine Attenuates Citrobacter rodentium-Induced Colitis by Regulating ATF6-Mediated Endoplasmic Reticulum Stress in Mice. *Mol. Nutr. Food Res.* **2021**, *65*, 2001065. [CrossRef]
26. Yu, Y.M.; Chang, W.C.; Liu, C.S.; Tsai, C.M. Effect of young barley leaf extract and adlay on plasma lipids and LDL oxidation in hyperlipidemic smokers. *Biol. Pharm. Bull.* **2004**, *27*, 802–805. [CrossRef]
27. Lopez, C.A.; Miller, B.M.; Rivera-Chávez, F.; Velazquez, E.M.; Byndloss, M.X.; Chávez-Arroyo, A.; Lokken, K.L.; Tsolis, R.M.; Winter, S.E.; Bäumler, A.J. Virulence factors enhance Citrobacter rodentium expansion through aerobic respiration. *Science* **2016**, *353*, 1249–1253. [CrossRef]
28. Bergstrom, K.S.B.; Kissoon-Singh, V.; Gibson, D.L.; Ma, C.; Montero, M.; Sham, H.P.; Ryz, N.; Huang, T.; Velcich, A.; Finlay, B.B.; et al. Muc2 protects against lethal infectious colitis by disassociating pathogenic and commensal bacteria from the colonic mucosa. *PLoS Pathog.* **2010**, *6*, e1000902. [CrossRef]
29. Mukhopadhya, I.; Hansen, R.; El-Omar, E.M.; Hold, G.L. IBD—What role do Proteobacteria play? *Nat. Rev. Gastroenterol. Hepatol.* **2012**, *9*, 219–230. [CrossRef]
30. Wu, H.; Xie, S.; Miao, J.; Li, Y.; Wang, Z.; Wang, M.; Yu, Q. Lactobacillus reuteri maintains intestinal epithelial regeneration and repairs damaged intestinal mucosa. *Gut Microbes* **2020**, *11*, 997–1014. [CrossRef]
31. Waki, N.; Kuwabara, Y.; Yoshikawa, Y.; Suganuma, H.; Koide, H.; Oku, N.; Ohashi, N. Amelioration of Citrobacter rodentium proliferation in early stage of infection in mice by pre-treatment with lactobacillus brevis KB290 and verification using in vivo bioluminescence imaging. *FEMS Microbiol. Lett.* **2017**, *364*. [CrossRef] [PubMed]
32. Ryu, S.H.; Park, J.H.; Choi, S.Y.; Jeon, H.Y.; Park, J.-I.; Kim, J.Y.; Ham, S.H.; Choi, Y.K. The probiotic Lactobacillus prevents citrobacter rodentium-induced murine colitis in a TLR2-dependent manner. *J. Microbiol. Biotechnol.* **2016**, *26*, 1333–1340. [CrossRef] [PubMed]
33. Kumar, A.; Anbazhagan, A.N.; Coffing, H.; Chatterjee, I.; Priyamvada, S.; Gujral, T.; Saksena, S.; Gill, R.K.; Alrefai, W.A.; Borthakur, A.; et al. Lactobacillus acidophilus counteracts inhibition of NHE3 and DRA expression and alleviates diarrheal phenotype in mice infected with Citrobacter rodentium. *Am. J. Physiol.-Gastrointest. Liver Physiol.* **2016**, *311*, G817–G826. [CrossRef] [PubMed]
34. Chen, C.-C.; Louie, S.; Shi, H.N.; Walker, W.A. Preinoculation with the probiotic Lactobacillus acidophilus early in life effectively inhibits murine Citrobacter rodentium colitis. *Pediatr. Res.* **2005**, *58*, 1185–1191. [CrossRef] [PubMed]
35. Sun, D.; Bai, R.; Zhou, W.; Yao, Z.; Liu, Y.; Tang, S.; Ge, X.; Luo, L.; Luo, C.; Hu, G.F.; et al. Angiogenin maintains gut microbe homeostasis by balancing α-Proteobacteria and Lachnospiraceae. *Gut* **2021**, *70*, 666–676. [CrossRef] [PubMed]
36. Alexander, K.L.; Zhao, Q.; Reif, M.; Rosenberg, A.F.; Mannon, P.J.; Duck, L.W.; Elson, C.O. Human Microbiota Flagellins Drive Adaptive Immune Responses in Crohn's Disease. *Gastroenterology* **2021**, *161*, 522–535. [CrossRef]
37. Brown, K.; Abbott, D.W.; Uwiera, R.R.E.; Inglis, G.D. Removal of the cecum affects intestinal fermentation, enteric bacterial community structure, and acute colitis in mice. *Gut Microbes* **2018**, *9*, 218–235. [CrossRef]
38. Milosavljevic, M.N.; Kostic, M.; Milovanovic, J.; Zaric, R.Z.; Stojadinovic, M.; Jankovic, S.M.; Stefanovic, S.M. Antimicrobial treatment of erysipelatoclostridium ramosum invasive infections: A systematic review. *Rev. Inst. Med. Trop. Sao Paulo* **2021**, *63*, e30. [CrossRef]
39. Nagayama, M.; Yano, T.; Atarashi, K.; Tanoue, T.; Sekiya, M.; Kobayashi, Y.; Sakamoto, H.; Miura, K.; Sunada, K.; Kawaguchi, T.; et al. TH1 cell-inducing Escherichia coli strain identified from the small intestinal mucosa of patients with Crohn's disease. *Gut Microbes* **2020**, *12*, 1788898. [CrossRef]
40. Li, D.; Wang, P.; Wang, P.; Hu, X.; Chen, F. Gut microbiota promotes production of aromatic metabolites through degradation of barley leaf fiber. *J. Nutr. Biochem.* **2018**, *58*, 49–58. [CrossRef]
41. Hwang, H.; Lee, S.R.; Yoon, J.; Moon, H.; Zhang, J.; Park, E.; Yoon, S.; Cho, J.A. Ferulic Acid as a Protective Antioxidant of Human Intestinal Epithelial Cells. *Antioxidants* **2022**, *11*, 1448. [CrossRef] [PubMed]
42. Jiminez, J.A.; Uwiera, T.C.; Abbott, D.W.; Uwiera, R.R.E.; Inglis, G.D. Butyrate Supplementation at High Concentrations Alters Enteric Bacterial Communities and Reduces Intestinal Inflammation in Mice Infected with Citrobacter rodentium. *mSphere* **2017**, *2*, e00243-17. [CrossRef] [PubMed]
43. Yang, W.; Yu, T.; Huang, X.; Bilotta, A.J.; Xu, L.; Lu, Y.; Sun, J.; Pan, F.; Zhou, J.; Zhang, W.; et al. Intestinal microbiota-derived short-chain fatty acids regulation of immune cell IL-22 production and gut immunity. *Nat. Commun.* **2020**, *11*, 4457. [CrossRef] [PubMed]
44. Singh, N.; Gurav, A.; Sivaprakasam, S.; Brady, E.; Padia, R.; Shi, H.; Thangaraju, M.; Prasad, P.D.; Manicassamy, S.; Munn, D.H.; et al. Activation of Gpr109a, receptor for niacin and the commensal metabolite butyrate, suppresses colonic inflammation and carcinogenesis. *Immunity* **2014**, *40*, 128–139. [CrossRef]

nutrients

MDPI

Article

Secondary Metabolites in the *Dendrobium heterocarpum* Methanolic Extract and Their Impacts on Viability and Lipid Storage of 3T3-L1 Pre-Adipocytes

Sakan Warinhomhoun [1,2,3], Hnin Ei Ei Khine [4], Boonchoo Sritularak [3,5], Kittisak Likhitwitayawuid [3], Tomofumi Miyamoto [6], Chiaki Tanaka [6,7], Chuchard Punsawad [1], Yanyong Punpreuk [8], Rungroch Sungthong [4] and Chatchai Chaotham [4,9,*]

1 School of Medicine, Walailak University, Nakhon Si Thammarat 80160, Thailand; sakan.cu@gmail.com (S.W.); chuchard.pu@wu.ac.th (C.P.)
2 Center of Excellence in Marijuana, Hemp, and Kratom, Walailak University, Nakhon Si Thammarat 80160, Thailand
3 Department of Pharmacognosy and Pharmaceutical Botany, Faculty of Pharmaceutical Sciences, Chulalongkorn University, Bangkok 10330, Thailand; boonchoo.sr@chula.ac.th (B.S.); kittisak.l@chula.ac.th (K.L.)
4 Department of Biochemistry and Microbiology, Faculty of Pharmaceutical Sciences, Chulalongkorn University, Bangkok 10330, Thailand; hnineieikhine12@gmail.com (H.E.E.K.); rungroch.s@chula.ac.th (R.S.)
5 Natural Products for Ageing and Chronic Diseases Research Unit, Faculty of Pharmaceutical Sciences, Chulalongkorn University, Bangkok 10330, Thailand
6 Graduate School of Pharmaceutical Sciences, Kyushu University, Fukuoka 812-8582, Japan; miyamoto@phar.kyushu-u.ac.jp (T.M.); ctanaka@wakayama-med.ac.jp (C.T.)
7 School of Pharmaceutical Sciences, Wakayama Medical University, Wakayama 640-8156, Japan
8 Department of Agriculture, Kasetsart University, Bangkok 10900, Thailand; cyyp01@hotmail.co.th
9 Preclinical Toxicity and Efficacy Assessment of Medicines and Chemicals Research Unit, Faculty of Pharmaceutical Sciences, Chulalongkorn University, Bangkok 10330, Thailand
* Correspondence: cchoatham@gmail.com

Citation: Warinhomhoun, S.; Khine, H.E.E.; Sritularak, B.; Likhitwitayawuid, K.; Miyamoto, T.; Tanaka, C.; Punsawad, C.; Punpreuk, Y.; Sungthong, R.; Chaotham, C. Secondary Metabolites in the *Dendrobium heterocarpum* Methanolic Extract and Their Impacts on Viability and Lipid Storage of 3T3-L1 Pre-Adipocytes. *Nutrients* **2022**, *14*, 2886. https://doi.org/10.3390/nu14142886

Academic Editors: Cristian Del Bo' and Susanne Klaus

Received: 11 May 2022
Accepted: 12 July 2022
Published: 14 July 2022

Abstract: Although many natural products have proven their potential to regulate obesity through the modulation of adipocyte biology, none of them has yet been approved for clinical use in obesity therapy. This work aims to isolate valuable secondary metabolites from an orchid species (*Dendrobium heterocarpum*) and evaluate their possible roles in the growth and differentiation of 3T3-L1 pre-adipocytes. Six compounds were isolated from the orchid's methanolic extracts and identified as amoenylin (**1**), methyl 3-(4-hydroxyphenyl) propionate (**2**), 3,4-dihydroxy-5,4′-dimethoxybibenzyl (**3**), dendrocandin B (**4**), dendrofalconerol A (**5**), and syringaresinol (**6**). Among these phytochemicals, compounds **2**, **3**, and **6** exhibited lower effects on the viability of 3T3-L1 cells, offering non-cytotoxic concentrations of $\lesssim 10$ μM. Compared to others tested, compound **3** was responsible for the maximum reduction of lipid storage in 3T3-L1 adipocytes ($IC_{50} = 6.30 \pm 0.10$ μM). A set of protein expression studies unveiled that compound **3** at non-cytotoxic doses could suppress the expression of some key transcription factors in adipocyte differentiation (i.e., PPARγ and C/EBPα). Furthermore, this compound could deactivate some proteins involved in the MAPK pathways (i.e., JNK, ERK, and p38). Our findings prove that *D. heterocarpum* is a promising source to explore bioactive molecules capable of modulating adipocytic growth and development, which can potentially be assessed and innovated further as pharmaceutical products to defeat obesity.

Keywords: *Dendrobium heterocarpum*; phytochemical; bibenzyl; adipocyte differentiation; obesity

1. Introduction

As a critical risk factor that correlates to various pathological conditions and chronic diseases, obesity turns out to be a public health concern worldwide. In 2020, the World Health Organization reported that 39% and 13% of adults (age > 18 years) were overweight and obese, respectively [1]. Irregular adipose tissue formation due to excess growth and development of adipocytic cells is a sign of obesity. This uneven biological process does not induce obesity only but also consequently provokes other diseases such as cancers, diabetes mellitus, cardiovascular diseases, chronic kidney diseases, and metabolic syndrome [2–4]. Therefore, modulating adipocytic growth and development by pharmaceutical products has become a considerable research hotspot for regulating obesity in these recent years [5].

Adipogenesis is a complex multistep process to convert pluripotent stem cells to mature adipocytes [6]. One of the critical steps in adipogenesis is the cellular lipogenesis triggered by some transcription factors (e.g., peroxisome proliferators activated receptor γ (PPARγ) and CCAAT/enhancer-binding protein α (C/EBPα)) during adipocyte differentiation [7]. Repressing the expression of these transcription factors can disrupt adipocyte differentiation and cellular lipid accumulation [8]. The other molecular events that occur instinctively in cells (including adipocytic cells) are the mitogen-activated protein kinase (MAPK) signaling pathways that involve several kinase cascades, mainly c-Jun N-terminal kinase (JNK), extracellular signal-regulated kinase (ERK), and stress-activated protein kinase (p38) [9,10]. Deactivation of these upstream regulators has been proposed to have some potential links to the suppressed adipocyte differentiation [11], which offers a promising molecular mechanism to modulate obesity.

Although some available anti-obesity drugs have been revealed for their success in weight loss, they still pose various adverse effects, such as cardiovascular toxicity, hallucinations, headache, anxiety, and hepatic toxicity [11]. In recent years, natural compounds with a high safety profile, particularly those derived from plants, have gained more attention due to their inhibitory activity in adipocyte differentiation and obesity development [12,13]. However, none of the discovered natural products have been approved for clinical use in obesity therapy. Moreover, only a few percentages of these compounds have been unveiled for their suppressive function in adipocyte differentiation and anti-obesity potential.

Dendrobium is one of the largest genera in the family *Orchidaceae*, consisting of approximately 1400 member species. More than 1100 species are predominant in Asia and Australia, of which approximately 150 species are omnipresent in Thailand [14,15]. *Dendrobium* plants are great sources for anti-diabetes mellitus [16–26], but their roles in the differentiation and cellular response of adipocytic cells are still scarce [20,26,27]. *Dendrobium heterocarpum* Wall. ex. Lindl, so-called "Ueang Si Tan (in Thai)", is another prevalent *Dendrobium* species in Thailand [28], while its phytoconstituents and pharmaceutical properties are still waiting to explore.

In this study, we aimed to profile secondary metabolites in the methanolic extracts derived from the whole *D. heterocarpum* plants and assess how the identified compounds exert their roles in the viability and differentiation of the mouse embryonic 3T3-L1 preadipocytes. The biological impacts of such phytochemicals on cellular lipid storage and some underlying molecular mechanisms potentially linked to obesity incidence and therapy were elucidated and discussed.

2. Materials and Methods

2.1. Plant Materials

The whole *D. heterocarpum* plant was purchased from Chatuchak market in Bangkok, Thailand (June 2019). The plant specimen was identified and confirmed by a curator, Mr. Yanyong Punpreuk, using the database available at the Botanical Garden Organization. A specimen voucher (BS-Dhet-012562) was generated and deposited at the department of Pharmacognosy and Pharmaceutical Botany, Faculty of Pharmaceutical Sciences, Chulalongkorn University, Thailand.

2.2. Extraction, Fractionation, and Phytochemical Elucidation

The dried plant material (3.6 kg) was extracted with MeOH (4 $\times$ 15 L), and the methanolic extract (300.6 g) was partitioned in a mixture of H_2O, EtOAc, and *n*-butanol to give aqueous (105.24 g), ethyl acetate (126.59 g), and *n*-butanol (112.78 g) extracts. The ethyl acetate extract was partitioned in hexane-20% H_2O/MeOH to obtain hexane (11.0 g) and methanolic (16.5 g) extracts. The methanolic extract was subsequently fractioned through several rounds of column chromatography and/or high-performance liquid chromatography (HPLC). The column chromatography was conducted with 70–320 μm of silica gel 60 (Merck Kieselgel 60, Darmstadt, Germany), 230–400 μm of silica gel 60 (Merck Kieselgel 60, Darmstadt, Germany), 40–63 μm of C-18 (Merck Kieselgel 60 RP-18, Darmstadt, Germany), or 25–100 μm of Sephadex LH-20 (GE Healthcare, Göteborg, Sweden). The semi-preparative HPLC was performed using a Shimadzu LC-20 AD system (Shimadzu, Kyoto, Japan) equipped with an autosampler (SIL-20 AC), a column oven (CTO-20A), a photodiode array detector (SPD-M20A), and liquid chromatography with a COSMOSIL C_{18} column (4.6 × 150 mm, 5 μm) (Nacalai Tesque Inc., Kyoto, Japan). For chemical structure elucidation, mass spectra were recorded on an electrospray ionization-quadrupole-quadrupole-time-of-flight-mass spectrometer (Bruker ESI-QqTOF-MS, Manchester, UK). ^{1}H nuclear magnetic resonance (NMR) (600 MHz) and ^{13}C NMR (125 MHz) spectra were recorded on an NMR spectrometer (Bruker Avance III^{TM} HD 600 MHz, Graduate School of Pharmaceutical Sciences, Kyushu University).

The methanolic extract was fractioned through a Sephadex LH-20 column and eluted with acetone to give six fractions (A–F). Fraction A (2.57 g) was separated by reverse-phase column chromatography (C-18, gradient mixture of MeOH-H_2O) to yield six fractions (A1–A6). Fraction A3 (24.5 mg) was fractioned on the reverse-phase column chromatography to yield four fractions (A3.1–A3.4). Compound **1** (2.7 mg) was obtained from fraction A3.2. Fraction B (6.85 g) was fractioned by a Sephadex LH-20 column and eluted with MeOH to give eight fractions (B1-B8). Under the same fractionation system, fraction B1 (2.04 g) was divided into three fractions, in which fraction B1.2 (1.81 g) was further purified on a semi-preparative HPLC column chromatography (C-18, gradient mixture of MeOH-H_2O) to gain compound **2** (6.7 mg). Fractions B3 (627.1 mg), B4 (817.5 mg), B5 (500.9 mg), and B6 (126.7 mg) were individually separated by a Sephadex LH-20 column eluted with MeOH and further purified on a column chromatography (silica gel, gradient mixture of hexane-acetone) to obtain compounds **3** (144.0 mg), **4** (6.7 mg), **5** (9.2 mg), and **6** (3.0 mg), respectively.

2.3. Cell Culture and Adipocyte Differentiation

The mouse embryonic 3T3-L1 pre-adipocytes purchased from the American Type Culture Collection (ATCC, Manassas, VA, USA) were cultured at 37 °C with a 5% CO_2 stream in a complete Dulbecco's modified Eagle medium (DMEM) (Gibco, Gaithersburg, MA, USA) that contained 10% fetal bovine serum (FBS; Gibco, Gaithersburg, MA, USA), 1000 units mL^{-1} of penicillin/streptomycin (Gibco, Gaithersburg, MA, USA), and 2 mM of L-glutamine (Gibco, Gaithersburg, MA, USA). For adipocyte differentiation, 3T3-L1 cells at 100% confluent were incubated at 37 °C with a 5% CO_2 stream for two days in a differentiation medium made of the complete DMEM plus 0.5 mM isobutyl methylxanthine (IBMX) (Sigma Aldrich, St. Louis, MO, USA), 1 μM of dexamethasone (Sigma Aldrich, St. Louis, MO, USA), and 5 μg mL^{-1} of insulin (Sigma Aldrich, St. Louis, MO, USA). Then, the medium was replaced with an insulin medium made of the complete DMEM plus 5 μg mL^{-1} of insulin, and cells were continuously incubated under the same conditions for another two days. The medium was replaced with the complete DMEM every two days to maintain adipocyte development. Cellular lipid droplets developed along the differentiation program could be observed under a light microscope (Nikon Ts2, Tokyo, Japan).

2.4. Cytotoxicity Assessments

The cytotoxic effect of elucidated compounds from *D. heterocarpum* on 3T3-L1 cell viability was evaluated using 3-(4,5-dimethylthiazol-2-yl)-2,5-diphenyl tetrazolium bromide (MTT) (Sigma Aldrich, St. Louis, MO, USA) assay. 3T3-L1 cells were seeded in 96-well plates at a density of 2 × 10^3 cells/well and incubated at 37 °C with a 5% CO_2 stream until reaching 100% confluent. Cells were then exposed to varying concentrations (5, 10, and

20 μM) of the tested compounds in the complete DMEM for 48 h. Before testing, every tested compound resuspended in absolute dimethyl sulfoxide (DMSO) (Sigma Aldrich, St. Louis, MO, USA) was prepared by dilution using the complete DMEM. The concentration of DMSO in each treatment was limited to <0.5% (*v*/*v*). DMSO at 0.5% (*v*/*v*) and 20 μM of oxyresveratrol (a known compound with the ability to suppress adipocyte differentiation [29]) were included in the tests as untreated vehicle and positive controls, respectively. After exposure for 48 h, the medium was removed, and the MTT solution (0.45 mg mL^{-1}), prepared by dissolving in phosphate-buffered saline (PBS), was added. The assay was incubated at 37 °C with a 5% CO_2 stream for 3 h. Then, the MTT solution was replaced with absolute DMSO to dissolve the purple formazan crystal that formed, which was further measured for absorbance at a 570 nm wavelength using a microplate reader (Anthros, Durham, NC, USA). The relative absorbance value between treated cells to vehicle control cells was presented as the percentage of cell viability. Non-cytotoxic concentrations of a tested compound were defined based on no significant difference in the percentages of cell viability derived from the treatments and the vehicle control.

3T3-L1 cell death after exposure to varying concentrations of the tested compounds was also determined by the nuclear staining technique. Hoechst 33342 (Sigma Aldrich, St. Louis, MO, USA) and propidium iodide (PI) (Sigma Aldrich, St. Louis, MO, USA) were used as staining dyes. After 48 h of treatment, the assay medium was removed from 3T3-L1 cells and replaced with PBS containing 2 μg mL^{-1} of Hoechst 33342 and 1 μg mL^{-1} of PI. Cells were stained for 30 min at 37 °C with a 5% CO_2 stream and observed/photographed under a fluorescence microscope (Olympus IX51 with DP70, Tokyo, Japan). Bright cells with a disseminated nucleus under both blue and red filters were referred to as dead cells.

2.5. Quantification of Cellular Lipid Content

Oil Red O staining was performed to determine the content of lipid droplets in differentiating 3T3-L1 cells. The adipocyte differentiation program as described previously was established, and cells were exposed to each tested compound at 5 μM for 48 h after obtaining 100% confluence. Cells exposed to 0.5% (*v*/*v*) DMSO and 20 μM of oxyresveratrol served as untreated vehicle and positive controls, respectively. Tested compounds were diluted in the differentiation medium before testing, in which the concentrations of DMSO were restricted to be lower than 0.5% (*v*/*v*). After the treatments, differentiating 3T3-L1 cells on day 8 of the differentiation program were fixed with 10% formalin and incubated with the Oil Red O staining solution (Sigma Aldrich, St. Louis, MO, USA) at an ambient temperature for 1 h. Cells were then washed thrice with distilled water and 60% (*v*/*v*) isopropanol (Sigma Aldrich, St. Louis, MO, USA) to remove the excess dye. Oil Red O-stained 3T3-L1 cells were observed using a light microscope (Nikon Ts2, Tokyo, Japan). The cellular Oil Red O content was extracted using absolute isopropanol and measured for absorbance at a 510 nm wavelength by the microplate reader. The percentage of Oil Red O-stained cells was calculated after normalization with the total cellular protein content determined by a BCA protein assay kit (Thermo Scientific, Rockford, CA, USA).

2.6. Western Blot Analysis

The impacts of a tested compound on the expression and activation of proteins that are linked to cellular responses in the early phase of adipocyte differentiation were assessed using Western blot analysis. The list of proteins analyzed in this study included adipogenic regulators (i.e., PPARγ and C/EBPα) and the MAPK signaling pathways-mediated proteins (i.e., JNK, ERK, and p38). The expression or activation of each target protein was normalized with β-Actin as a reference cellular protein. The activation of JNK, ERK, and p38 was assessed using the phosphorylation ratio between those phosphorylated (i.e., p-ERK, p-JNK, and p-p38) and the sum of those phosphorylated and non-phosphorylated.

3T3-L1 pre-adipocytes at 100% confluence were exposed for 48 h to a tested compound at non-cytotoxic concentrations prepared in the differentiation medium. Cells were then washed with PBS (pH 7.4) and used for protein extraction by incubation in an ice-cold radioimmunoprecipitation assay buffer (Thermo Scientific, Rockford, CA, USA) supplemented with the protease inhibitor cocktail (Roche Applied Science, Indianapolis, IN, USA) for 45 min. Derived cell lysates were centrifuged at 4 °C, 6440× *g* for 15 min, and the supernatant was collected. The BCA protein assay kit was conducted to quantify the total protein content in the supernatant.

The same amount (30 μg) of protein from each treatment was loaded and separated onto 10% (*w*/*v*) sodium dodecyl sulfate-polyacrylamide (Bio-Lad Laboratories, Hercules, CA, USA) gel electrophoresis. The separated proteins were then transferred onto nitrocellulose membranes (Bio-Rad Laboratories, Hercules, CA, USA), which were blocked with 5% skim milk (Sigma Aldrich, St. Louis, MO, USA) in tris-buffered saline with 0.1% tween 20 (TBST) (pH 7.2). The membranes were then immunoblotted with specific primary antibodies, including PPARγ (1:1000), C/EBPα (1:1000), ERK (1:1000), p-ERK (1:1000), JNK (1:1000), p-JNK (1:1000), p38 (1:1000), p-p38 (1:1000), and β-Actin (1:1000) (Cell Signaling Technol ogy, Danvers, MA, USA) at 4 °C for overnight. After washing thrice (~7 min) with TBST, the membranes were incubated for 2 h with the specific horseradish peroxidase-linked secondary antibody. The target protein bands and their intensities on the membranes were visualized, captured, and quantified under UV light after adding chemiluminescent substrates (Bio-Rad Laboratories, Hercules, CA, USA), using Chemiluminescent ImageQuant LAS 4000 (GE Healthcare Bio-Sciences AB, Uppsala, Sweden).

2.7. Statistical Analysis

All data were obtained from three independent experiments and reported as means ± standard deviations. Statistical analysis was performed with a one-way analysis of variant (ANOVA) using GraphPad Prism 9.3.1 (GraphPad Software Inc., San Diego, CA, USA). Differences in means at $p < 0.05$ were considered statically significant.

3. Results

3.1. Secondary Metabolites in the Methanolic Extracts of *D. heterocarpum*

The chemical structures of the six compounds (**1**–**6**) extracted from *D. heterocarpum* (Figure 1) were elucidated using their spectroscopic data in comparison with previously reported structures [30–35]. The description of each compound was listed below.

Compound **1** (Figure 1) was identified as amoenylin [30]: Yellow–brown amorphous solid; HR-ESIMS: *m/z* 311.1260 $[M + Na]^+$ calculated for 311.1259 suggesting $C_{17}H_{20}O_4Na$. ^{1}H-NMR (300 MHz, acetone-d_6) δ: 5.36 (H-4, s), 3.79 (3H, s, OMe-4′), 3.84 (6H, s, OMe-3, OMe-5), 6.35 (2H, s, H-2, H-6), 6.82 (2H, d, J = 9.0 Hz, H-3′, H-5′), 7.07 (2H, d, J = 8.4 Hz, H-2′, H-6′), 2.83 (4H, m, H-α, H-α′); ^{13}C-NMR (75 MHz, acetone-d_6) δ: 37.2 (C-α′), 38.3 (C-α), 55.2 (OMe-4′), 56.2 (OMe-3, OMe-5), 105.0 (C-1, C-6), 113.6 (C-3′, C-5′), 129.3 (C-2′, C-6′), 132.4 (C-4), 132.6 (C-1′), 133.4 (C-1), 147.0 (C-3, C-5), 157.9 (C-4′).

Compound **2** (Figure 1) was identified as methyl 3-(4-hydroxyphenyl) propionate [31]: White powder; HR-ESIMS: *m/z* 203.0665 $[M + Na]^+$ calculated for 203.0684 suggesting $C_{10}H_{12}O_3Na$. ^{1}H-NMR (300 MHz, acetone-d_6) δ: 2.53 (2H, t, J = 7.8 Hz, H-2), 2.79 (2H, t, J = 7.8 Hz, H-3), 3.58 (3H, s, OMe-1), 6.73 (2H, d, J = 8.4 Hz, H-3′, H-5′), 7.03 (2H, d, J = 8.4 Hz, H-2′, H-6′); ^{13}C-NMR (75 MHz, acetone-d_6) δ: 29.8 (C-3), 35.6 (C-2), 52.3 (OMe-1), 115.1 (C-3′, C-5′), 129.1 (C-2′, C-6′), 131.4 (C-1′), 155.7 (C-4′) 172.6 (C-1).

Compound **3** (Figure 1) was identified as 3,4-dihydroxy-5,4′-dimethoxybibenzyl [32]: Brown amorphous solid; HR-ESIMS: *m/z* 297.1127 $[M + Na]^+$ calculated for 297.1102 suggesting $C_{16}H_{18}O_4Na$. ^{1}H-NMR (300 MHz, acetone-d_6) δ: 2.53 (2H, t, J = 7.8 Hz, H-2), 2.79 (2H, t, J = 7.8 Hz, H-3), 3.58 (3H, s, OMe-1), 6.73 (2H, d, J = 8.4 Hz, H-3′, H-5′), 7.03 (2H, d, J = 8.4 Hz, H-2′, H-6′); ^{13} C-NMR (75 MHz, acetone-d_6) δ: 29.8 (C-3), 35.6 (C-2), 52.3 (OMe-1), 115.1 (C-3′, C-5′), 129.1 (C-2′, C-6′), 131.4 (C-1′), 155.7 (C-4′), 172.6 (C-1).

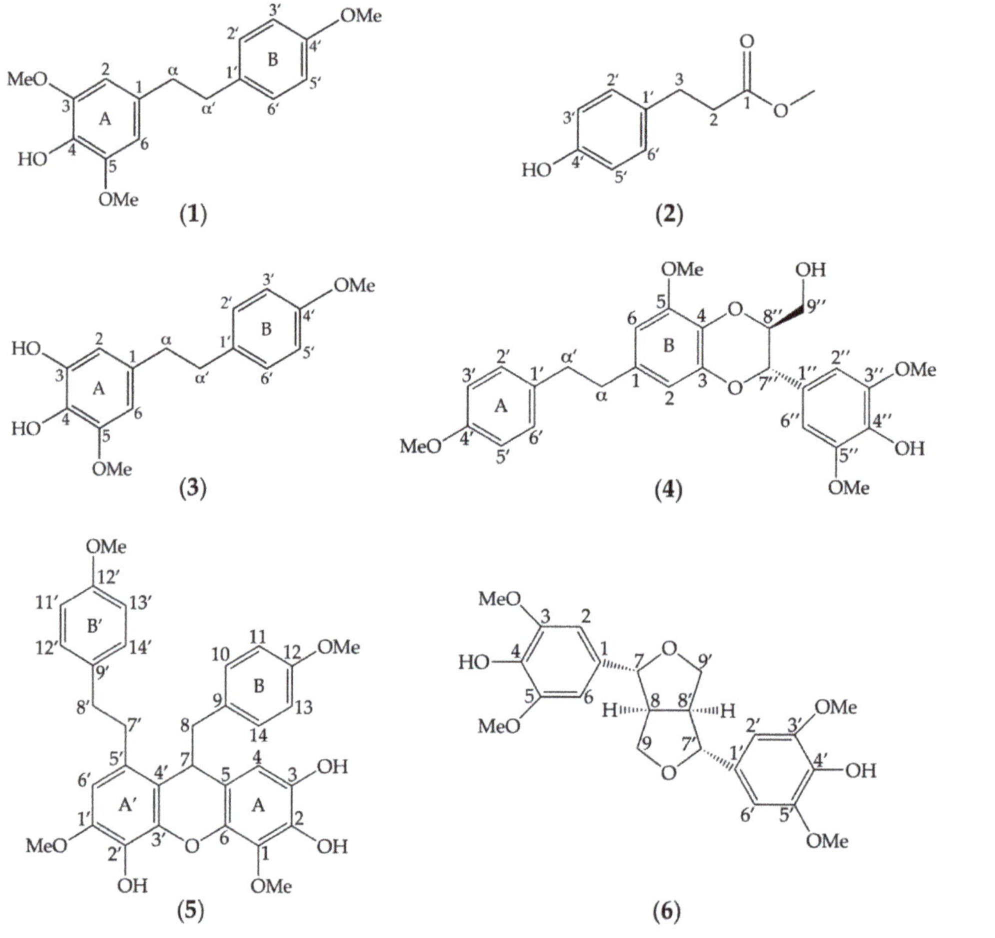

Figure 1. Chemical structures of secondary metabolites eluted from the methanolic extracts of *Dendrobium heterocarpum*. Six compounds (**1–6**) were assigned, and their structures were determined based on spectroscopic data in comparison with previously reported structures such as amoenylin (**1**), methyl 3-(4-hydroxyphenyl) propionate (**2**), 3,4-dihydroxy-5,4´-dimethoxybibenzyl (**3**), dendrocandin B (**4**), dendrofalconerol A (**5**), and syringaresinol (**6**).

Compound **4** (Figure 1) was identified as dendrocandin B [33]: White powder; HR-ESIMS: *m/z* 505.1850 [M + Na]$^+$ calculated for 505.1838 suggesting $C_{27}H_{30}O_8Na$. ^{1}H-NMR (300 MHz, acetone-d_6) δ: 2.84 (4H, m, H-α, H-α′), 3.54 (2H, m, H-9″), 3.80 (3H, m, OMe-4′), 3.83 (2H, m, H-9″), 3.87 (3H, s, OMe-5), 3.99 (6H, s, OMe-3″, OMe-5″), 4.02 (2H, m, H-8″), 4.96 (2H, d, J = 8.1 Hz, H-7″), 6.33 (2H, d, J = 1.8 Hz, H-6), 6.53 (2H, d, J = 1.8 Hz, H-2), 6.69 (2H, s, H-2″, H-6″), 7.11 (4H, d, H-2′, H-6′); ^{13}C-NMR (75 MHz, acetone-d_6) δ: 37.2 (C-α′), 38. 2 (C-α), 55.5 (OMe-4′), 56.3 (OMe-5), 56.6 (OMe-3″, OMe-5″), 105.0 (C-1, C-6), 113.6 (C-3′, C-5′), 129.3 (C-2′, C-6′), 132.4 (C-4), 61.8 (C-9″), 76.7 (C-7″), 78.5 (C-8″), 104.4 (C-6), 105.1 (C-2″,C-6″), 109.8 (C-2), 127.6 (C-1″), 129.6 (C-2′,C-6′), 131.2 (C-4), 134.0 (C-1′), 144.4 (C-3), 147.5 (C-3″, C-5″), 158.1 (C-4′).

Compound **5** (Figure 1) was identified as dendrofalconerol A [34]: Brown amorphous solid; HR-ESIMS: *m/z* 567.2003, [M + Na]$^+$ calculated for $C_{32}H_{32}O_8Na$; 567.1994, suggesting $C_{32}H_{32}O_8$. ^{1}H NMR (300 MHz, acetone-d_6) δ: 2.65-2.71 (2H, m, H8), 2.74-2.83 (2H, m, H-8), 2.75-2.84 (2H, m, H-8′), 2.78-2.86 (2H, m, H-7′), 2.80-2.86 (2H, m, H-8′), 2.87-2.90 (2H, m,

H-7′), 3.70 (3H, s, MeO-12), 3.73 (3H, s, MeO-12′), 3.81 (3H, s, MeO-1′), 3.89 (3H, s, MeO-1), 4.09 (1H, dd, *J* = 5.7, 6.9 Hz, H-7), 6.13 (1H, s, H-4), 6.60 (2H, d, *J* = 8.4 Hz, H-10, H-14), 6.65 (2H, s, H-6′), 6.68 (2H, d, *J* = 8.4 Hz, H-11, H-13), 6.82 (2H, d, *J* = 8.4 Hz, H-11′, H-13′), 7.13 (2H, d, *J* = 8.5 Hz, H-10′, H-14′); ^{13}C NMR (75 MHz, acetone-d_6) δ: 34.4 (C-7′), 37.5 (C-8′), 39.6 (C-7), 45.3 (C-8), 55.3 (MeO-12), 55.4 (MeO-12′), 56.2 (MeO-1′), 61.1 (MeO-1), 109.7 (C-4), 108.4 (C-6′), 113.9 (C-11, C-13), 117.8 (C- 5), 119.0 (C-4′), 129.5 (C-5′), 130.1 (C-10′, C-14′), 131.3 (C-10, C-14), 131.5 (C-9), 134.0 (C-2′), 134.6 (C-9′), 136.4 (C- 1), 137.3 (C-2), 139.9 (C-6), 142.3 (C-3′), 147.1 (C-1′), 158.9 (C-12′), 159.1 (C-12).

Compound **6** (Figure 1) was identified as syringaresinol [35]: Yellow–brown amorphous solid; HR-ESIMS: *m/z* 441.1527, [M + Na]$^+$ calculated for $C_{22}H_{26}O_8Na$; 441.1527, suggesting $C_{22}H_{26}O_8$. ^{1}H NMR (300 MHz, acetone-d_6) δ: 3.07 (2H, m, H-8, H-8′), 3.81 (12H, s, MeO-3, MeO-5, MeO-3′, MeO-5′), 4.21 (4H, m, H-9, H-9′), 4.65 (4H, m, H-7, H-7′), 6.67 (8H, s, H-2, H-2′, H-6, H-6′); ^{13}C NMR (75 MHz, acetone-d_6) δ: 54.3 (C-6, C-6′), 54.3 (C-8, C-8′), 55.7 (MeO-3, MeO-5), 55.7 (MeO-3′, MeO-5′), 71.4 (C-9, C-9′), 85.8 (C-7, C-7′), 103.5 (C-2, C-2′), 132.2 (C-1, C-1′), 147.7 (C-3, C-3′), 147.7 (C-5, C-5′), 157.8 (C-5, C-5′).

3.2. Roles of Identified Phytochemicals in the Growth and Development of Adipocytic Cells

Six identified compounds (**1**–**6**) were tested for their cytotoxic effects on 3T3-L1 pre-adipocytes (Figure 2a). The lowest percentage of cell viability (61.32 ± 1.1%) was observed after treating 3T3-L1 cells with compounds **4** and **5** at 20 μM. Even the lowest tested concentration (5 μM) of these compounds still posed cytotoxic effects on 3T3-L1 cells, supported by the significant decrease in cell viability compared to untreated vehicle controls (Figure 2a). Compounds **2**, **3**, and **6** demonstrated similar cytotoxicity, giving the same range of non-cytotoxic concentrations at $\lesssim$10 μM. The biological impacts of the six compounds at the same concentration (5 μM) on the cellular lipid storage in 3T3-L1 cells were also assessed with the Oil Red O staining technique (Figure 2b,c). In comparison with the untreated vehicle control, the decrease in the cellular lipid storage of 3T3-L1 cells was apparently observed after treating the cells with compounds **1**, **2**, and **3** (Figure 2b). A statistical comparison proved that compounds **1**, **2**, **3**, and **6** could significantly decrease the percentage of Oil Red O-stained 3T3-L1 cells (Figure 2c), while compound **3** posed the maximum reduction (51.73 ± 2.31%) compared to the others and controls. Besides, compound **3** at 5 μM could suppress the cellular lipid accumulation at the same level as that observed with a positive control made of 20 μM of oxyresveratrol (Figure 2c).

Compound **3**, with the greatest inhibitory activity against the lipid storage in 3T3-L1 cells, was investigated further for its possible molecular mechanisms that regulate adipocyte differentiation. The cytotoxic effect of compound **3** at different tested concentrations (Figure 2a) was confirmed with a nuclear straining approach (Figure 3a). It was explicit that some dead cells stained with Hoechst 33342 were observed after testing with compound **3** at 20 μM (Figure 3a). Non-cytotoxic concentrations (2.5, 5, and 10 μM) of this compound were tested for the capability to suppress cellular lipid storage in 3T3-L1 cells using the Oil Red O staining method (Figure 3b,c). The cellular lipid content was significantly decreased when the tested concentration of compound **3** increased (Figure 3c). Based on the percentage of Oil Red O-stained (Figure 3c), the half-maximal inhibitory concentration (IC_{50}) to suppress cellular lipid accumulation of compound **3** was computed and determined at 6.30 ± 0.10 μM. This IC_{50} value was approximately 3-fold lower than that of the positive control made of oxyresveratrol (20.20 ± 1.20 μM).

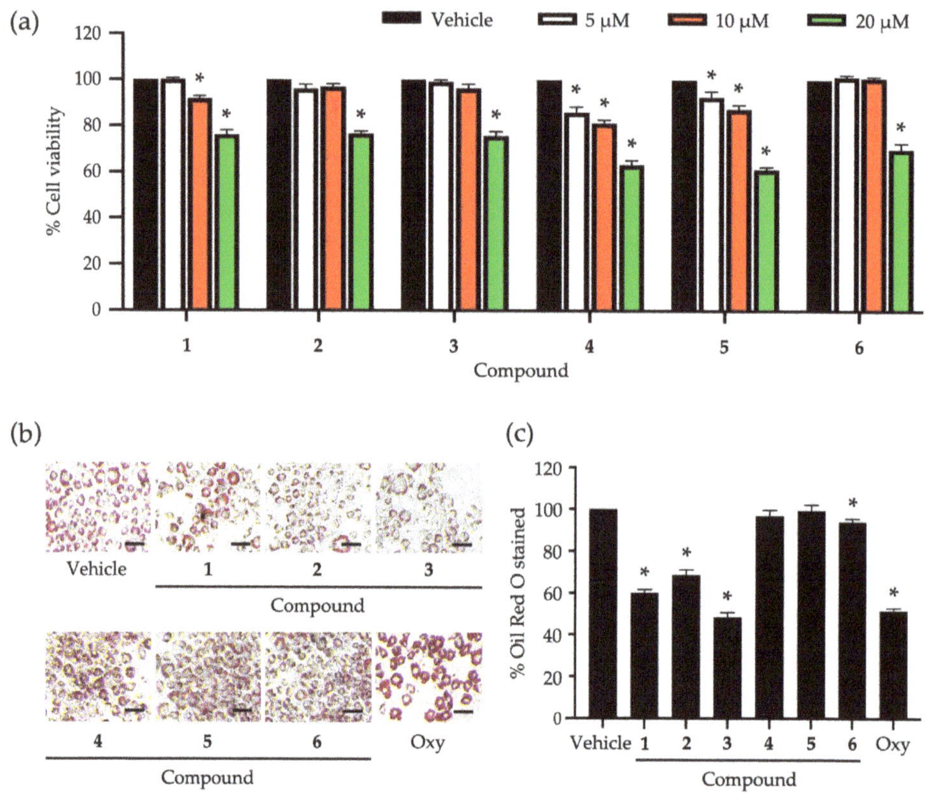

Figure 2. Roles of secondary metabolites eluted from *Dendrobium heterocarpum* in the viability and lipid storage of adipocytic cells. The percentage of cell viability was accounted for after treating 3T3-L1 cells with varying tested concentrations (5, 10, and 20 μM) of compounds **1–6** for 48 h (**a**). The impact of compounds **1–6** at the concentration of 5 μM on the lipid storage of 3T3-L1 cells was assessed using the Oil Red O staining method (**b**); scale bars = 20 μm. The percentage of Oil Red O stained was computed after the normalization with the total cellular protein (**c**). The graphical results demonstrate means ± SDs derived from three independent experiments. The asterisk (*) represents a statistical difference at $p < 0.05$ between the treatment and untreated vehicle control made of 0.5% (*v*/*v*) dimethyl sulfoxide, assessed by one-way ANOVA with Tukey's post hoc test. A positive control made of 20 μM of oxyresveratrol (Oxy) might be included in some experiments.

To elucidate certain underlying molecular mechanisms of compound **3** in adipocyte differentiation, the expression levels of adipogenic regulators (i.e., PPARγ and C/EBPα) were tracked at early adipocyte differentiation using the Western blot analysis (Figure 4). The band intensities of both transcription factors were gradually decreased when the tested concentrations of compound **3** increased (Figure 4a). This observation was consistent with the expression levels of these transcription factors relative to β-Actin as a reference protein (Figure 4b,c). Compound **3** at 10 μM offered the maximum suppression of both transcription factors compared to the treatments at lower concentrations and the untreated vehicle control. The decreased expression levels of PPARγ and C/EBPα also correlated with the decrease in lipid accumulation of 3T3-L1 cells treated with this compound (Figure 3c).

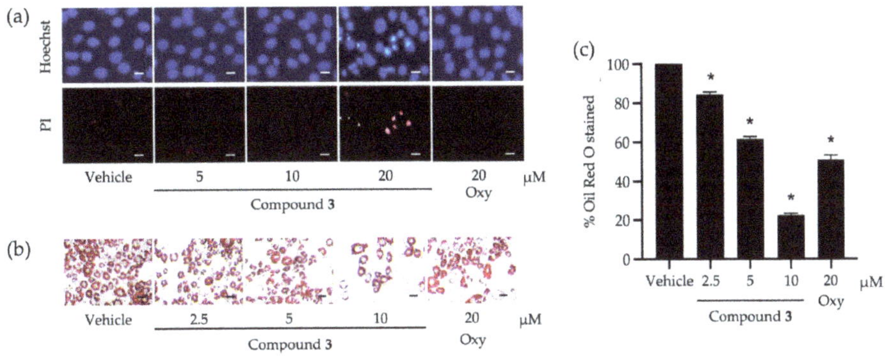

Figure 3. Roles of compound **3** derived from *Dendrobium heterocarpum* in the viability and lipid storage of adipocytic cells. Dead 3T3-L1 cells caused by compound **3** tested at different concentrations (5, 10, and 20 μM) were confirmed by a nuclear staining method using Hoechst 33342 (Hoechst) and propidium iodide (PI) (**a**). The impact of compound **3** at non-cytotoxic concentrations (2.5, 5, and 10 μM) on the lipid storage of 3T3-L1 cells was assessed using the Oil Red O staining method (**b**); scale bars = 10 μm. The percentage of Oil Red O-stained was computed after the normalization with the total cellular protein (**c**). The graphical results demonstrate means ± SDs derived from three independent experiments. The asterisk (*) represents a statistical difference at $p < 0.05$ between the treatment and untreated vehicle control made of 0.5% (*v*/*v*) dimethyl sulfoxide, assessed by one-way ANOVA with Tukey's post hoc test. A positive control made of 20 μM of oxyresveratrol (Oxy) might be included in some experiments.

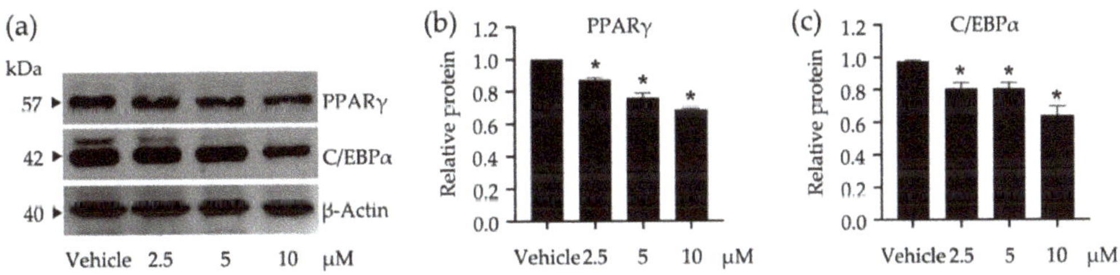

Figure 4. Impacts of compound **3** on the peroxisome proliferators activated receptor γ (PPARγ) and CCAAT/enhancer-binding protein α (C/EBPα) expression in 3T3-L1 cells. The band intensities of these transcription factors were visualized by Western blotting assay (**a**) and used for calculating their expression relative to β-Actin as a reference protein (**b**,**c**). Non-cytotoxic concentrations (2.5, 5, and 10 μM) of compound **3** were used in this study. The graphical results demonstrate means ± SDs derived from three independent experiments. The asterisk (*) represents a statistical difference at $p < 0.05$ between the treatment and untreated vehicle control made of 0.5% (*v*/*v*) dimethyl sulfoxide, assessed by one-way ANOVA with Tukey's post hoc test.

As the MAPK signaling pathways exert their significant roles in regulating several cellular responses such as proliferation, differentiation, survival, and apoptosis, the impact of a tested compound on the modulation of these pathways may offer a promising way to monitor adipocyte differentiation. Compound **3** was tested for its role in the activation of JNK/ERK/p38-mediated MAPK signaling pathways (Figure 5). The activation of these kinases was tracked through phosphorylation using Western blot analysis (Figure 5a), in which the phosphorylated ratio for each protein was accounted for after treating 3T3-L1 cells with varying concentrations of compound **3** (Figure 5b–d). In comparison to untreated vehicle controls, compound **3** at 10 μM significantly suppressed the phosphorylation of every tested protein, while this compound at 2.5 μM did not affect the phosphorylation of any tested proteins. JNK and p38 phosphorylation were more susceptible to compound **3** than that observed with ERK (Figure 5b–d).

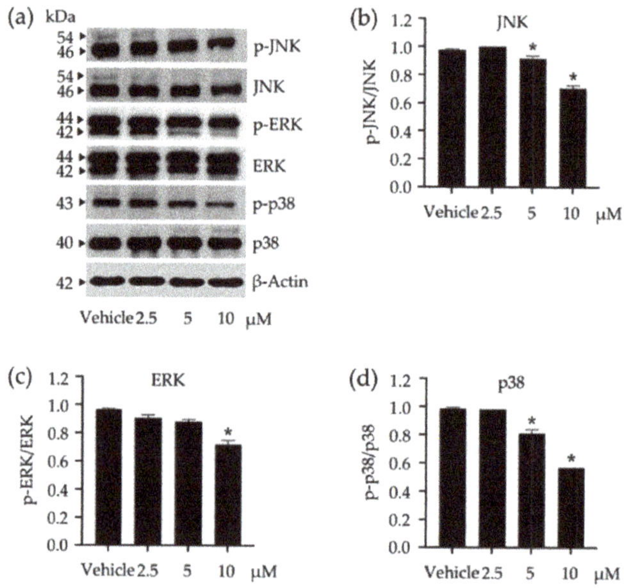

Figure 5. Impacts of compound **3** on the activation of c-Jun N-terminal kinase (JNK), extracellular signal-regulated kinase (ERK), and stress-activated protein kinase (p38) in 3T3-L1 cells. The band intensities of these kinases were visualized by Western blotting assay (**a**) and used for calculating their expression relative to β-Actin as a reference protein. The expression ratio between phosphorylated analog (p-JNK, p-ERK, and p-p38) and the summary (phosphorylated + non-phosphorylated) of each kinase was accounted for to describe the activation level (**b–d**). Non-cytotoxic concentrations (2.5, 5, and 10 μM) of compound **3** were used in this study. The graphical results demonstrate means ± SDs derived from three independent experiments. The asterisk (*) represents a statistical difference at $p < 0.05$ between the treatment and untreated vehicle control made of 0.5% (v/v) dimethyl sulfoxide, assessed by one-way ANOVA with Tukey's post hoc test.

4. Discussion

The genus *Dendrobium* has been reported as a rich source of valuable metabolites with potent bioactivity to suppress adipocyte differentiation—a critical biological process to indicate obesity [20,26,27,36]. Our study also proved that *D. heterocarpum*, a member of the genus *Dendrobium* and a widespread Thai orchid species, houses a variety of phytochemicals with inhibitory activity against adipocyte development. Six compounds (**1–6**) were profiled in *D. heterocarpum* methanolic extracts. An orchid bibenzyl, amoenylin (compound **1**), was firstly elucidated from the methanolic extract of *Dendrobium amoenum* [30]. This compound seemed to be a common metabolite found in the genus *Dendrobium*, while its bioactivity to suppress cellular lipid accumulation was firstly unveiled here. Methyl 3-(4-hydroxyphenyl) propionate (compound **2**) is a prevalent plant secondary metabolite previously reported to be found in the ethyl acetate extract of *Amyris texana* leaves and the methanolic extract of an orchid species, *Bulbophyllum retusiusculum* [31]. In addition to its recognized plant growth-related function as a biological nitrification inhibitor [37], this phenylpropanoid was firstly reported here with less of a cytotoxic effect on 3T3-L1 pre-adipocytes (compared to compounds **1**, **4**, and **5** eluted in this work) and the capability to reduce lipid storage of these cells.

Compound **3**, 3,4-dihydroxy-5,4′-dimethoxybibenzyl, is another bibenzyl derived from *D. heterocarpum*, which showed the greatest bioactivity to inhibit lipid accumulation in 3T3-L1 cells. This orchid bibenzyl was firstly elucidated from *Dendrobium moniliforme* with a given name, moniliformine (unapproved name) [32], which was also isolated later from *Dendrobium harveyanum* [20] and *Dendrobium officinale* [38]. This compound posed some potent bioactivities to impede the invasive mechanisms of non-small-cell lung cancer cells and sensitize the apoptosis of these cancer cells [39,40], but did not show a cytotoxic effect on the HeLa cervical cancer cell line [38]. In 2008, the sister molecule of compound **3**, dendrocandin A, was isolated from *Dendrobium candidum* [33]. The authors also found dendrocandin B (compound **4**) from this *Dendrobium* species [33], which was also elucidated in this study. Compound **4** was formerly reported to have inhibitory activity against diverse human cancer cells [38,41], but did not hamper α-glucosidase and pancreatic lipase activities [20]. Lacking such an enzyme-inhibiting function of compound **4** might be linked to the absence of its bioactivity to suppress lipid accumulation in 3T3-L1 cells tested in this study.

Dendrofalconerol A (compound **5**) was firstly isolated from *Dendrobium falconeri* [34] and further found in diverse *Dendrobium* species [17,20,41]. This bisbibenzyl was recognized for its anticancer potential [42,43] and its strong capability to inhibit α-glucosidase and pancreatic lipase activities [17,20]. However, based on our findings, compound **5** did not exhibit the inhibitory effect on lipid accumulation in 3T3-L1 cells, which was contrary to its sister molecule, dendrofalconerol B [20]. These results prove that the varied biological functions of chemicals rely significantly on their molecular structures with different radical groups and isomeric analogs. This molecular structure-dependent bioactivity was also observed with compounds **1** and **3**. These two compounds are bibenzyls that have only one different radical group by replacing a methoxy group at the C-3 on the A ring in compound **1** with a hydroxyl group in compound **3**. The replacement led to the decreased cytotoxic effect on 3T3-L1 cells (Figure 2a) and the increased capability to inhibit lipid accumulation in these cells (Figure 2c). This methoxy/hydroxyl replacement in a parent compound has been proposed to link with the shift in various bioactivities, including lipogenesis and lipolysis during adipocyte differentiation [44–47]. Compound **6**, syringaresinol, was isolated from *Tripterygium wilfordii*, and it was structurally elucidated for the first time in 1976 [48]. Syringaresinol is a member of lignans, which has been increasingly reported to be found in several *Dendrobium* species [35,49–51]. This lignan has been proven recently to have no function to promote glucose uptake in 3T3L-1 cells [52]. We found here that syringaresinol slightly affected lipid storage in 3T3-L1 cells compared to compounds **1–3** (Figure 2c), and even the influencing level was significantly different from that observed with the untreated vehicle control.

Adipose tissues are essential parts of the body and have some critical functions to maintain body temperature, storage energy, and protect viscera. The formation of adipose tissue requires regular growth and the development of adipocytes. Adipocyte differentiation is a key step to convert pre-adipocytes to mature adipocytes that fully function in cellular lipid metabolism and adipose tissue development [53,54]. Excess adipocyte differentiation can cause irregularities in adipose tissue formation and body energy homeostasis, leading to the incidence of obesity and other metabolism-related diseases [55]. Hence, adipocyte differentiation is a promising platform to understand the emergence of obesity and search for a pharmacotherapeutic way to prevent or monitor this disease. There are many molecular mechanisms taking place during adipocyte differentiation and regulating or maintaining this biological process [12]. Compound **3** already proved that it could strongly suppress the lipid accumulation in 3T3-L1 cells. One of the underlying molecular mechanisms is that compound **3** could downregulate a set of crucial transcription factors (i.e., PPARγ and C/EBPα) triggered during the early differentiation of 3T3-L1 cells. These transcription factors play a significant role in adipocyte differentiation by regulating lipogenesis and cellular lipid storage. A study reported that *Pparg*-knocked out mice had impaired lipid metabolisms and were resistant to high-fat diet-induced obesity [56]. Suppressed adipocyte differentiation and impaired lipid metabolisms were unveiled from *C/ebpa*-deleted 3T3-L1 cells and mice [57].

It was also proved here that compound **3** could suppress the activation of the MAPK signaling pathways (i.e., JNK, ERK, and p38) in 3T3-L1 cells. To date, the biological roles of these biomarkers in adipocyte differentiation are still rarely known and require further investigation. Some studies describe that JNK is implicated in the development of obesity-related insulin resistance, and ERK is required for cellular proliferation and differentiation [9,11,58]. The decrease in JNK and ERK phosphorylation observed in this study might not be a direct result of compound **3**. A study unveiled that the phosphorylation of JNK and ERK in 3T3-L1 cells was significantly activated after treating these cells with the chemokine (C-X-C motif) ligand 3 (CXCL3) [59]. The authors also found that the expression of C/EBPβ and δ (other adipogenic regulators) induced by CXCL3 was downregulated if JNK and ERK were inhibited by their specific inhibitors [59]. Hence, CXCL3 as a crucial adipokine in adipocyte differentiation might be negatively affected by compound **3**, which consequently caused the decrease in JNK and ERK phosphorylation. ERK has been reviewed, with a pivotal role in cell cycle progression and cell proliferation [9]. Many studies revealed that the decrease in ERK phosphorylation was responsible for the increase in apoptosis, autophagy, and metastasis of diverse cancer cells [60] and related references therein. Suppressing this transcription factor might restrain pre-adipocytes at the G0 phase of the mitotic clonal expansion, which is supposed to further limit hyperplasia in adipose tissue formation. The phosphorylation of p38—a stress-activated protein kinase—was also decreased in the presence of compound **3**. A study reported that high-fat diet-induced obesity was significantly diminished in mice administered the p38 inhibitor [61]. These p38 inhibitor-fed mice also had reduced macrophage infiltration into white adipose tissues and a lower level of tumor necrosis factor α, compared to untreated control mice. Thus, the decrease in p38 phosphorylation might offer a possible link to the modulation of adipocyte differentiation, which would further mitigate obesity.

Adipocyte differentiation involves the sequentially coordinated changes in hormonal sensitivity and a gene expression-accompanied morphological shift from fibroblasts to round-shaped cells. There are several genes and proteins triggered during adipocyte differentiation and maturation, some of which play a pivotal role in cellular lipid metabolism, known as adipogenic effectors (e.g., fatty acid synthase, perilipin 1, lipoprotein lipase, and adiponectin). Further investigations of how a natural molecule exerts its biological impact on the expression of these adipogenic effectors are also required to provide significant insights into its anti-adipogenic activity. The protein expression analysis in our study only describes the biological impact of a tested compound on the downstream central dogma of molecular biology. Additional studies at the transcription level (upstream) are also essential to elucidate the underlying molecular mechanisms of the tested molecule in-depth.

5. Conclusions

The *D. heterocarpum* methanolic extract houses diverse secondary metabolites. Among them, six compounds were identified, including amoenylin (**1**), methyl 3-(4-hydroxyphenyl) propionate (**2**), 3,4-dihydroxy-5,4′-dimethoxybibenzyl (**3**), dendrocandin B (**4**), dendrofalconerol A (**5**), and syringaresinol (**6**). These compounds had varying impacts on the viability and lipid accumulation of 3T3-L1 pre-adipocytes. Compound **3** exhibited the greatest reduction in 3T3-L1 lipid storage compared to the other compounds. This decreased cellular lipogenesis was proven to be linked to the downregulation of some key adipogenic regulators (i.e., PPARγ and C/EBPα) and the suppression of JNK/ERK/p38-mediated MAPK signaling pathways. To this end, *D. heterocarpum* proved its potential to be a natural source for discovering pharmaceutical products with promising bioactivities to modulate adipocyte differentiation. Compound **3** would be a potent candidate for further investigation and development as a phytomedicine for preventing and curing obesity.

Author Contributions: Conceptualization, C.C.; methodology, S.W., H.E.E.K., B.S., K.L., T.M. and C.T.; software, R.S. and C.C.; validation, B.S., K.L., T.M., C.T., C.P., Y.P. and C.C.; investigation, S.W. and H.E.E.K.; resources, B.S., K.L. and C.C.; data curation, S.W. and C.C.; writing—original draft preparation, S.W., H.E.E.K. and C.C.; writing—review and editing, R.S.; supervision, C.C.; funding acquisition, S.W. and C.C. All authors have read and agreed to the published version of the manuscript.

Funding: This research was funded by Walailak University (grant number: WU-IRG-64-076) and the Faculty of Pharmaceutical Sciences, Chulalongkorn University (grant number: Phar2564-RG008).

Institutional Review Board Statement: Not applicable.

Informed Consent Statement: Not applicable.

Data Availability Statement: Data are contained within the article.

Acknowledgments: We are grateful to all financial supports and research facilities from Walailak University and the Faculty of Pharmaceutical Sciences, Chulalongkorn University. H.E.E.K. has funding support from the Scholarship for International Graduate Students in ASEAN and Non-ASEAN Countries, Chulalongkorn University. R.S. is supported by the Second Century Fund (C2F), Chulalongkorn University.

Conflicts of Interest: The authors have no conflict of interest to declare.

References

1. Ambele, M.A.; Dhanraj, P.; Giles, R.; Pepper, M. Adipogenesis: A complex interplay of multiple molecular determinants and pathways. *Int. J. Mol. Sci.* **2020**, *21*, 4283. [CrossRef] [PubMed]
2. De Lorenzo, A.; Romano, L.; Di Renzo, L.; Di Lorenzo, N.; Cenname, G.; Gualtieri, P. Obesity: A preventable, treatable, but relapsing disease. *Nutrition* **2020**, *71*, 110615. [CrossRef] [PubMed]
3. Ma, S.; Xi, B.; Yang, L.; Sun, J.; Zhao, M.; Bovet, P. Trends in the prevalence of overweight, obesity, and abdominal obesity among Chinese adults between 1993 and 2015. *Int. J. Obes.* **2021**, *45*, 427–437. [CrossRef] [PubMed]
4. Mark, D.H. Deaths attributable to obesity. *JAMA.* **2005**, *293*, 1918–1919. [CrossRef] [PubMed]
5. Lee, H.-W.; Rhee, D.-K.; Kim, B.-O.; Pyo, S. Inhibitory effect of sinigrin on adipocyte differentiation in 3T3-L1 cells: Involvement of AMPK and MAPK pathways. *Biomed. Pharmacother.* **2018**, *102*, 670–680. [CrossRef] [PubMed]
6. Jakab, J.; Miškić, B.; Miškić, S.; Juranić, B.; Ćosić, V.; Schwarz, D.; Včev, A. Adipogenesis as a potential anti-obesity target: A review of pharmacological treatment and natural products. *Diabetes Metab. Syndr. Obes.* **2021**, *14*, 67–83. [CrossRef]
7. Cowherd, R.M.; Lyle, R.E.; McGehee, R.E., Jr. Molecular regulation of adipocyte differentiation. *Semin. Cell Dev. Biol.* **1999**, *10*, 3–10. [CrossRef]
8. Rayalam, S.; Yang, J.-E.; Ambati, S.; Della-Fera, M.A.; Baile, C.A. Resveratrol induces apoptosis and inhibits adipogenesis in 3T3-L1 adipocytes. *Phytother. Res.* **2008**, *22*, 1367–1371. [CrossRef]
9. Bost, F.; Aouadi, M.; Caron, L.; Binétruy, B. The role of MAPKs in adipocyte differentiation and obesity. *Biochimie* **2005**, *87*, 51–56. [CrossRef]
10. Prusty, D.; Park, B.-H.; Davis, K.E.; Farmer, R.S. Activation of MEK/ERK signaling promotes adipogenesis by enhancing peroxisome proliferator-activated receptor γ (PPARγ) and C/EBPα gene expression during the differentiation of 3T3-L1 preadipocytes. *J. Biol. Chem.* **2002**, *277*, 46226–46232. [CrossRef]
11. Lim, S.H.; Lee, H.S.; Han, H.-K.; Choi, C.-I. Saikosaponin A and D inhibit adipogenesis via the AMPK and MAPK signaling pathways in 3T3-L1 adipocytes. *Int. J. Mol. Sci.* **2021**, *22*, 11409. [CrossRef] [PubMed]
12. Guru, A.; Issac, P.K.; Velayutham, M.; Saraswathi, N.T.; Arshad, A.; Arockiaraj, J. Molecular mechanism of down-regulating adipogenic transcription factors in 3T3-L1 adipocyte cells by bioactive anti-adipogenic compounds. *Mol. Biol. Rep.* **2021**, *48*, 743–761. [CrossRef] [PubMed]
13. Lao, W.; Tan, Y.; Jin, X.; Xiao, L.; Kim, J.J.Y.; Qu, X. Comparison of cytotoxicity and the anti-adipogenic effect of green tea polyphenols with epigallocatechin-3-gallate in 3T3-L1 preadipocytes. *Am. J. Chin. Med.* **2015**, *43*, 1177–1190. [CrossRef] [PubMed]
14. Lam, Y.; Ng, T.B.; Yao, R.M.; Shi, J.; Xu, K.; Cho Wing Sze, S.; Zhang, K.Y. Evaluation of chemical constituents and important mechanism of pharmacological biology in *Dendrobium* plants. *Evid.-Based Compl. Alt.* **2015**, *2015*, 841752. [CrossRef] [PubMed]
15. Xiaohua, J.; Singchi, C.; Yibo, L. Taxonomic revision of *Dendrobium moniliforme* complex (Orchidaceae). *Sci. Hortic.* **2009**, *120*, 143–145. [CrossRef]
16. Inthongkaew, P.; Chatsumpun, N.; Supasuteekul, C.; Kitisripanya, T.; Putalun, W.; Likhitwitayawuid, K.; Sritularak, B. α-Glucosidase and pancreatic lipase inhibitory activities and glucose uptake stimulatory effect of phenolic compounds from *Dendrobium formosum*. *Rev. Bras. Farmacogn.* **2017**, *27*, 480–487. [CrossRef]
17. Limpanit, R.; Chuanasa, T.; Likhitwitayawuid, K.; Jongbunpresert, V.; Sritularak, B. α-Glucosidase inhibitors from *Dendrobium tortile*. *Rec. Nat. Prod.* **2016**, *10*, 609.

18. Liu, Y.; Yang, L.; Zhang, Y.; Liu, X.; Wu, Z.; Gilbert, R.G.; Deng, B.; Wang, K. *Dendrobium officinale* polysaccharide ameliorates diabetic hepatic glucose metabolism via glucagon-mediated signaling pathways and modifying liver-glycogen structure. *J. Ethnopharmacol.* **2020**, *248*, 112308. [CrossRef]
19. Lu, Y.; Kuang, M.; Hu, G.-P.; Wu, R.-B.; Wang, J.; Liu, L.; Lin, Y.-C. Loddigesiinols G–J: α-Glucosidase inhibitors from *Dendrobium loddigesii*. *Molecules* **2014**, *19*, 8544–8555. [CrossRef]
20. Maitreesophone, P.; Khine, H.E.E.; Quiel Lasam Nealiga, J.; Kongkatitham, V.; Panuthai, P.; Chaotham, C.; Likhitwitayawuid, K.; Sritularak, B. α-Glucosidase and pancreatic lipase inhibitory effects and anti-adipogenic activity of dendrofalconerol B, a bisbibenzyl from *Dendrobium harveyanum*. *S. Afr. J. Bot.* **2022**, *146*, 187–195. [CrossRef]
21. Na Ranong, S.; Likhitwitayawuid, K.; Mekboonsonglarp, W.; Sritularak, B. New dihydrophenanthrenes from *Dendrobium infundibulum*. *Nat. Prod. Res.* **2019**, *33*, 420–426. [CrossRef] [PubMed]
22. San, H.T.; Boonsnongcheep, P.; Putalun, W.; Mekboonsonglarp, W.; Sritularak, B.; Likhitwitayawuid, K. α-Glucosidase inhibitory and glucose uptake stimulatory effects of phenolic compounds from *Dendrobium christyanum*. *Nat. Prod. Commun.* **2020**, *15*, 1934578X20913453.
23. Sarakulwattana, C.; Mekboonsonglarp, W.; Likhitwitayawuid, K.; Rojsitthisak, P.; Sritularak, B. New bisbibenzyl and phenanthrene derivatives from *Dendrobium scabrilingue* and their α-glucosidase inhibitory activity. *Nat. Prod. Res.* **2020**, *34*, 1694–1701. [CrossRef] [PubMed]
24. Sun, J.; Zhang, F.; Yang, M.; Zhang, J.; Chen, J.; Zhan, R.; Li, L.; Chen, Y. Isolation of α-glucosidase inhibitors including a new flavonol glycoside from *Dendrobium devonianum*. *Nat. Prod. Res.* **2014**, *28*, 1900–1905. [CrossRef]
25. Thant, M.T.; Chatsupun, N.; Mekboonsonglarp, W.; Sritularak, B.; Likhitwitayawuid, K. New fluorene derivatives from *Dendrobium gibsonii* and their α-glucosidase inhibitory activity. *Molecules* **2020**, *25*, 4931. [CrossRef] [PubMed]
26. Thant, M.T.; Khine, H.E.E.; Quiel Lasam Nealiga, J.; Chatsumpun, N.; Chaotham, C.; Sritularak, B.; Likhitwitayawuid, K. α-Glucosidase inhibitory activity and anti-adipogenic effect of compounds from *Dendrobium delacourii*. *Molecules* **2022**, *27*, 1156. [CrossRef] [PubMed]
27. Li, X.-W.; Huang, M.; Lo, K.; Chen, W.-L.; He, Y.-Y.; Xu, Y.; Zheng, H.; Hu, H.; Wang, J. Anti-diabetic effect of a shihunine-rich extract of *Dendrobium loddigesii* on 3T3-L1 cells and db/db mice by up-regulating AMPK–GLUT4–PPARα. *Molecules* **2019**, *24*, 2673. [CrossRef] [PubMed]
28. Vaddhanaphuti, N.B. *A Field Guide to the Wild Orchids of Thailand*, 4th ed.; Silkworm Books: Chiang Mai, Thailand, 2005; p. 117.
29. Tan, H.Y.; Iris, M.Y.; Li, E.T.; Wang, M. Inhibitory effects of oxyresveratrol and cyanomaclurin on adipogenesis of 3T3-L1 cells. *J. Funct. Foods* **2015**, *15*, 207–216. [CrossRef]
30. Majumder, P.L.; Guha, S.; Sen, P. Bibenzyl derivatives from the orchid *Dendrobium amoenum*. *Phytochemistry* **1999**, *52*, 1365–1369. [CrossRef]
31. Fang, Y.-S.; Yang, M.-H.; Cai, L.; Wang, J.-P.; Yin, T.-P.; Yu, J.; Ding, Z.-T. New phenylpropanoids from *Bulbophyllum retusiusculum*. *Arch. Pharm. Res.* **2018**, *41*, 1074–1081. [CrossRef]
32. Zhi-Ming, B.; Zheng-Tao, W.; Luo-Shan, X. Chemical constituents of *Dendrobium moniliforme*. *Acta Bot. Sin.* **2004**, *46*, 124–126.
33. Li, Y.; Wang, C.-L.; Guo, S.-X.; Yang, J.-S.; Xiao, P.-G. Two new compounds from *Dendrobium candidum*. *Chem. Pharm. Bull.* **2008**, *56*, 1477–1479. [CrossRef] [PubMed]
34. Sritularak, B.; Likhitwitayawuid, K. New bisbibenzyls from *Dendrobium falconeri*. *Helv. Chim. Acta* **2009**, *92*, 740–744. [CrossRef]
35. Sritularak, B.; Duangrak, N.; Likhitwitayawuid, K. A new bibenzyl from *Dendrobium secundum*. *Z. Naturforsch. C* **2011**, *66*, 205–208. [CrossRef]
36. Qu, J.; Tan, S.; Xie, X.; Wu, W.; Zhu, H.; Li, H.; Liao, X.; Wang, J.; Zhou, Z.-A.; Huang, S. *Dendrobium Officinale* polysaccharide attenuates insulin resistance and abnormal lipid metabolism in obese mice. *Front. Pharmacol.* **2021**, *12*, 659626. [CrossRef]
37. Zakir, H.A.; Subbarao, G.V.; Pearse, S.J.; Gopalakrishnan, S.; Ito, O.; Ishikawa, T.; Kawano, N.; Nakahara, K.; Yoshihashi, T.; Ono, H.; et al. Detection, isolation and characterization of a root-exuded compound, methyl 3-(4-hydroxyphenyl) propionate, responsible for biological nitrification inhibition by sorghum (*Sorghum bicolor*). *New Phytol.* **2008**, *180*, 442–451. [CrossRef]
38. Ren, G.; Deng, W.Z.; Xie, Y.F.; Wu, C.H.; Li, W.Y.; Xiao, C.Y.; Chen, Y.L. Bibenzyl derivatives from leaves of *Dendrobium officinale*. *Nat. Prod. Commun.* **2020**, *15*, 1934578X20908678. [CrossRef]
39. Putri, H.E.; Nutho, B.; Rungrotmongkol, T.; Sritularak, B.; Vinayanuwattikun, C.; Chanvorachote, P. Bibenzyl analogue DS-1 inhibits MDM2-mediated p53 degradation and sensitizes apoptosis in lung cancer cells. *Phytomedicine* **2021**, *85*, 153534. [CrossRef]
40. Putri, H.E.; Sritularak, B.; Chanvorachote, P. DS-1 inhibits migration and invasion of non-small-cell lung cancer cells through suppression of epithelial to mesenchymal transition and integrin β1/FAK signaling. *Anticancer Res.* **2021**, *41*, 2913–2923. [CrossRef]
41. Mittraphab, A.; Muangnoi, C.; Likhitwitayawuid, K.; Rojsitthisak, P.; Sritularak, B. A new bibenzyl-phenanthrene derivative from *Dendrobium signatum* and its cytotoxic activity. *Nat. Prod. Commun.* **2016**, *11*, 1934578X1601100526. [CrossRef]
42. Pengpaeng, P.; Sritularak, B.; Chanvorachote, P. Dendrofalconerol A sensitizes anoikis and inhibits migration in lung cancer cells. *J. Nat. Med.* **2015**, *69*, 178–190. [CrossRef] [PubMed]
43. Pengpaeng, P.; Sritularak, B.; Chanvorachote, P. Dendrofalconerol A suppresses migrating cancer cells via EMT and integrin proteins. *Anticancer Res.* **2015**, *35*, 201–205. [PubMed]
44. Kayser, O.; Kolodziej, H. Antibacterial activity of simple coumarins: Structural requirements for biological activity. *Z. Naturforsch. C* **1999**, *54*, 169–174. [CrossRef] [PubMed]

45. Zhang, H.; Yang, F.; Qi, J.; Song, X.C.; Hu, Z.F.; Zhu, D.N.; Yu, B.Y. Homoisoflavonoids from the fibrous roots of *Polygonatum odoratum* with glucose uptake-stimulatory activity in 3T3-L1 adipocytes. *J. Nat. Prod.* **2010**, *73*, 548–552. [CrossRef]
46. Yang, Y.; Yang, X.; Xu, B.; Zeng, G.; Tan, J.; He, X.; Hu, C.; Zhou, Y. Chemical constituents of *Morus alba* L. and their inhibitory effect on 3T3-L1 preadipocyte proliferation and differentiation. *Fitoterapia* **2014**, *98*, 222–227. [CrossRef]
47. Nishina, A.; Ukiya, M.; Fukatsu, M.; Koketsu, M.; Ninomiya, M.; Sato, D.; Yamamoto, J.; Kobayashi-Hattori, K.; Okubo, T.; Tokuoka, H.; et al. Effects of various 5, 7-dihydroxyflavone analogs on adipogenesis in 3T3-L1 cells. *Biol. Pharm. Bull.* **2015**, *38*, 1794–1800. [CrossRef]
48. Bryan, R.F.; Fallon, L. Crystal structure of syringaresinol. *J. Chem. Soc. Perkin Trans.* **1976**, *2*, 341–345. [CrossRef]
49. Chen, Y.G.; Yu, H.; Lian, X. Isolation of stilbenoids and lignans from *Dendrobium hongdie*. *Trop. J. Pharm. Res.* **2015**, *14*, 2055–2059. [CrossRef]
50. Hu, J.M.; Chen, J.J.; Yu, H.; Zhao, Y.X.; Zhou, J. Two novel bibenzyls from *Dendrobium trigonopus*. *J. Asian Nat. Prod. Res.* **2008**, *10*, 647–651. [CrossRef]
51. Zhou, J. Chemical constituents of *Dendrobium officinale*. *Chin. Tradit. Herb. Drugs* **2015**, *24*, 1292–1295.
52. Lv, H.W.; Li, Y.X.; Luo, M.; Qi, J.M.; Fu, Z.F.; Zhang, H.J.; Guo, Y.Q.; Chu, C.; Li, H.B.; Yan, J.Z. Two new nor-lignans from *Selaginella pulvinata* (Hook. & Grev.) Maxim and their antihyperglycemic activities. *Nat. Prod. Res.* **2021**, *36*, 279–286.
53. Poulos, S.P.; Dodson, M.V.; Hausman, G.J. Cell line models for differentiation: Preadipocytes and adipocytes. *Exp. Biol. Med.* **2010**, *235*, 1185–1193. [CrossRef] [PubMed]
54. Morrison, S.; McGee, S.L. 3T3-L1 adipocytes display phenotypic characteristics of multiple adipocyte lineages. *Adipocyte* **2015**, *4*, 295–302. [CrossRef] [PubMed]
55. Li, Y.; Rong, Y.; Bao, L.; Nie, B.; Ren, G.; Zheng, C.; Amin, R.; Arnold, R.; Jeganathan, R.B.; Huggins, K.W. Suppression of adipocyte differentiation and lipid accumulation by stearidonic acid (SDA) in 3T3-L1 cells. *Lipids Health Dis.* **2017**, *16*, 181. [CrossRef]
56. Jones, J.R.; Barrick, C.; Kim, K.A.; Lindner, J.; Blondeau, B.; Fujimoto, Y.; Shiota, M.; Kesterson, R.A.; Kahn, B.B.; Magnuson, M.A. Deletion of PPARγ in adipose tissues of mice protects against high fat diet-induced obesity and insulin resistance. *Proc. Natl. Acad. Sci. USA* **2005**, *102*, 6207–6212. [CrossRef]
57. Darlington, G.J.; Wang, N.; Hanson, R.W. C/EBPα: A critical regulator of genes governing integrative metabolic processes. *Curr. Opin. Genet. Dev.* **1995**, *5*, 565–570. [CrossRef]
58. Tarantino, G.; Caputi, A. JNKs, insulin resistance and inflammation: A possible link between NAFLD and coronary artery disease. *World J. Gastroenterol.* **2011**, *17*, 3785–3794. [CrossRef]
59. Kusuyama, J.; Komorizono, A.; Bandow, K.; Ohnishi, T.; Matsuguchi, T. CXCL3 positively regulates adipogenic differentiation. *J. Lipid Res.* **2016**, *57*, 1806–1820. [CrossRef]
60. He, L.; Su, Q.; Bai, L.; Li, M.; Liu, J.; Liu, X.; Zhang, C.; Jiang, Z.; He, J.; Shi, J.; et al. Recent research progress on natural small molecule bibenzyls and its derivatives in *Dendrobium* species. *Eur. J. Med. Chem.* **2020**, *204*, 112530. [CrossRef]
61. Maekawa, T.; Jin, W.; Ishii, S. The role of ATF-2 family transcription factors in adipocyte differentiation: Antiobesity effects of p38 inhibitors. *Mol. Cell. Biol.* **2010**, *30*, 613–625. [CrossRef]

nutrients

MDPI

Article

Anti-Obesity Effects of Traditional and Commercial Kochujang in Overweight and Obese Adults: A Randomized Controlled Trial

A Lum Han [1,*], Su-Ji Jeong [2], Myeong-Seon Ryu [2], Hee-Jong Yang [2], Do-Youn Jeong [2], Do-Sim Park [3] and Hee Kyung Lee [1]

1 Department of Family Medicine, Wonkwang University Hospital, Iksan 54538, Korea; yebbnkim@gmail.com
2 Microbial Institute for Fermentation Industry, Sunchang 56048, Korea; yo217@naver.com (S.-J.J.); rms6223@naver.com (M.-S.R.); godfiltss@naver.com (H.-J.Y.); jdy2534@korea.kr (D.-Y.J.)
3 Department of Laboratory Medicine, Wonkwang University Hospital, Iksan 54538, Korea; emailds@hanmail.net
* Correspondence: qibosarang@naver.com or vivian@wku.ac.kr; Tel.: +82-063-859-1300

Abstract: Kochujang shows anti-obesity effects in cell and animal models. Kochujang is traditionally prepared via slow fermentation or commercially using *Aspergillus oryzae*. We analyze the anti-obesity effects of two types of Kochujang in overweight and obese adults. The analyses included the following groups: traditional Kochujang containing either a high-dose (HTK; $n = 19$), or a low-dose of beneficial microbes (LTK; $n = 18$), and commercial Kochujang (CK; $n = 17$). Waist circumference decreased significantly in the HTK and CK groups. Total cholesterol, low-density lipoprotein cholesterol, high-density lipoprotein cholesterol, and triglyceride levels decreased in the HTK and LTK groups. Visceral fat is significantly reduced in the HTK group. The population of beneficial microorganisms in stool samples increased in all groups. Consumption of Kochujang reduces visceral fat content and improves the lipid profile, which can be enhanced by enrichment with beneficial microbes. These results suggest that Kochujang has the potential for application in obesity prevention.

Keywords: capsaicin; phytochemicals; fermentation; anti-obesity; kochujang

Citation: Han, A.L.; Jeong, S.-J.; Ryu, M.-S.; Yang, H.-J.; Jeong, D.-Y.; Park, D.-S.; Lee, H.K. Anti-Obesity Effects of Traditional and Commercial Kochujang in Overweight and Obese Adults: A Randomized Controlled Trial. *Nutrients* **2022**, *14*, 2783. https://doi.org/10.3390/nu14142783

Academic Editor: Anna Gramza-Michałowska

Received: 9 June 2022
Accepted: 2 July 2022
Published: 6 July 2022

1. Introduction

The prevalence of overweight and obesity worldwide is increasing, with nearly one-third of the world's population now classified as overweight or obese [1]. Obesity adversely affects almost every function of the body, contributing to several disease states such as diabetes, cardiovascular disease, several types of cancer, various musculoskeletal disorders, and poor mental health [1]. While obesity is on the rise worldwide, the anti-obesity effect of Kochujang is attracting attention. Kochujang is one of Korea's representative sauces with a combination of flavors. Kochujang has a savory taste owing to proteins derived from soybeans; sweetness from carbohydrates obtained from glutinous rice, non-glutinous rice, and barley rice; a spicy taste from red pepper powder; asalty taste from salt used for seasoning [2]. The fermentation process of food is associated with several health benefits, such as the degradation of toxic substances, improvement of digestion, and the production of vitamins that increase the nutritional value [3]. The weight loss- and obesity-reducing effects of Kochujang have been reported in several studies. Kochujang contains approximately 20–30% red pepper powder, rich in capsaicin, which exhibits anti-obesity effects [4]. The capsaicin present in red pepper powder can improve energy metabolism, lower cholesterol levels, improve blood lipid profiles, and break down fat tissues [5]. The use of red pepper powder has been shown to alter thermogenesis and appetite [6]. Other studies have reported changes in appetite and energy balance upon ingestion of capsaicin [7,8]. This change may occur via changes in the hypothalamic anorexia and anorexia neuropeptides [9]. Meju, the main component of Kochujang, is a fermented soybean product that shows anti-obesity and anti-atherosclerotic activities in obese adults [10]. Kochujang can be prepared using traditional methods or via commercial mass production. Generally, traditional Kochujang is prepared by fermenting Meju, salt, and red pepper powder together with glutinous rice and malt [11]. In contrast, for commercial Kochujang preparation, koji products are manufactured by culturing fungi such as *Aspergillus oryzae* on rice grains or flakes [11]. As all of the main ingredients of Kochujang have anti-obesity effects, Kochujang itself may show anti-obesity effects. However, to date, clinical trials comparing the post-use effects of traditional and commercial Kochujang in human participants are lacking. In this randomized, double-blind clinical trial, we tested the hypothesis that traditional and commercial Kochujang supplementation can reduce body fat content and improve blood lipid profiles in overweight adults. Further, the difference in anti-obesity effects between traditional and commercial Kochujang was investigated. The effects of traditional and commercial Kochujang supplementation on microbial changes in the stool were also evaluated.

2. Materials and Methods

2.1. Study Design

This study was a 6-week randomized, double-blind clinical trial (registration number: KCT0006899). The participants were asked to visit the research center three times. After 3 weeks of Kochujang supplementation, the participants visited the research center to check vital signs/side effects, and adherence to dosing. Efficacy indicators such as body weight, abdominal fat computed tomography (CT), adipokine levels, lipid profiles, and blood tests were evaluated at the first and last visits. The following Kochujang pills were provided to the participants: a pill prepared using traditional Kochujang with a high content of beneficial microorganisms (HTK), a pill prepared using traditional Kochujang containing a low content of beneficial microorganisms (LTK), and a pill prepared using commercial Kochujang (CK). Enrolled participants were to be randomly assigned to one of three groups receiving HTK pills (n = 20), LTK pills (n = 20), or CK pills (n = 20). The number of screened participants was 62 and the dropout rate was 20%. In total, 48 participants (16 participants in each intake group) completed the study.

The following method was used to conduct the double-blind randomized trial: a screening number was provided to participants who provided written consent to participate in the study. The screening number consisted of a total of four digits, beginning with S followed by three digits (for example, S001). Participants who passed the screening test were assigned a study subject number. The study subject number was randomly assigned in order. The study subject number began with GCJ-R and had a set rule with the last two digits. The numbers indicated the order of participants in the study and ranged from 01 to 62. The subject number assigned to each subject was used as a subject identification code to identify the subject until the end of the trial. Participants were asked to continue their normal diet and not consume other nutraceuticals or dietary supplements. Nutrient intake levels and activity levels were measured before and after the intervention period. At the beginning of the trial, all participants were instructed to maintain a usual diet and level of physical activity. The study design is presented in Figure 1.

2.2. Participants

In total, 62 volunteers aged 19–70 years with a body mass index (BMI) of $\geq$23 kg/m^2 were recruited and randomly assigned to three groups. Exclusion criteria were as follows: >10% change in body weight in the past 3 months; cardiovascular diseases such as arrhythmia, heart failure, myocardial infarction, and use of a pacemaker; allergy or hypersensitivity to any component of the test product; colon diseases such as Crohn's disease; history of gastrointestinal surgery (such as appendix or intestinal surgery); participation in another clinical trial within the past 2 months; liver dysfunction or acute/chronic renal disease; use of antipsychotic drug therapy within the past 2 months; laboratory test abnormalities as assessed by the investigator; psychological state abnormalities; history of alcohol or drug abuse; pregnancy or breastfeeding. All 62 volunteers provided consent prior to treatment. Participants under treatment with additional medicines for personal conditions during Kochujang supplementation and those who intended to stop Kochujang supplementation were excluded. The protocol was approved by the institutional review board of Wonkwang University Hospital (IRB approval no. 2021-04-046).

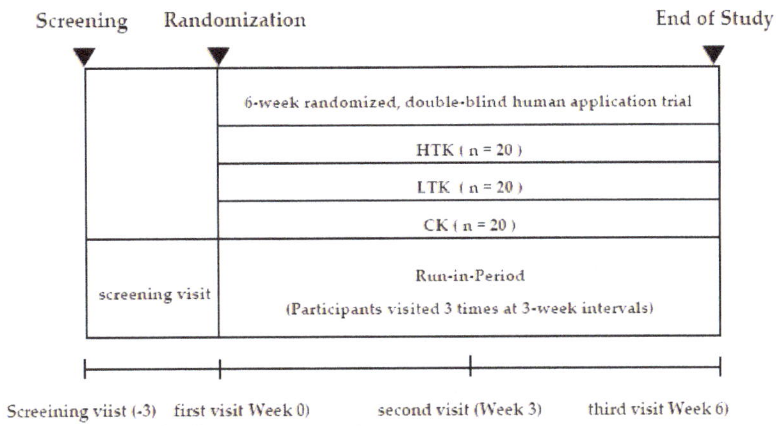

Figure 1. Study design.

Participants consumed Kochujang pills thrice a day after every meal, and the total amount of Kochujang pills taken per day was 25.3 g (19 g/d as Kochujang powder). In the 2014 National Health and Nutrition Survey, the average daily intake of red pepper paste for Koreans was 10.75 g and the extreme intake was 3686 g [12]. Therefore, in this study, 19 g of Kochujang powder was set as the daily intake.

2.3. Preparation of Kochujang Products

The preparation of Kochujang varies with region and household. Generally, Kochujang is prepared by mixing ingredients such as glutinous rice, red pepper powder, malt, and soybean paste powder in the fermentation process. Meju, which is one of the major ingredients, is prepared between August and September for fermentation prior to cold weather. Generally, Kochujang is prepared during the cold season between October and March. Prepared Kochujang can be consumed immediately; however, it is allowed to ripen for a few months for improved quality and flavor. The preparation process of traditional Kochujang is shown in Figure 2.

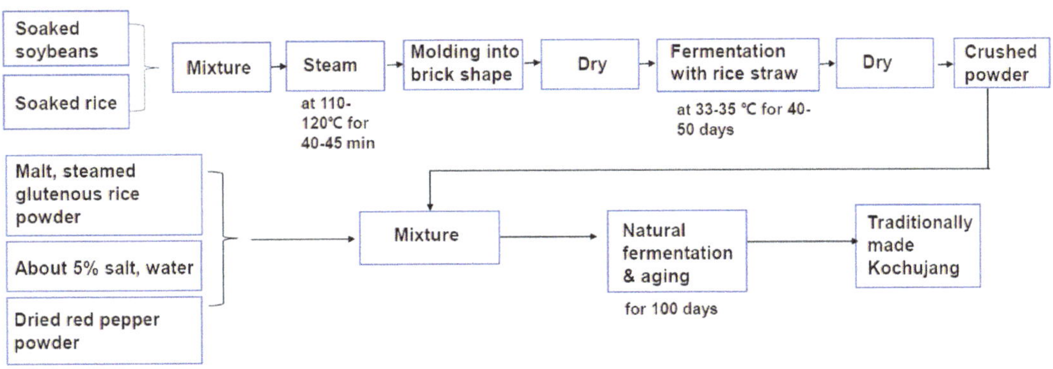

Figure 2. Schematic representation of the traditional process for Kochujang preparation.

Kochujang in pill form was prepared by mixing the Kochujang sample (Microbial Institute for Fermentation Industry, Sunchang, Korea) with deionized water to reduce the influence of high sugar content on removing moisture, and then the sample was freeze-dried. Dried samples were ground and then mixed with microcrystalline cellulose and magnesium stearate (Microbial Institute for Fermentation Industry) in defined proportions as shown in Table 1.

Table 1. Composition of Kochujang pills.

Ingredient	HTK		LTK		CK	
	Content (g)	Ratio (%)	Content (g)	Ratio (%)	Content (g)	Ratio (%)
Freeze-dried Kochujang powder	19.0	75	19.0	75	19.0	75
Microcrystalline cellulose	5.1	20	5.1	20	5.1	20
Magnesium stearate	1.2	5	1.2	5	1.2	5
Total	25.3	100	25.3	100	25.3	100

HTK, traditional Kochujang containing a high-dose of beneficial microbes; LTK, traditional Kochujang containing a low-dose of effective microbes; CK, commercially prepared Kochujang.

2.4. *Safety Assessment*

The health of the participants was assessed via screening tests, including electrocardiogram, urinalysis, hematology, and blood chemistry tests. These included analyses of white and red blood cell counts; hemoglobin, hematocrit, and platelet counts; total protein, albumin, alanine aminotransferase (ALT), aspartate aminotransferase (AST), blood urea nitrogen (BUN), and creatinine levels. Pulse and blood pressure were measured at each visit after a 10-min break. Participants were asked to report any side effects or changes in lifestyle or eating patterns during Kochujang supplementation and to report medication compliance.

2.5. *Biochemical Analyses, Abdomen Fat CT Analysis, and Lifestyle Survey*

A dietary intake survey was performed based on the Meal Recording Act; the research participants received a dietary log and recorded all food they consumed as accurately as possible [13]. Research participants completed a physical activity survey questionnaire based on the Global Physical Activity Questionnaire during their visits [14]. To analyze changes in the intestinal microbiome, $\geq$1 g feces was frozen using a stool collection kit after the first visit and 6 weeks later. The stool samples before and after Kochujang supplementation were collected from each subject using MICROBE & ME stool collecting kit (Macrogen, Seoul, South Korea). For CT scanning of abdominal fat, CT scans (Hispeed CT/e; General Electric, Boston, MA, USA) were taken with participants in the supine position. Total abdominal fat area, subcutaneous fat area, and visceral fat area were measured and expressed as mm2. For blood tests, the participants fasted for >12 h before blood collection. The total cholesterol (TC), low-density lipoprotein cholesterol (LDL-C), high-density lipoprotein cholesterol (HDL-C), triglyceride (TG), non-HDL-cholesterol (NonHDL-C), ALT, AST, gamma-glutamyl transferase, BUN, creatinine, glucose, and high sensitivity C-reactive protein levels were measured using the Hitachi 7600 automatic analyzer (Hitachi, Tokyo, Japan). Serum leptin and plasma adiponectin levels were measured via radioimmunoassay using a human leptin kit (Linco Research, St. Charles, MO, USA).

2.6. *Statistical Analyses*

All statistical analyses were performed using PASW statistics 23 (previously SPSS statistics) (SPSS version 23.0; IMP SPSS, Chicago, IL, USA). All data are expressed as the mean $\pm$ standard error or percentages (%) for categorical variables. Values of $p < 0.05$ were considered significant.

The sample size was determined to achieve 80% statistical power with an alpha of 0.05. The sample size for each group was determined by allowing a dropout rate of 20%. Efficacy parameters were analyzed in the per-protocol group, and safety parameters were analyzed in the intention-to-treat group. A chi-square test was performed to determine baseline differences in the frequencies of categorized variables between groups. Student's paired t-test was performed to assess differences between groups before and after the 6-week intervention period. A linear mixed-effects model was applied to the repeated measures data for each continuous outcome variable and data. An expert analyzed the 24-h dietary intake data using Can-Pro 3.0 software (Korean Nutrition Society, Seoul, South Korea).

3. Results

3.1. Participants

A total of 62 participants were recruited; four were excluded during the screening process, two did not meet the inclusion criteria, and two declined participation. The remaining 58 patients were randomly assigned to the following three groups: HTK, LTK, and CK. All participants completed the 6-week study, except for four who withdrew consent. The final analysis included a total of 54 participants in the HTK (n = 19), LTK (n = 18), and CK (n = 17) groups (Figure 1). Four individuals dropped out, two withdrew consent, one required treatment for a hand fracture, and one required treatment for dermatosis. This was not related to Kochujang supplementation, and none of the participants complained of side effects.

3.2. Anthropometric Parameters

The general characteristics of the participants are shown in Table 2. Cross-analysis was performed for sex, alcohol consumption, and smoking; analysis of variance was performed for age, weight, and BMI. There were no significant differences in baseline characteristics among the three groups, such as age, sex, weight, height, alcohol consumption, smoking, initial weight, and initial BMI. No significant changes in caloric intake were found within or between groups in the dietary intake survey ($p > 0.05$). No significant changes in the metabolic equivalent of the task were found within or between groups in the physical activity survey ($p > 0.05$).

Table 2. General characteristics of participants.

Value	Group			
	HTK (n = 19)	LTK (n = 18)	CK (n = 17)	p
Sex (M/F)	6/13	9/11	5/14	0.447
Alcohol consumption (n)	11 (57.9)	10 (50.0)	8 (42.1)	0.623
Smoking (n)	1 (5.3)	3 (15.0)	1 (5.3)	0.455
Age (years)	41.58 ± 9.19	41.1 ± 10.08	37.16 ± 11.21	0.345
Weight (kg)	70.99 ± 10.21	71.05 ± 9.16	71.44 ± 12.58	0.991
BMI (kg/m^2)	26.02 ± 2.32	25.87 ± 2.08	26.03 ± 2.71	0.975

The HTK, traditional Kochujang containing a high-dose of beneficial microbes; LTK, traditional Kochujang containing a low-dose of effective microbes; CK, commercially prepared Kochujang; M, male; F, female; BMI, body mass index. Values are presented as the mean ± standard deviation or number (percentage). A significant difference between three groups, p-value for independent t-test. A significant difference between three groups, p-value for independent Chi-square test.

3.3. Efficacy Evaluation

Waist circumference (WC) decreased significantly after Kochujang supplementation in the HTK and CK groups. Hip circumference (HC) and waist-hip ratio (WHR) decreased in all three groups; however, the reduction was only significant in the CK group. TC, LDL-C, HDL-C, TG, and NonHDL-C levels decreased in the HTK and LTK groups; however, the reduction was significant in only the HTK group (Figure 3). Visceral fat area, subcutaneous fat area, and visceral fat/subcutaneous fat ratio (V/S) decreased in the HTK and LTK groups; however, the decrease in the visceral fat area was significant only in the HTK group (Figure 4). In all three groups, there was no significant change in BMI, body fat mass (BFM), or percent of body fat (PBF) before and after treatment (Table 3).

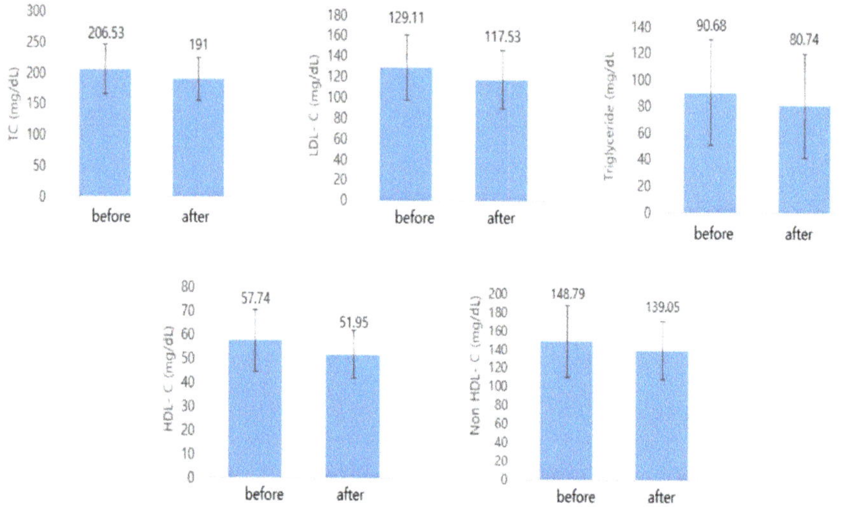

Figure 3. Lipid profile before and after Kochujang treatment in the HTK group. HTK, traditional Kochujang containing a high dose of beneficial microbes. TC, total cholesterol; LDL-C, low-density lipoprotein cholesterol; HDL-C, high-density lipoprotein cholesterol; NonHDL-C, non-HDL-C.

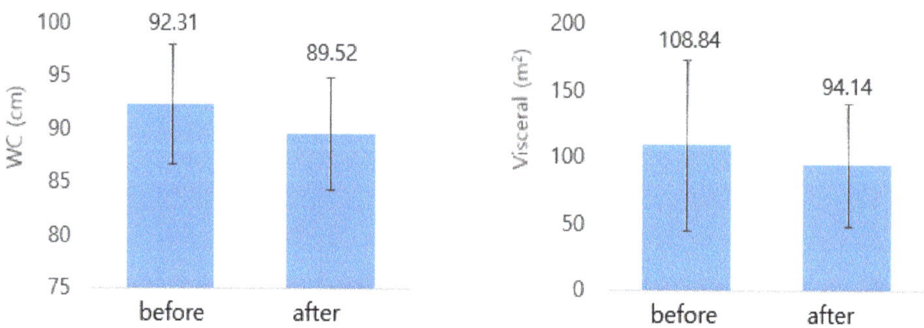

Figure 4. WC and visceral fat area before and after Kochujang treatment in the HTK group. HTK, traditional Kochujang containing a high dose of beneficial microbes; WC, waist circumference.

3.4. Microbiome Analysis

The presence of beneficial, harmful, and other microorganisms was evaluated in stool samples. After Kochujang supplementation, the population of beneficial microorganisms increased in all three groups, and that of harmful microorganisms decreased in the HTK and CK groups; however, this was not significant. Beneficial microorganisms included *Lactobacillus* spp., *Bifidobacterium* spp., *Lactococcus lactis*, *Enterococcus faecium*, and Bacteroides. Harmful bacteria include *Clostridium perfringens*, *Bacteroides eggerthii*, *Sutterella stercoricanis*, *Ruminococcus torques*, *Parabacteroides merdae*, and *Parabacteroides distasonis* (Table 4).

Table 3. Efficacy comparison between the three groups.

Value	Group								
	HTK (n = 19)			LTK (n = 18)			CK (n = 17)		
	Before	**After**	p	**Before**	**After**	p	**Before**	**After**	p
HC (cm)	101.46 ± 5.64	100.5 ± 4.99	0.099	100.84 ± 4.64	99.48 ± 4.01	0.057	101.66 ± 6.88	100.38 ± 6.73	0.022
WHR	0.91 ± 0.03	0.89 ± 0.03	0.053	0.91 ± 0.04	0.91 ± 0.03	0.956	0.91 ± 0.05	0.89 ± 0.05	0.024
TC (mg/dL)	206.53 ± 39.91	191.00 ± 34.95	0.011	219.06 ± 33.73	218.56 ± 38.31	0.901	205.59 ± 35.21	205 ± 36.92	0.928
LDL-C (mg/dL)	129.11 ± 32.07	117.53 ± 28.50	0.020	138.94 ± 31.16	138.44 ± 35.68	0.912	123.24 ± 30.51	124.41 ± 32.07	0.822
HDL-C (mg/dL)	57.74 ± 13.12	51.95 ± 10.10	0.003	56.28 ± 13.23	55.17 ± 14.87	0.420	62.29 ± 12.39	60.24 ± 9.93	0.152
Triglyceride (mg/dL)	90.68 ± 40.19	80.74 ± 39.20	0.003	118.89 ± 58.96	131.5 ± 60.78	0.389	99.35 ± 78.48	95.47 ± 100.43	0.638
Non HDL-C (mg/dL)	148.79 ± 38.40	139.05 ± 31.44	0.050	162.78 ± 32.57	163.39 ± 37.34	0.869	143.29 ± 33.75	144.76 ± 37.88	0.796
WC (cm)	92.31 ± 5.66	89.52 ± 5.33	0.006	91.94 ± 5.67	90.76 ± 5.56	0.248	92.34 ± 9.21	89.19 ± 8.58	0.002
Adiponectin (μg/mL)	21.48 ± 7.12	20.75 ± 5.53	0.497	20.71 ± 10.84	23.22 ± 11.93	0.099	29.69 ± 10.44	27.59 ± 14.22	0.367
Leptin (ng/mL)	17,109.06 ± 14,015.12	20,527 ± 20,871.21	0.286	12,258.87 ± 10,118.63	15,728.07 ± 8290.3	0.141	23,299 ± 25,617.05	21,998.07 ± 21,734.74	0.764
VF (m^2)	108.84 ± 63.81	94.14 ± 46.29	0.021	122.18 ± 61.54	119.15 ± 63.47	0.272	90.79 ± 60.55	94.95 ± 58.19	0.086
SF (m^2)	202.66 ± 64.94	204.32 ± 57.27	0.789	184.94 ± 57.09	182.43 ± 60.39	0.520	218.91 ± 101.03	219.41 ± 83.6	0.949
V/S	0.58 ± 0.48	0.54 ± 0.37	0.403	0.69 ± 0.36	0.69 ± 0.39	0.894	0.45 ± 0.29	0.46 ± 0.28	0.732
GGT (IU/L)	22.11 ± 14.9	21.42 ± 10.95	0.682	25.78 ± 17.45	26.5 ± 19.05	0.622	22.18 ± 18.56	21.76 ± 19.24	0.744
AST (IU/L)	22.05 ± 5.3	22.74 ± 4.85	0.495	22.78 ± 7.97	24.44 ± 10	0.182	20.59 ± 6.96	22 ± 7.09	0.148
ALT (IU/L)	20.79 ± 8.69	21.11 ± 7.79	0.811	27.61 ± 23.86	28.67 ± 28.73	0.668	23.18 ± 14.69	23 ± 17.48	0.928
BUN (mg/dL)	12.63 ± 3.00	13.68 ± 3.43	0.030	13.67 ± 4.7	13.17 ± 4.41	0.480	12.41 ± 3.22	12.18 ± 4.45	0.797

Table 3. *Cont.*

Value	Group								
	HTK (n = 19)			LTK (n = 18)			CK (n = 17)		
	Before	After	p	Before	After	p	Before	After	p
Cr (mg/dL)	0.75 ± 0.17	0.77 ± 0.18	0.108	0.77 ± 0.17	0.78 ± 0.18	0.472	0.71 ± 0.14	0.71 ± 0.16	0.921
Glucose (mg/dL)	105.68 ± 10.59	103.42 ± 8.01	0.250	106.28 ± 10.85	103.89 ± 10.35	0.112	103.94 ± 8.22	101.12 ± 9.77	0.185
hs-CRP (mg/dL)	1.76 ± 3.69	0.80 ± 0.42	0.255	2.59 ± 5.91	1.3 ± 1.16	0.362	1.32 ± 1.93	0.94 ± 0.73	0.421
BMI (kg/m^2)	26.02 ± 2.32	26.20 ± 2.43	0.201	25.87 ± 2.08	25.97 ± 2.1	0.462	26.03 ± 2.71	26.25 ± 2.63	0.127
Weight (kg)	70.99 ± 10.21	71.44 ± 10.24	0.228	71.05 ± 9.16	71.26 ± 8.94	0.564	71.44 ± 12.58	72.09 ± 12.56	0.095
BFM (kg)	22.74 ± 5.02	23.29 ± 4.79	0.105	23.14 ± 5.04	23.23 ± 5.06	0.798	24.2 ± 8.17	24.52 ± 7.92	0.201
PBF (%)	32.31 ± 6.82	32.87 ± 6.48	0.118	32.76 ± 6.73	32.73 ± 6.65	0.933	33.83 ± 9.14	34.02 ± 9.04	0.460

HTK, traditional Kochujang containing a high-dose of beneficial microbes; LTK, traditional Kochujang containing a low-dose of effective microbes; CK, commercially prepared Kochujang; HC, Hip circumference; WC, waist circumference; BMI, body mass index, WHR, waist-hip ratio; TC, total cholesterol, LDL-C, low-density lipoprotein cholesterol; HDL-C, high-density lipoprotein cholesterol; TG, triglyceride, NonHDL-C, NonHDL-cholesterol; VF, visceral fat; SF, subcutaneous fat; V/S, visceral fat/subcutaneous fat ratio; GGT, gamma-GT; AST, aspartate transaminase; ALT, alanine transaminase; BUN, blood urea nitrogen; Cr, creatinine, glucose, hs-CRP, high-sensitivity C-reactive protein; BFM, body fat mass; PBF, percent of body fat. Values are presented as the mean ± standard deviation. A paired t-test was performed to analyze the statistical significance of each variable before and after administration of Kochujang.

Table 4. Comparison of microbiome changes in stools between three groups.

Value	Group								
	HTK (n = 19)			LTK (n = 18)			CK (n = 17)		
	Before (%)	After (%)	p	Before (%)	After (%)	p	Before (%)	After (%)	p
Beneficial Bacteria	26.82 ± 11.26	31.04 ± 11.73	0.128	25.61 ± 6.53	29.79 ± 11.44	0.206	27.59 ± 11.72	28.05 ± 9.03	0.898
Harmful Bacteria	5.03 ± 10.01	4.11 ± 7.95	0.157	3.00 ± 2.91	3.16 ± 5.59	0.846	2.88 ± 2.27	2.34 ± 1.34	0.294
Others	37.11 ± 16.70	64.85 ± 10.55	<0.0001	41.59 ± 13.99	67.05 ± 14.71	<0.0001	41.48 ± 15.31	69.61 ± 9.30	<0.0001

HTK, traditional Kochujang containing a high-dose of beneficial microbes; LTK, traditional Kochujang containing a low-dose of effective microbes; CK, commercially prepared Kochujang. Values are presented as the mean ± standard deviation. Comparison of microbiome changes in stools after Kochujang administration between three groups.

4. Discussion

Epidemiological studies have shown that the incidence of obesity is increasing worldwide and in Korea [15]. Fermented foods have been suggested to serve as a source of healthy nutrients such as dietary fiber, antioxidants, phytochemicals, and minerals [16]. Among various fermented products, Kochujang is a traditional Korean sauce that can be used to prepare various food products or consumed for its health benefits [17]. The anti-obesity effect of Kochujang has been proven in several animal experiments [4,18]. In the present study, we presented evidence that Kochujang consumption may reduce visceral fat content and improve blood lipid profiles in overweight and obese patients. We also found that the population of beneficial bacteria may increase in the stool microbiome upon Kochujang supplementation.

Kochujang can be prepared via two approaches. One approach involves the fermentation of soybean paste (fermented soybean), glutinous rice, and red pepper powder by bacteria (Bacillus subtilis and *Bacillus licheniformis*) or yeast (*Saccharomyces cerevisiae* and *Zygosaccharomyces rouxii*) [19]. This traditional homemade Kochujang preparation method requires at least 6 months of fermentation [19]. The quality of Meju added to Kochujang greatly depends on the microorganisms used in the fermentation process [19]. The second method of Kochujang preparation involves the commercial method, in which steamed rice inoculated with yeast (*A. oryzae*) is used. One of the advantages of using koji is the short period required for fermentation, which usually ranges from 2 weeks to 1 month [19].

Traditional Kochujang contains more carbohydrate sources than those present in other traditional sauces and may contain a large population of various bacteria [20]. Since the presence of beneficial bacteria such as *Bacillus* spp. in Kochujang varies with fermentation conditions and component composition [20], the health effects may differ between traditional Kochujang containing varying amounts of microorganisms and commercial Kochujang. Therefore, we divided the traditional Kochujang administration group into two groups and used Kochujang inoculated with a high content of an effective strain and Kochujang inoculated with a low content. The anti-obesity effects of two types of traditional Kochujang and commercially available Kochujang were compared.

A previous study investigated the anti-obesity effect of Kochujang according to the polymorphism of the obesity-related gene, peroxisome proliferator activator receptor c (PPARc2) in overweight/obese subjects [21]. Study participants with BMI $\geq$ 23 or a high waist/hip ratio (WHR; $\geq$0.90 for men and $\geq$0.85 for women) received a placebo or Kochujang for 12 weeks. The result showed that neither Kochujang nor placebo had any effect on anthropometric measures such as BMI, WC, BP, VF area, and SF area after the intervention [21]. However, in our study, WC and VF areas decreased after Kochujang treatment with a high content of effective microorganisms added. This may be a result of the increase in the anti-obesity effect of the high-content effective microorganism.

Traditional Kochujang and commercial Kochujang reduced weight gain in a dietary intake experiment with rats [22]. The study investigated the effect of Kochujang on the reduction of body weight, adipose tissue, and serum lipid levels in male Sprague-Dawley rats fed a high-fat diet containing 10% of unfermented traditional Kochujang, fermented traditional Kochujang, or fermented commercial Kochujang. The changes in lipid profile, body weight, and weight of each organ and adipose tissue were measured. The results showed that Kochujang had an anti-obesity effect; however, the non-fermented Kochujang group had no significant effect on organ fat reduction, weight loss, cholesterol, or triglycerides [22].

Perirenal fat and hepatic total lipids were decreased in Kochujang-fed rats. In a human study with overweight participants, the group that was fed Kochujang for 12 weeks showed no change in body weight and WHR compared to the group that was fed the placebo; however, they did show a significant decrease in visceral fat. In addition, serum concentrations of triglycerides and apolipoprotein B were also decreased in the Kochujang-fed group compared to that of the placebo group [23]. In rats fed a high-fat diet, Kochujang supplementation reduces the increase in body fat content by 30%, which has been found to be mediated via an increase in energy expenditure and activity of brown adipocytes [24]. In a similar rat experiment, it was reported that the addition of capsaicin has an anti-obesity effect that reduces the increase in body fat content by 30%, which is attributed to an increase in mitochondrial activity and beta-adrenergic activation of brown adipocytes [25]. However, they reported that capsaicin supplementation does not improve cholesterol, TGL, and HDL-C profiles [25]. The increase in energy expenditure induced by capsaicin occurs via the secretion of catecholamine from the adrenal medulla [26]. Substance P acts as a cooperating factor in the cholinergic regulation of adrenal catecholamine release [26].

Fermentation improves the metabolic function of active ingredients by converting inactive forms present in food to active mediators [27]. It has been reported that daidzein and genistein levels are considerably increased upon fermentation of soybeans [27]. The increased daidzein and genistein content in fermented soybeans is attributed to the contribution of microbial enzymes to hydrolysis during fermentation [28]. During sufficient fermentation, various enzymes are produced by yeast and lactic acid bacteria [28].

Traditional Kochujang composition varies with the province of Korea in production, carbohydrate content, and bacteria present in the fermentation environment [20]. Since Kochujang has a lower salinity and higher carbohydrate content than soybean paste, the growth of harmful bacteria may increase [29]. Traditional Kochujang of high quality should be well managed and contain a large number of beneficial bacteria and <0.5% harmful bacteria [29]. Fecal transplantation studies provide convincing evidence for an association between the microbiome and blood lipids. Indeed, many metabolic phenotypes, including lipid metabolism, obesity, glucose response, and insulin sensitivity, can be transmitted to the host via fecal transplantation. Metagenomic sequencing (MGS) profiling yielded an average of 3.1 Gb of sequencing data for each sample, aligned to 18.6 million bacterial genes. Different microorganisms have varying effects on lipids, further demonstrating that triglycerides with different sized fatty acid moieties are associated with different microorganisms [30]. Although the exact mechanisms of microbiome types and human lipid metabolism are still being studied, they show potential as individualized therapeutics in the future.

The inoculation of Kochujang with beneficial bacteria ensures that the bacterial population is controlled and has specific functions. Kochujang inoculated with *Bacillus* spp. has anti-cerebrovascular disease effects, and the properties differ depending on the province of Korea it is prepared in [20]. Traditional Kochujang has also been found to show a difference in functionality depending on the abundance of microorganisms [20].

We found that Kochujang supplementation had anti-obesity effects. The visceral fat content decreased only in the HTK group. A previous study evaluated the abdominal fat CT scans and blood lipid profiles after administering Kochujang to overweight Korean adults for 12 weeks. They found that although the body weight and WHR did not change, there was a significant reduction in the visceral fat area [23]. In addition, serum TG and apolipoprotein B levels decreased in the Kochujang-administered group compared with those in the control group [23]. These results can be attributed to bioactive compounds such as isoflavone aglycones, peptides in fermented soy, and capsaicin present in red pepper powder via the activity of the obesity-linked gene and peroxisome proliferator activator receptor γ (PPARγ2) [21].

Adiponectin and leptin are adipokines that considerably influence obesity-related metabolic diseases by modulating fat metabolism, energy homeostasis, and insulin sensitivity. Adiponectin and leptin levels are inversely related to the progression of obesity, insulin resistance, and atherosclerosis [31]. In our study, there was no significant change in the levels of leptin and adiponectin in any of the three groups. Previously, leptin secretion was measured upon adding Kochujang and garlic-added Kochujang to cultured 3T3-L1 adipocytes [32]; leptin secretion was found to decrease by 34% and 48%, respectively, compared with that in control adipocytes. mRNA expression levels of obesity-related genes such as tumor necrosis factor-alpha, PPARγ, CCAAT/enhancer-binding protein alpha, and sterol regulatory element-binding transcription factor 1 (SREBP1c) have also been found to decrease in cells treated with Kochujang and garlic-added Kochujang compared with those in the control group [32]. In another study, when 3T3-L1 adipocytes were treated with Kochujang extract, adipocyte size and leptin secretion were found to decrease [33]. The study reported that Kochujang suppresses lipogenesis via downregulation of obesity-related genes SREBP-1c and PPAR-quin and stimulates lipolysis due to increased hormone-sensitive lipase activity [33]. Among healthy Korean participants, a study found that the adiponectin level and risk of metabolic syndrome were reduced with an increase in kimchi consumption [34].

Owing to the clinical nature of this study, it cannot provide information on the exact mechanism of the chemical compounds and their interactions responsible for the biological activity of Kochujang. However, based on the results of previous studies, the anti-obesity effect of Kochujang, which contains a high number of beneficial microorganisms, may be due to the transformation of components such as lipolytic enzymes, certain end-products of Kochujang, and capsaicin into a more active form or some combination of these. Additionally, isomaltooligosaccharide (0.8~6.5%) and small peptides are produced, which have a beneficial effect on fat metabolism [33].

Kochujang ingestion can cause digestive symptoms such as heartburn. Some studies suggest that capsaicin, one of its components, may increase the risk of gastric adenocarcinoma in mice. However, there are studies showing that after adjusting for related factors, there is no relationship with gastric adenocarcinoma; thus, there is no evidence of any negative effects of Kochujang at this point [35].

A major limitation of our study is the small sample size, which limits the generalizability of our results to other populations of overweight or obese individuals. Additionally, changes in certain obesity-related indicators were not significant. During the course of the trial, the food intake and activity levels of the participants were not controlled because we relied on self-reported data.

5. Conclusions

In conclusion, the anti-obesity effect of traditional Kochujang may be greater than that of commercial Kochujang, which can be further increased with an increased number of beneficial bacteria present. Traditional Kochujang incorporated into the diet can prevent obesity. Despite the limitations, our study is meaningful in that it observed the anti-obesity effect of Kochujang in human participants with respect to various indicators. The study also compared the number of beneficial bacteria between traditional and commercial Kochujang. Future studies should address the above limitations.

Author Contributions: Conceptualization, D.-Y.J. and H.-J.Y.; methodology, D.-Y.J. and H.-J.Y.; formal analysis, A.L.H. and S.-J.J.; investigation, A.L.H., D.-S.P., M.-S.R. and H.-J.Y.; writing—review and editing, A.L.H.; Graphic abstract, H.K.L. All authors have read and agreed to the published version of the manuscript.

Funding: This research was funded by the Ministry of Agriculture, Food and Rural Affairs and the Korea Agro-Fisheries and Food Trade Corporation.

Institutional Review Board Statement: The study was conducted in accordance with the Declaration of Helsinki and approved by the Institutional Review Board (or Ethics Committee) of Wonkwang University Hospital (IRB approval no. 2021-04-046).

Informed Consent Statement: Informed consent was obtained from all subjects involved in the study.

Data Availability Statement: Raw data can be provided upon request.

Conflicts of Interest: The authors declare no conflict of interest.

References

1. Chooi, Y.C.; Ding, C.; Magkos, F. The Epidemiology of Obesity. *Metabolism* **2019**, *92*, 6–10. [CrossRef] [PubMed]
2. Kwon, D.Y.; Jang, D.J.; Yang, H.J.; Chung, K.R. History of Korean Gochu, Gochujang, and Kimchi. *J. Ethn. Foods* **2014**, *1*, 3–7. [CrossRef]

3. Anlier, N.; Gökcen, B.B.; Sezgin, A.C. Health Benefits of Fermented Foods. *Crit. Rev. Food Sci. Nutr.* **2019**, *59*, 506–527. [CrossRef] [PubMed]
4. Shin, H.W.; Jang, E.S.; Moon, B.S.; Lee, J.J.; Lee, D.E.; Lee, C.H.; Shin, C.S. Anti-obesity Effects of Gochujang Products Prepared Using Rice Koji and Soybean Meju in Rats. *J. Food Sci. Technol.* **2016**, *53*, 1004–1013. [CrossRef]
5. Kawada, T.; Hagihara, K.I.; Iwai, K. Effects of Capsaicin on Lipid Metabolism in Rats Fed a High Fat Diet. *J. Nutr.* **1986**, *116*, 1272–1278. [CrossRef]
6. Ludy, M.J.; Mattes, R.D. The Effects of Hedonically Acceptable Red Pepper Doses on Thermogenesis and Appetite. *Physiol. Behav.* **2011**, *102*, 251–258. [CrossRef]
7. Janssens, P.L.; Hursel, R.; Westerterp-Plantenga, M.S. Capsaicin Increases Sensation of Fullness in Energy Balance, and Decreases Desire to Eat After Dinner in Negative Energy Balance. *Appetite* **2014**, *77*, 44–49. [CrossRef]
8. Rigamonti, A.E.; Casnici, C.; Marelli, O.; De Col, A.; Tamini, S.; Lucchetti, E.; Tringali, G.; De Micheli, R.; Abbruzzese, L.; Bortolotti, M.; et al. Acute Administration of Capsaicin Increases Resting Energy Expenditure in Young Obese Subjects Without Affecting Energy Intake, Appetite, and Circulating Levels of Orexigenic/Anorexigenic Peptides. *Nutr. Res.* **2018**, *52*, 71–79. [CrossRef]
9. Baboota, R.K.; Murtaza, N.; Jagtap, S.; Singh, D.P.; Karmase, A.; Kaur, J.; Bhutani, K.K.; Boparai, R.K.; Premkumar, L.S.; Kondepudi, K.K.; et al. Capsaicin-Induced Transcriptional Changes in Hypothalamus and Alterations in Gut Microbial Count in High Fat Diet Fed Mice. *J. Nutr. Biochem.* **2014**, *25*, 893–902. [CrossRef]
10. Soh, J.R.; Shin, D.H.; Kwon, D.Y.; Cha, Y.S. Effect of Cheonggukjang Supplementation Upon Hepatic Acyl-CoA Synthase, Carnitine Palmitoyltransferase I, Acyl-CoA Oxidase and Uncoupling Protein 2 mRNA Levels in C57BL/6J Mice Fed with High Fat Diet. *Genes Nutr.* **2008**, *2*, 365–369. [CrossRef]
11. Nam, Y.D.; Park, S.L.; Lim, S.I. Microbial Composition of the Korean Traditional Food "Kochujang" Analyzed by a Massive Sequencing Technique. *J. Food Sci.* **2012**, *77*, M250–M256. [CrossRef] [PubMed]
12. Kim, Y. The Korea National Health and Nutrition Examination Survey (KNHANES): Current Status and Challenges. *Epidemiol. Health* **2014**, *36*, e2014002. [CrossRef] [PubMed]
13. Shim, J.S.; Oh, K.; Kim, H.C. Dietary Assessment Methods in Epidemiologic Studies. *Epidemiol. Health* **2014**, *36*, e2014009. [CrossRef]
14. Armstrong, T.; Bull, F. Development of the World Health Organization Global Physical Activity Questionnaire (GPAQ). *J. Public Health* **2006**, *14*, 66–70. [CrossRef]
15. Ranasinghe, P.; Mathangasinghe, Y.; Jayawardena, R.; Hills, A.P.; Misra, A. Prevalence and Trends of Metabolic Syndrome among Adults in the Asia-Pacific Region: A Systematic Review. *BMC Public Health* **2017**, *17*, 101. [CrossRef] [PubMed]
16. Septembre-Malaterre, A.; Remize, F.; Poucheret, P. Fruits and Vegetables, as a Source of Nutritional Compounds and Phytochemicals: Changes in Bioactive Compounds during Lactic Fermentation. *Food Res. Int.* **2018**, *104*, 86–99. [CrossRef] [PubMed]
17. Kwon, D.Y.; Chung, K.R.; Yang, H.J.; Jang, D.J. Gochujang (Korean Red Pepper Paste): A Korean Ethnic Sauce, Its Role and History. *J. Ethn. Foods* **2015**, *2*, 29–35. [CrossRef]
18. Son, H.K.; Shin, H.W.; Jang, E.S.; Moon, B.S.; Lee, C.H.; Lee, J.J. Gochujang Prepared Using Rice and Wheat Koji Partially Alleviates High-Fat Diet-Induced Obesity in Rats. *Food Sci. Nutr.* **2020**, *8*, 1562–1574. [CrossRef]
19. Shin, D.H.; Kim, D.H.; Choi, U.; Lim, D.K.; Lim, M.S. Studies on the Physicochemical Characteristics of Traditional Kochujang. *Korean J. Food Sci. Technol.* **1996**, *28*, 157–161.
20. Ha, G.; Yang, H.J.; Ryu, M.S.; Jeong, S.J.; Jeong, D.Y.; Park, S. Bacterial Community and Anti-cerebrovascular Disease-Related Bacillus species Isolated From Traditionally Made Kochujang From Different Provinces of Korea. *Microorganisms* **2021**, *9*, 2238. [CrossRef]
21. Lee, Y.; Cha, Y.S.; Park, Y.; Lee, M. PPARã2 C1431T Polymorphism Interacts With the Antiobesogenic Effects of Kochujang, a Korean Fermented, Soybean-Based Red Pepper Paste, in Overweight/Obese Subjects: A 12-Week, Double-Blind Randomized Clinical Trial. *J. Med. Food* **2017**, *20*, 610–617. [CrossRef] [PubMed]
22. Rhee, S.H.; Kong, K.R.; Jung, K.O.; Park, K.Y. Decreasing Effect of Kochujang on Body Weight and Lipid Levels of Adipose Tissues and Serum in Rats Fed a High-Fat Diet. *J. Korean Soc. Food Sci. Nutr.* **2003**, *32*, 882–886. [CrossRef]
23. Cha, Y.S.; Kim, S.R.; Yang, J.A.; Back, H.I.; Kim, M.G.; Jung, S.J.; Song, W.O.; Chae, S.W. Kochujang, Fermented Soybean-Based Red Pepper Paste, Decreases Visceral Fat and Improves Blood Lipid Profiles in Overweight Adults. *Nutr. Metab.* **2013**, *10*, 24. [CrossRef]
24. Choo, J.J. Antiobesity Effects of Kochujang in Rats Fed on a High-Fat Diet. *Korean J. Nutr.* **2000**, *33*, 787–793.
25. Kawada, T.; Watanabe, T.; Takaishi, T.; Tanaka, T.; Iwai, K. Capsaicin-Induced Beta-Adrenergic Action on Energy Metabolism in Rats: Influence of Capsaicin on Oxygen Consumption, the Respiratory Quotient, and Substrate Utilization. *Proc. Soc. Exp. Biol. Med.* **1986**, *183*, 250–256. [CrossRef]
26. Watanabe, T.; Kawada, T.; Iwai, K. Enhancement by Capsaicin of Energy Metabolism in Rats through Secretion of Catecholamine From Adrenal Medulla. *Agric. Biol. Chem.* **1987**, *51*, 75–79. [CrossRef]
27. Chiou, R.Y.; Cheng, S.L. Isoflavone Transformation during Soybean Koji Preparation and Subsequent Miso Fermentation Supplemented with Ethanol and NaCl. *J. Agric. Food Chem.* **2001**, *49*, 3656–3660. [CrossRef]

28. Okabe, Y.; Shimazu, T.; Tanimoto, H. Higher Bioavailability of Isoflavones After a Single Ingestion of Aglycone-Rich Fermented Soybeans Compared with Glucoside-Rich Non-fermented Soybeans in Japanese Postmenopausal Women. *J. Sci. Food Agric.* **2011**, *91*, 658–663. [CrossRef]
29. Shin, D.H.; Jeong, D.Y. The Present Status of Kochujang (Fermented Hot Pepper-Soybean Paste) in Korea and Its Future. *J. Korean Soc. Food Sci. Nutr.* **2004**, *11*, 238–245.
30. Wang, Z.; Koonen, D.; Hofker, M.; Fu, J. Gut microbiome and lipid metabolism: From associations to mechanisms. *Curr. Opin. Lipidol.* **2016**, *27*, 216–224. [CrossRef]
31. Scherer, P.E. Adipose Tissue: From Lipid Storage Compartment to Endocrine Organ. *Diabetes* **2006**, *55*, 1537–1545. [CrossRef] [PubMed]
32. Kong, C.S.; Park, K.Y. Anti-obesity Effect of Garlic-Added Kochujang in 3T3-L1 Adipocytes. *Prev. Nutr. Food Sci.* **2008**, *13*, 66–70. [CrossRef]
33. Ahn, I.S.; Do, M.S.; Kim, S.O.; Jung, H.S.; Kim, Y.I.; Kim, H.J.; Park, K.Y. Antiobesity Effect of Kochujang (Korean Fermented Red Pepper Paste) Extract in 3T3-L1 Adipocytes. *J. Med. Food* **2006**, *9*, 15–21. [CrossRef]
34. Oh, I.M.; Joung, H.J.; Oh, S.W.; Yoon, Y.S.; Yoo, K.H.; Park, J.E.; Park, J.S.; Jang, E.J.; Park, S.J.; Park, S.W.; et al. Relationship of Combined Consumption of Rice and Kimchi, Korean Traditional Diet and the Risk of Metabolic Syndrome in Healthy Korean Volunteers. *J. Clin. Nutr.* **2013**, *5*, 110–116. [CrossRef]
35. Toth, B.; Rogan, E.; Walker, B. Tumorigenicity and mutagenicity studies with capsaicin of hot peppers. *Anticancer Res.* **1984**, *4*, 117–119.

nutrients

Article

Phytochemical Combination (*p*-Synephrine, *p*-Octopamine Hydrochloride, and Hispidulin) for Improving Obesity in Obese Mice Induced by High-Fat Diet

Dahae Lee [1], Ji Hwan Lee [1], Byoung Ha Kim [2], Sanghyun Lee [3], Dong-Wook Kim [4,*] and Ki Sung Kang [1,*]

1 College of Korean Medicine, Gachon University, Seongnam 13120, Korea; pjsldh@gachon.ac.kr (D.L.); kleert26@naver.com (J.H.L.)
2 D. Nature Co., Ltd., Seongnam 13174, Korea; mot37@d-nature.co.kr
3 Department of Plant Science and Technology, Chung-Ang University, Anseong 17546, Korea; slee@cau.ac.kr
4 College of Pharmacy, Wonkwang University, Iksan 54538, Korea
* Correspondence: pharmengin1@wku.ac.kr (D.-W.K.); kkang@gachon.ac.kr (K.S.K.); Tel.: +82-43-229-7984 (D.-W.K.); +82-31-750-5402 (K.S.K.)

Abstract: Obesity treatment efficiency can be increased by targeting both central and peripheral pathways. In a previous study, we identified two natural compounds (hispidulin and *p*-synephrine) that affect adipocyte differentiation. We tested whether obesity treatment efficiency may be improved by adding an appetite-controlling agent to the treatment in the present study. Alkaloids, such as *p*-octopamine, are adrenergic agonists and are thus used as dietary supplements to achieve weight loss. Here, we assessed anti-obesity effects of a mixture of *p*-synephrine, *p*-octopamine HCl, and hispidulin (SOH) on murine preadipocyte cells and on mice receiving a high-fat diet (HFD). SOH showed stronger inhibition of the formation of red-stained lipid droplets than co-treatment with hispidulin and *p*-synephrine. Moreover, SOH reduced the expression of adipogenic marker proteins, including CCAAT/enhancer-binding protein alpha, CCAAT/enhancer-binding protein beta, and peroxisome proliferator-activated receptor gamma. In the HFD-induced obesity model, body weight and dietary intake were lower in mice treated with SOH than in the controls. Additionally, liver weight and the levels of alanine aminotransferase and total cholesterol were lower in SOH-treated mice than in the controls. In conclusion, our results suggest that consumption of SOH may be a potential alternative strategy to counteract obesity.

Keywords: hispidulin; *p*-synephrine; *p*-octopamine hydrochloride; adipocytes; adipogenesis

Citation: Lee, D.; Lee, J.H.; Kim, B.H.; Lee, S.; Kim, D.-W.; Kang, K.S. Phytochemical Combination (*p*-Synephrine, *p*-Octopamine Hydrochloride, and Hispidulin) for Improving Obesity in Obese Mice Induced by High-Fat Diet. *Nutrients* **2022**, *14*, 2164. https://doi.org/10.3390/nu14102164

Academic Editors: Satoshi Nagaoka and Lindsay Brown

Received: 4 April 2022
Accepted: 19 May 2022
Published: 23 May 2022

1. Introduction

Obesity is the underlying cause of various metabolic disorders such as hypertension, hyperglycemia, and diabetes. Therefore, it has become a global medical issue [1]. Imbalances between energy intake and consumption can lead to excessive fat accumulation and functional and morphological changes in adipocytes, which is characteristic of obesity [2,3]. Typically, obesity develops due to an increased number and size of adipocytes. Therefore, controlling the differentiation of preadipocytes into mature adipocytes (termed adipogenesis) is important in the treatment of obesity [4]. Although the precise molecular mechanisms underlying obesity remain unclear, recent studies indicated that CCAAT/enhancer-binding protein alpha (C/EBPα), C/EBPβ, and peroxisome proliferator-activated receptor gamma (PPARγ) are important regulators released during adipogenesis [5,6].

Several approaches such as inhibition of dietary fat absorption, appetite control, insulin and leptin revival, inhibition of fat synthesis, and increased fat mobilization and burning have been used to develop obesity treatments. The FDA-approved combination drug Qsymia®, which is a combination of the stimulant phentermine and the anti-seizure drug topiramate, highlights the synergistic effect of the individual compounds. As this drug is associated with undesirable side effects, such as increased blood pressure, anxiety, and dizziness, recent studies attempted to develop safe anti-obesity drugs derived from natural products [7,8]. Of note, compounds isolated from natural products for the prevention and treatment of obesity are frequently used as mixtures because such mixtures may exert synergistic effects as a result of different mechanisms of action [9]. For instance, combined treatment with naringin, hesperidin, and *p*-synephrine results in weight loss in human patients [10]. In our previous study, combinations of phytochemicals (*p*-synephrine and hispidulin) produced anti-adipogenic effects in a murine preadipocyte cell line (3T3-L1), and *p*-synephrine and hispidulin exhibited different mechanisms of action [11]. Several studies confirmed low or negligible toxicity of these compounds in animal models [12,13]; however, the anti-obesity effects of co-treatment with *p*-synephrine and hispidulin in animal models are still unclear. Hispidulin has been isolated from *Grindelia argentina* and *Arrabidaea chica* [14]. *p*-Synephrine is abundant in *Citrus aurantium*, and it is biosynthesized through N-methylation of *p*-octopamine [15,16].

We reviewed the literature and found that obesity treatment efficiency may be improved by synchronously targeting central and peripheral pathways using natural products. We hypothesized that *p*-octopamine was a suitable natural ingredient that may exert effects similar to those of Qsymia phentermine. Alkaloids, such as *p*-octopamine, exhibit adrenergic agonist activity and are used as dietary supplements to achieve weight loss. *p*-Octopamine is a hydroxylated phenylethylamine that occurs naturally in animals, including insects. In particular, *p*-octopamine is abundant in *Citrus limon* and *Citrus aurantium*. In obese Zucker rats, intraperitoneal administration of *p*-octopamine results in weight loss without adverse effects [17]. *p*-Octopamine hydrochloride (*p*-octopamine HCl) is used for the treatment of circulatory disorders and hypotensive regulation [15,18,19]; however, its potential anti-obesity effects remain unclear. Here, we assessed antidiabetic effects of a mixture of *p*-synephrine, *p*-octopamine HCl, and hispidulin (SOH) in 3T3-L1 adipocytes and high-fat diet (HFD)-fed mice. We hypothesized that SOH treatment would significantly reduce adipogenesis in 3T3-L1 adipocytes and reduce the body weight of HFD-fed mice.

2. Materials and Methods

2.1. Cell Culture and Adipogenic Differentiation

3T3-L1 mouse preadipocyte cells were obtained from the American Type Culture Collection (Manassas, VA, USA). The 3T3-L1 cells were cultured in Dulbecco's modified Eagle's medium (Cellgro, Manassas, VA, USA) containing antibiotics, i.e., 1% penicillin/streptomycin (P/S; Gaithersburg, MD, USA), and 10% bovine calf serum (Gaithersburg, MD, USA). The 3T3-L1 preadipocytes were seeded in a 24-well plate at 4×10^4 cells/well, and after two days, the medium was replaced by adipogenic differentiation medium containing 1-methyl-3-isobutylxanthine (Sigma-Aldrich, St. Louis, MO, USA), 1% P/S, 0.4 μg/mL dexamethasone (Sigma-Aldrich), 10% fetal bovine serum (FBS; Gaithersburg), and 5 μg/mL insulin (Sigma-Aldrich). After incubation for two days, the cultivation medium was replaced by adipogenic differentiation medium containing 1% P/S, 10% FBS, and 5 μg/mL insulin, which was exchanged every two days. On differentiation day 6, the medium was replaced with adipogenic differentiation medium containing 10% FBS and 1% P/S, followed by cultivation for two days. Hispidulin, *p*-synephrine, and *p*-octopamine HCl were added individually or in combination to the culture medium during adipogenic differentiation. *p*-Synephrine (≥98%), and *p*-octopamine HCl (≥95%) were purchased from Sigma-Aldrich. Hispidulin (≥98%) was obtained from Natural Product Institute of Science and Technology (www.nist.re.kr), Anseong, Korea.

2.2. Cell Viability

Viability of 3T3-L1 cells was measured using a water-soluble tetrazolium salt-1-based colorimetric assay kit (Ez-Cytox Cell Viability Assay Kit; Daeil Lab Service, Seoul, South Korea). 3T3-L1 cells were seeded in a 96-well plate at 1×10^4 cells/well. Hispidulin, *p*-synephrine, and/or *p*-octopamine HCl were then individually added at different concentrations (5, 10, 20, or 40 μM, each) for 8 days, 10 μL EZ-Cytox reagent was added, and absorbance was measured after 40 min using a PowerWave XS microplate reader (Bio-Tek Instruments, Winooski, VT, USA) at 490 nm.

2.3. Oil Red O Staining

After cell differentiation, differentiated cells were fixed using 4% paraformaldehyde solution (Sigma-Aldrich) for 1 h. Fixed cells were stained using Oil Red O solution (ORO; Sigma-Aldrich) for 1 h. After removal of the ORO solution, the cells were washed using distilled water, and images were recorded. Accumulated intracellular ORO was completely eluted using 100% isopropanol and was quantified through spectrophotometric absorbance using a PowerWave XS microplate reader (Bio-Tek Instruments) at 540 nm.

2.4. Western Blotting Analysis

Equal amounts of cellular lysates were evaluated by 10% sodium dodecyl sulfate–polyacrylamide gel and transferred by electroblotting to polyvinylidene difluoride membranes (Pall Corporation, Washington, DC, USA). The proteins were incubated with the primary antibodies, including PPARγ (53 kDa), C/EBPα (42 kDa), C/EBPβ (46 kDa), and GAPDH (37 kDa), overnight at 4 °C, and then incubated with an appropriate secondary antibody for 1 h at room temperature. All antibodies were purchased from Cell Signaling (Boston, MA, USA). Immune complexes were developed with enhanced chemiluminescence reagent (GE Healthcare UK Limited, Buckinghamshire, UK). Reactions were visualized using a chemiluminescence system (FUSION Solo; PEQLAB Biotechnologie GmbH, Erlangen, Germany).

2.5. Study Animals and Experimental Design

Male C57/BL6 mice (five weeks old upon receipt, Daehan Biolink, Chungbuk, Korea) were used for the experiments, after one week of acclimation. During acclimation and experimental periods, four mice were kept per cage at constant temperature (20–25 °C) and humidity (30–35%) and under a 12/12-h light/dark cycle. The mice had ad libitum access to feed and water. Four treatment groups (comprising eight individuals each) were used: (1) normal diet control group (ND); ND and administration of sodium carboxyl methyl cellulose (CMC) solution. (2) HFD control group: HFD after administration of the CMC solution; (3) positive group: HFD and administration of atorvastatin (ATS; 30 mg/kg); and (4) experimental group: HFD and administration of SOH (i.e., a 1:1:1 mixture of hispidulin, synephrine, and *p*-octopamine HCl at 20 mg/kg). The high fat diet (HFD) used food with 60% of energy from lipids, 18% of energy from protein and 21% of energy from carbohydrate (D12492 Research Diets, New Brunswick, NJ, USA). The food intake and body weight during the experimental period were measured weekly. All treatment compounds were dissolved in CMC solution and were orally administered at 5 mL/kg body weight. The blood was collected after the animals were sacrificed at the end of the experiment. The plasma was separated from collected blood and stored at −20 °C until analysis. The study was approved by institutional animal care and use committees of Gachon University (GU1-2022-IA0027-00). The experiment procedures were conducted based on the guidelines of the Declaration of Helsinki.

2.6. Blood Biochemistry

At the end of the experimental period, blood samples were collected from the inferior vena cava after fasting for 24 h and were placed in test tubes containing 0.18 M ethylenediaminetetraacetic acid (EDTA). The plasma was centrifuged in collected blood for 15 min at 1200 RCF and then stored at −20 °C until analysis. The levels of creatinine, alanine aminotransferase (ALT), and total cholesterol were measured using respective commercial kits (GENIA, Seong-Nam, Korea).

2.7. Statistical Analyses

Differences between treatments were determined using one-way analyses of variance and multiple comparisons with Bonferroni correction. Statistical significance is reported at $p < 0.05$. All analyses were performed using SPSS Statistics ver. 19.0 (SPSS Inc., Chicago, IL, USA).

3. Results

3.1. Inhibitory Effects of Hispidulin, p-Synephrine, and p-Octopamine HCl on Adipogenesis in 3T3-L1 Preadipocytes

Before initiating adipogenic differentiation in 3T3-L1 preadipocytes, no cytotoxicity was observed in the presence or absence of hispidulin, *p*-synephrine, and *p*-octopamine HCl using the Ez-Cytox cell viability assay kit. After 24 h of incubation, hispidulin, *p*-synephrine, and *p*-octopamine HCl, at concentrations of 40 μM, each, did not affect the viability of 3T3-L1 preadipocytes (Figure 1A–C). ORO staining showed that treatment with hispidulin and *p*-synephrine inhibited the differentiation of 3T3-L1 preadipocytes into mature adipocytes. However, *p*-octopamine HCl did not inhibit the formation of red-stained lipid droplets, which was, however, slightly inhibited by treatment with 20 μM and 40 μM hispidulin (53.06% ± 2.95% and 43.92% ± 2.54% reduction, respectively). Similarly, formation of red-stained lipid droplets was slightly inhibited by treatment with 20 μM and 40 μM *p*-synephrine (48.37% ± 3.58% and 43.54% ± 0.84% reduction, respectively (Figure 1D–G).

3.2. Inhibitory Effects of SOH on Adipogenesis in 3T3-L1 Preadipocytes

As shown in Figure 2A–D, a mixture of *p*-synephrine, *p*-octopamine HCl, and hispidulin, at concentrations of 40 μM, each, did not affect the viability of 3T3-L1 preadipocytes. As shown in Figure 2E–I, co-treatment with 5, 10, 20, and 40 μM hispidulin and 5, 10, 20, and 40 μM *p*-synephrine resulted in stronger inhibition of red-stained lipid droplet formation than treatment with hispidulin or *p*-synephrine alone. Cells treated with equal concentrations of hispidulin and *p*-synephrine (5, 10, 20, or 40 μM, each) showed significant inhibition (53.46% ± 0.98%, 54.08% ± 0.93%, 37.99% ± 0.56%, and 37.38% ± 0.93% reduction, respectively) of the formation of red-stained lipid droplets. Co-treatment with 5, 10, 20, and 40 μM *p*-synephrine and 5, 10, 20, and 40 μM *p*-octopamine HCl resulted in stronger inhibition of red-stained lipid droplet formation than treatment with *p*-octopamine HCl alone. Red-stained lipid droplet formation was slightly inhibited by treatment with equal concentrations (20 and 40 μM, each) of *p*-synephrine and *p*-octopamine HCl (57.91% ± 2.75% and 48.69% ± 0.71% reduction, respectively). Co-treatment with 5, 10, 20, and 40 μM hispidulin and 5, 10, 20, and 40 μM *p*-octopamine HCl resulted in stronger inhibition of red-stained lipid droplet formation than treatment with *p*-octopamine HCl alone. Red-stained lipid droplet formation was slightly inhibited by treatment with equal concentrations (20 and 40 μM, each) of hispidulin and *p*-octopamine HCl (53.33% ± 1.33% and 48.14% ± 0.97% reduction, respectively). SOH treatment resulted in greater inhibition of the formation of red-stained lipid droplets than treatment with hispidulin, *p*-synephrine, or *p*-octopamine HCl alone. Cells treated with equal concentrations of hispidulin, *p*-synephrine, and *p*-octopamine HCl (5, 10, 20, and 40 μM, each) showed significant inhibition (50.38% ± 2.79%, 50.67% ± 1.08%, 40.32% ± 0.87%, and 36.11% ± 1.16% reduction, respectively) of red-stained lipid droplet formation.

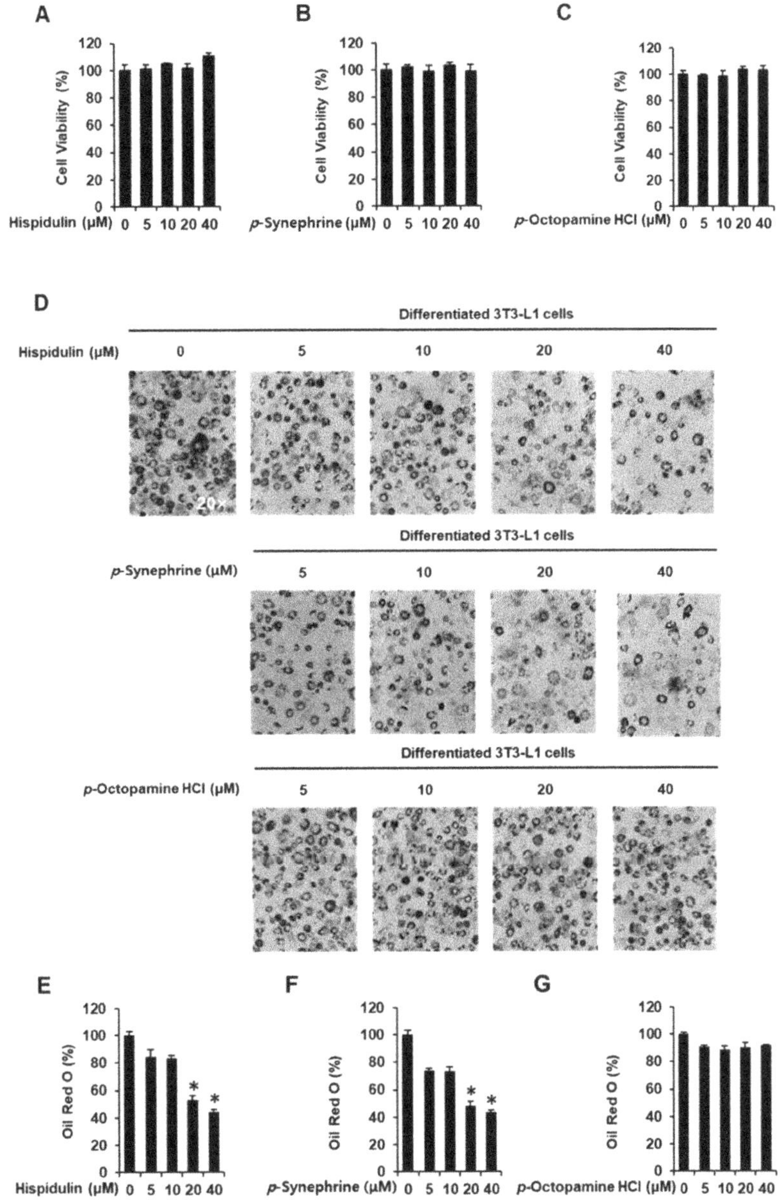

Figure 1. Inhibitory effects of hispidulin, *p*-synephrine, and *p*-octopamine HCl on adipogenesis in 3T3L-1 preadipocytes. Effect of (**A**) hispidulin, (**B**) *p*-synephrine, and (**C**) *p*-octopamine HCl on the viability of 3T3L-1 cells for 8 days using an Ez-Cytox cell viability assay. (**D**) Images of Oil Red O (ORO) staining of differentiated 3T3L-1 cells were produced using an inverted microscope at 20-fold magnification on day 8 after treatment with hispidulin, *p*-synephrine, and *p*-octopamine HCl. (**E–G**) Quantification of ORO staining as percentage of untreated control (three independent replicates; * $p < 0.05$, Kruskal–Wallis nonparametric test). Data are shown as means ± standard error of the mean (SEM).

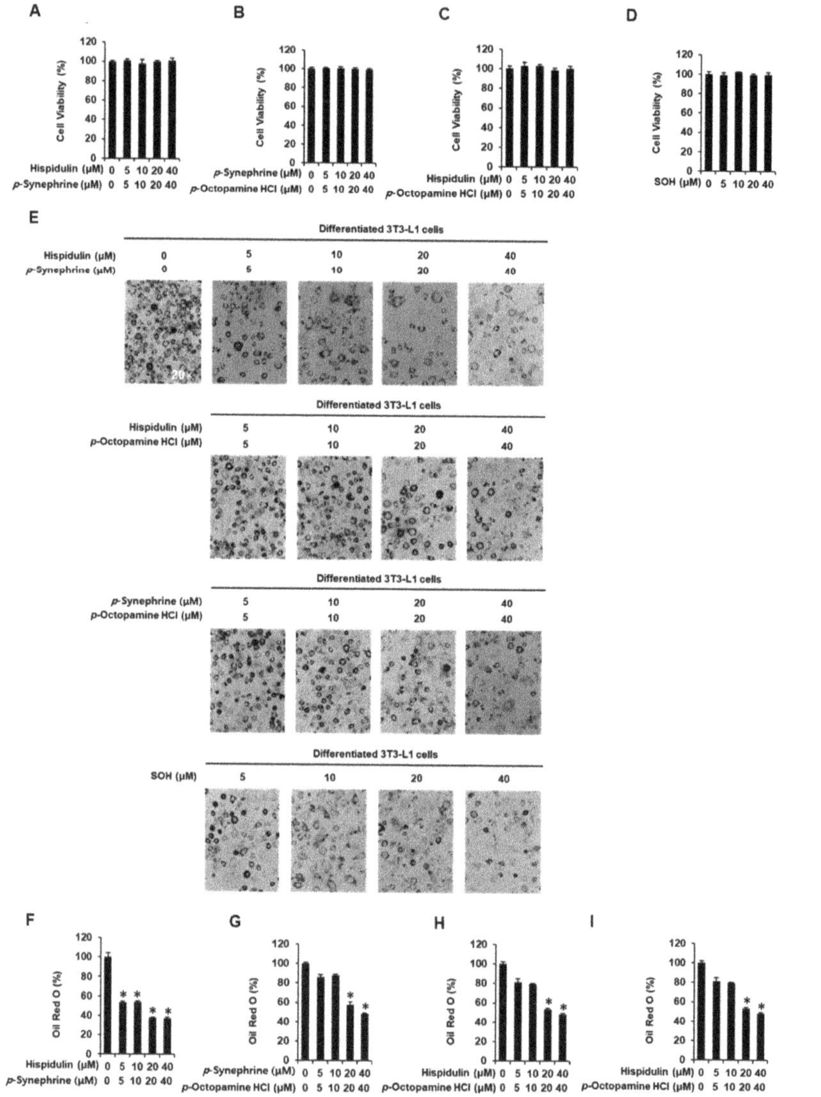

Figure 2. Inhibitory effects of a mixture of *p*-synephrine, *p*-octopamine HCl, and hispidulin (SOH) on adipogenesis of 3T3L-1 preadipocytes. Effect of (**A**) a mixture of hispidulin and *p*-synephrine, (**B**) a mixture of *p*-synephrine and *p*-octopamine HCl, (**C**) a mixture of hispidulin and *p*-octopamine HCl, and (**D**) SOH on the viability of 3T3L-1 cells for 8 days using an Ez-Cytox cell viability assay. (**E**) Images of ORO staining of differentiated 3T3L-1 cells recorded using an inverted microscope at 20-fold magnification on day 8 after treatment with a mixture of hispidulin and *p*-synephrine, a mixture of *p*-synephrine and *p*-octopamine HCl, a mixture of hispidulin and *p*-octopamine HCl, and SOH. (**F–I**) Quantification of ORO staining expressed as percentage of untreated controls (three independent replicates; * $p < 0.05$, Kruskal–Wallis nonparametric test). Data are shown as means ± SEM.

3.3. *Inhibitory Effects of SOH on the Expression of Proteins Involved in Adipogenesis of Differentiated 3T3L-1 Cells*

Western blotting was performed to examine the expression of adipogenic marker proteins including PPARγ, C/EBPα, and C/EBPβ in 3T3L-1 cells. Treatment with SOH inhibited the expression of PPARγ, C/EBPα, and C/EBPβ (Figure 3). This suggests that SOH was effective in decreasing adipogenic marker proteins during the eight-day adipocyte differentiation period.

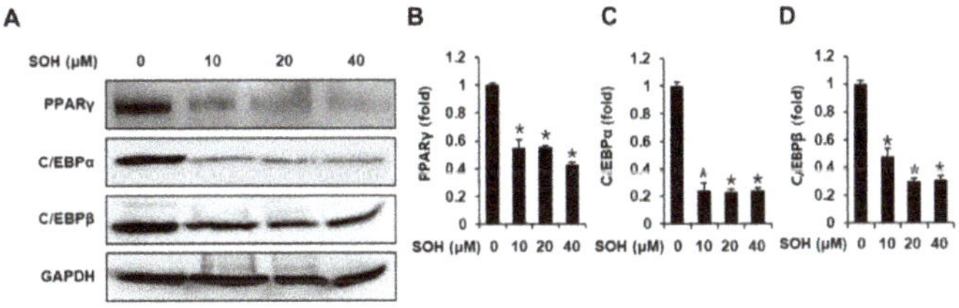

Figure 3. Inhibitory effects of SOH on the expression of adipogenesis-related proteins in differentiated 3T3L-1 cells. (**A**) Protein expression of peroxisome proliferator-activated receptor gamma (PPARγ), CCAAT/enhancer-binding protein alpha (C/EBPα), CCAAT/enhancer-binding protein beta (C/EBPβ), and glyceraldehyde 3-phosphate dehydrogenase (GAPDH) in differentiated 3T3L-1 cells on day 8 after treatment with SOH. (**B–D**) Analysis of the ratios of band intensities of PPARγ, C/EBPα, and C/EBPβ in treated cells compared with those of untreated differentiated 3T3L-1 cells (three independent replicates, * $p < 0.05$, Kruskal-Wallis nonparametric test). Data are shown as means ± SEM.

3.4. *Effects of SOH on Body Weight and Food Intake in HFD-Induced Obese Mice*

Body weight of HFD mice was higher than that of mice in the other groups at eight weeks. The weight of HFD mice was 34.7 ± 0.65 g, which was approximately 4.5 g higher than that of the ND control mice. Those mice treated with SOH had a lower body weight (29.1 ± 0.97 g) than HFD mice and ND mice. Mice treated with ATS had a lower body weight (31.9 ± 0.41 g) than the HFD group (Figure 4A). Dietary intake showed a tendency to increase in all groups during the experimental period. The HFD group showed high dietary intake during all experimental periods, except for the first week. The intake of ATS-treated mice showed a similar tendency to the HFD mice without improvement effect. However, SOH-treated mice showed a tendency towards lower intake than HFD mice during all of the experimental period, except the first week. In the last week of the experiment, the food intake of the SOH-treated group was 9.9 ± 0.8 g, which was a approximately 5% decrease in food intake compared with HFD group (10.4 ± 0.78 g) (Figure 4B).

3.5. *Effects of SOH on Blood Biochemistry in HFD-Induced Obese Mice*

Blood creatinine concentrations were 0.31 ± 0.001 mg/dL in the ND group, 0.30 ± 0.001 mg/dL in the HFD group, 0.29 ± 0.001 mg/dL in the ATS group, and 0.32 ± 0.001 mg/dL in the SOH group. The creatinine levels did not differ significantly between groups (Figure 5A). Blood ALT concentrations were 23.20 ± 1.86 IU/L in the HFD group, which showed slightly higher tendency, not statistically significant, than the 18.67 ± 1.76 IU/L in the ND group; it was 18.0 ± 1.54 IU/L in the SOH group, and ALT levels were approximately 22% lower, compared to the HFD group. ALT levels in the ATS group were 37.0 ± 3.69 IU/L, thus significantly higher than in the HFD group, by approximately 59% (Figure 5B). Total blood cholesterol was 191.50 ± 5.45 mg/dL in HFD mice, which was significantly higher than in ND mice (134.33 ± 5.81 mg/dL); it was 148.40 ± 10.68 mg/dL in SOH mice, which was significantly lower than in HFD mice, by approximately 22%. Cholesterol concentrations were 180.80 ± 5.68 mg/dL in ATS mice, which was slightly lower than in HFD mice by approximately 5% (Figure 5C).

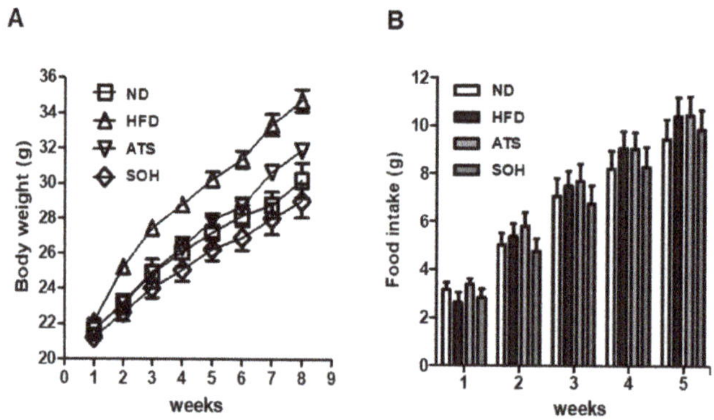

Figure 4. Body weight and food intake. Mice were fed a high-fat diet (HFD) and normal food from one week of age. Body weight and food intake were recorded weekly. (**A**) body weight, (**B**) food intake. ND: normal food control; HFD: control; ATS: HFD + atorvastatin 30 mg/kg oral administration; SOH: HFD + SOH 20 mg/kg (hispidulin, synephrine, and *p*-octopamine HCl in a 1:1:1 mixture), oral administration. Data are shown as means ± SEM. A one-way ANOVA and Bonferroni post-hoc analysis were used to compare normal and HFD, and HFD with each group.

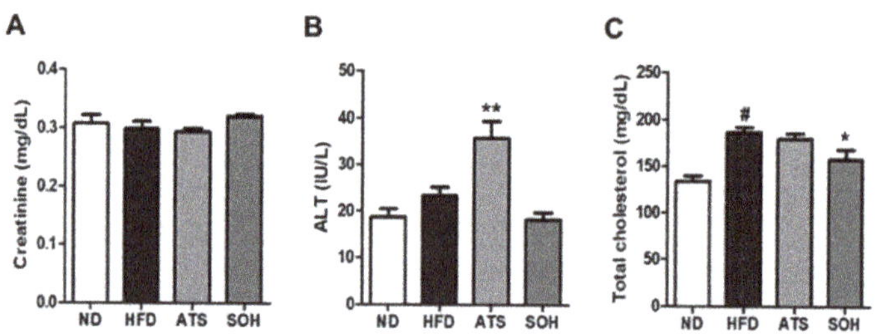

Figure 5. Blood analysis. Blood analysis was performed after killing the mice after 8 weeks. (**A**) Creatinine, (**B**) alanine aminotransferase (ALT), and (**C**) total cholesterol. ND: normal food control, HFD: HFD control; ATS: HFD + atorvastatin 30 mg/kg oral administration; SOH: HFD + SOH 20 mg/kg (hispidulin, synephrine, *p*-octopamine HCl in a 1:1:1 mixture) oral administration. Data are shown as means ± SEM. A one-way ANOVA and Bonferroni post-hoc analysis were used to compare normal and HFD ($^{\#}$ $p < 0.05$) and HFD with each group (* $p < 0.05$, ** $p < 0.001$).

4. Discussion

In the present study, we assessed the inhibitory effects of SOH on adipogenesis. Underlying molecular mechanisms of SOH in 3T3-L1 cells were assessed by analyzing its effect on the expression of pro-adipogenic proteins. In addition, its anti-obesity effects in an HFD-induced obesity mouse model were evaluated. SOH reduced adipogenesis by inhibiting the expression of pro-adipogenic C/EBPα, C/EBPβ, and PPARγ in 3T3-L1 cells, leading to decreased dietary intake and body weight in mice with HFD-induced obesity. Adipogenesis and lipid accumulation are associated with the development of obesity. A reduction in obesity is associated with the inhibition of adipogenesis and lipid accumulation in adipocytes, and a reduction in the number of adipocytes [20]. Treatment with SOH resulted in stronger inhibition of the formation of red-stained lipid droplets than treatment with hispidulin, *p*-synephrine, or *p*-octopamine HCl alone. In particular, treatment with SOH (5 μM) resulted in slightly stronger inhibition of the formation of red-stained lipid droplets than co-treatment with 5 μM hispidulin and 5 μM *p*-synephrine.

Adipogenesis is regulated by transcription factors such as C/EBPα, C/EBPβ, and PPARγ. During the eight-day adipocyte differentiation period, C/EBPβ is immediately activated and elicits expression of C/EBPα and PPARγ as master regulators. Expression of C/EBPα and PPARγ is vital for the progression of the late stages of adipogenesis. Moreover, PPARγ activation is associated with lipid accumulation [21]. In the present study, treatment with SOH resulted in stronger inhibition of the expression of C/EBPα, C/EBPβ, and PPARγ, compared to co-treatment with hispidulin plus *p*-synephrine. Thus, our results indicate that SOH significantly reduced lipid accumulation and adipogenesis by inhibiting adipogenic transcription factors during the eight-day adipocyte differentiation period.

We used an HFD-induced obesity mouse model to evaluate the anti-obesity activity of SOH. At the end of the experiment, the body weights of HFD mice treated with SOH were significantly lower than those of the HFD controls. Consistent with the decrease in body weight, the SOH-treated group showed a decrease in food intake, compared with the HFD group. These results indicate that the SOH-induced decrease in body weight was due to reduced food intake. We examined changes in the blood levels of creatinine (to assess kidney function), ALT (indicating liver function), and total cholesterol (as an obesity factor), and no changes in blood creatinine levels after the administration of ATS or SOH were observed, suggesting no kidney toxicity. However, blood ALT levels were significantly increased in ATS-treated mice, suggesting liver toxicity. In general, elevated ALT as an indicator of hepatic function is associated with the development of hepatic steatosis [22]. Obesity is a characteristic of metabolic syndrome that causes cardiovascular disease and diabetes due to the accumulation of visceral fat in the body [23]. Recently, normal weight obesity, or skinny obesity, which has high body fat despite normal body weight, has also been known [24]. Total cholesterol level is known as one of the obesity factors because it is higher in obese and/or overweight people compared with normal weight subjects [25]. In our experiment, SOH-treated mice showed a tendency to suppress weight gain and to decrease dietary intake compared to the HFD group. In addition, the levels of total cholesterol and ALT decreased or tended to decrease compared with those in HFD mice.

Thus, our results indicate that SOH treatment may alleviate the development of hepatic steatosis. We demonstrated that SOH plays an important role in adipogenesis and lipid accumulation by inhibiting C/EBPα, C/EBPβ, and PPARγ expression in 3T3-L1 cells. Moreover, SOH administration mitigated HFD-induced body-weight gain, increased dietary intake, and increased the levels of ALT and total cholesterol. Therefore, we suggest that SOH exerted anti-adipogenic effects in 3T3-L1 cells and obesity effects in HFD-fed mice. These results suggest that administration of SOH may be a potential strategy to reduce obesity.

5. Conclusions

SOH treatment attenuated the differentiation and lipid accumulation of 3T3-L1 preadipocytes by inhibiting the expression of adipogenic marker proteins, including C/EBPα, C/EBPβ, and PPARγ. In addition, administration of SOH mitigated HFD-induced body-weight gain, increased dietary intake, and increased the levels of ALT and total cholesterol in mice. Our results suggest that anti-obesity effects of SOH result from a decrease in adipogenesis, and that SOH has a positive effect by reducing body-weight gain in obese mice. Thus, administration of SOH may be a promising alternative strategy for the management of obesity-associated metabolic disorders.

Author Contributions: K.S.K., D.L. and D.-W.K. conceived and designed the experiments; D.L., J.H.L., B.H.K. and S.L. performed the experiments; D.L. and J.H.L. analyzed the data; D.-W.K. and K.S.K. contributed reagents, materials, and analytical tools; and D.L., J.H.L. and K.S.K. wrote the manuscript. All authors have read and agreed to the published version of the manuscript.

Funding: This research was supported by the Basic Science Research Program through the National Research Foundation of Korea, funded by the Ministry of Education (2020M3A9E410438012, NRF-2021M3H9A1097888).

Institutional Review Board Statement: Not applicable.

Informed Consent Statement: Not applicable.

Data Availability Statement: Data is contained within the article.

Conflicts of Interest: The authors declare no conflict of interest.

Abbreviations

C/EBPα	adenosine monophosphate
PPARγ	peroxisome proliferator-activated receptor gamma
p-octopamine HCl	*p*-Octopamine hydrochloride
SOH	a mixture of *p*-synephrine, *p*-octopamine HCl, and hispidulin
P/S	penicillin/streptomycin
FBS	fetal bovine serum
ORO	Oil Red O solution
ND	normal diet control group
CMC	sodium carboxyl methyl cellulose
HFD	high-fat diet
ATS	atorvastatin
EDTA	ethylenediaminetetraacetic acid
ALT	alanine aminotransferase
SEM	standard error of the mean

References

1. Maric, C.; Hall, J.E. Obesity, metabolic syndrome and diabetic nephropathy. *Contrib. Nephrol.* **2011**, *170*, 28–35. [PubMed]
2. Prentice, A.; Jebb, S. Energy intake/physical activity interactions in the homeostasis of body weight regulation. *Nutr. Rev.* **2004**, *62*, S98–S104. [CrossRef] [PubMed]
3. Soria, A.; D'Alessandro, M.E.; Lombardo, Y.B. Duration of feeding on a sucrose-rich diet determines metabolic and morphological changes in rat adipocytes. *J. Appl. Physiol.* **2001**, *91*, 2109–2116. [CrossRef] [PubMed]
4. Hausman, D.; DiGirolamo, M.; Bartness, T.; Hausman, G.; Martin, R. The biology of white adipocyte proliferation. *Obes. Rev.* **2001**, *2*, 239–254. [CrossRef]
5. Wu, Z.; Rosen, E.D.; Brun, R.; Hauser, S.; Adelmant, G.; Troy, A.E.; McKeon, C.; Darlington, G.J.; Spiegelman, B.M. Cross-regulation of C/EBPα and PPARγ controls the transcriptional pathway of adipogenesis and insulin sensitivity. *Mol. Cell* **1999**, *3*, 151–158. [CrossRef]
6. Koh, Y.-K.; Lee, M.-Y.; Kim, J.-W.; Kim, M.; Moon, J.-S.; Lee, Y.-J.; Ahn, Y.-H.; Kim, K.-S. Lipin1 is a key factor for the maturation and maintenance of adipocytes in the regulatory network with CCAAT/enhancer-binding protein α and peroxisome proliferator-activated receptor γ2. *J. Biol. Chem.* **2008**, *283*, 34896–34906. [CrossRef] [PubMed]
7. Shin, J.H.; Gadde, K.M. Clinical utility of phentermine/topiramate (Qsymia™) combination for the treatment of obesity. *Diabetes Metab. Syndr. Obes. Targets Ther.* **2013**, *6*, 131–139.
8. Vermaak, I.; Viljoen, A.M.; Hamman, J.H. Natural products in anti-obesity therapy. *Nat. Prod. Rep.* **2011**, *28*, 1493–1533. [CrossRef]
9. Roh, C.; Jung, U.; Jo, S.-K. Screening of anti-obesity agent from herbal mixtures. *Molecules* **2012**, *17*, 3630–3638. [CrossRef]
10. Stohs, S.J.; Preuss, H.G.; Keith, S.C.; Keith, P.L.; Miller, H.; Kaats, G.R. Effects of *p*-synephrine alone and in combination with selected bioflavonoids on resting metabolism, blood pressure, heart rate and self-reported mood changes. *Int. J. Med. Sci.* **2011**, *8*, 295. [CrossRef]
11. Lee, D.; Kwak, H.J.; Kim, B.H.; Kim, S.H.; Kim, D.-W.; Kang, K.S. Combined anti-adipogenic effects of hispidulin and p-synephrine on 3T3-L1 adipocytes. *Biomolecules* **2021**, *11*, 1764. [CrossRef] [PubMed]
12. Ferrandiz, M.; Bustos, G.; Paya, M.; Gunasegaran, R.; Alcaraz, M. Hispidulin protection against hepatotoxicity induced by bromobenzene in mice. *Life Sci.* **1994**, *55*, PL145–PL150. [CrossRef]
13. Kim, K.-W.; Kim, H.-D.; Jung, J.-S.; Woo, R.-S.; Kim, H.-S.; Suh, H.-W.; Kim, Y.-H.; Song, D.-K. Characterization of antidepressant-like effects of p-synephrine stereoisomers. *Naunyn Schmiedebergs Arch. Pharmacol.* **2001**, *364*, 21–26. [CrossRef] [PubMed]
14. Patel, K.; Patel, D.K. Medicinal importance, pharmacological activities, and analytical aspects of hispidulin: A concise report. *J. Tradit. Complement. Med.* **2017**, *7*, 360–366. [CrossRef] [PubMed]
15. Stohs, S.J. Physiological functions and pharmacological and toxicological effects of *p*-octopamine. *Drug Chem. Toxicol.* **2015**, *38*, 106–112. [CrossRef] [PubMed]
16. Stohs, S.J.; Preuss, H.G.; Shara, M. The safety of Citrus aurantium (bitter orange) and its primary protoalkaloid *p*-synephrine. *Phytother. Res.* **2011**, *25*, 1421–1428. [CrossRef]
17. Bour, S.; Visentin, V.; Prévot, D.; Carpéné, C. Moderate weight-lowering effect of octopamine treatment in obese Zucker rats. *J. Physiol. Biochem.* **2003**, *59*, 175–182. [CrossRef]
18. Mestrovic, N. Uber die behandlung hypotoner zustande bei chirugischen patienten. *Z. Allg.* **1972**, *48*, 1301–1303.
19. Stucke, W. Die behandlung hypotoner kreislaufstorungen mit Norphen retard. *Z. Allg.* **1972**, *48*, 240–243.

20. Madsen, L.; Petersen, R.K.; Kristiansen, K. Regulation of adipocyte differentiation and function by polyunsaturated fatty acids. *Biochim. Biophys. Acta Mol. Basis Dis.* **2005**, *1740*, 266–286. [CrossRef]
21. Guo, L.; Li, X.; Tang, Q.-Q. Transcriptional regulation of adipocyte differentiation: A central role for CCAAT/enhancer-binding protein (C/EBP) β. *J. Biol. Chem.* **2015**, *290*, 755–761. [CrossRef] [PubMed]
22. de Andrade, R.S.B.; de Carvalho França, L.F.; Pessoa, L.D.S.; de Almeida Arrais Landim, B.; Rodrigues, A.A.; Alves, E.H.P.; Lenardo, D.D.; Nascimento, H.M.S.; de Melo Sousa, F.B.; Barbosa, A.L.D.R. High-fat diet aggravates the liver disease caused by periodontitis in rats. *J. Periodontol.* **2019**, *90*, 1023–1031. [CrossRef] [PubMed]
23. Alberti, K.G.; Eckel, R.H.; Grundy, S.M.; Zimmet, P.Z.; Cleeman, J.I.; Donato, K.A.; Fruchart, J.C.; James, W.P.; Loria, C.M.; Smith, S.C., Jr. Harmonizing the metabolic syndrome: A joint interim statement of the international diabetes federation task force on epidemiology and prevention; National Heart, Lung, and Blood Institute; American Heart Association; World Heart Federation; International Atherosclerosis Society; and International Association for the Study of Obesity. *Circulation* **2009**, *120*, 1640–1645. [PubMed]
24. Kapoor, N.; Feingold, K.R.; Anawalt, B.; Boyce, A.; Chrousos, G.; de Herder, W.W.; Dhatariya, K.; Dungan, K.; Hershman, J.M.; Hofland, J.; et al. Thin Fat Obesity: The Tropical Phenotype of Obesity. In *Endotext [Internet]*; MDText.com, Inc.: South Dartmouth, MA, USA, 2021.
25. Szczygielska, A.; Widomska, S.; Jaraszkiewicz, M.; Knera, P.; Muc, K. Blood lipids profile in obese or overweight patients. *Ann. Univ. Mariae Curie Sklodowska Med.* **2003**, *58*, 343–349.

nutrients

MDPI

Article

Distinct AMPK-Mediated FAS/HSL Pathway Is Implicated in the Alleviating Effect of Nuciferine on Obesity and Hepatic Steatosis in HFD-Fed Mice

Hanyuan Xu, Xiaorui Lyu, Xiaonan Guo, Hongbo Yang, Lian Duan, Huijuan Zhu, Hui Pan, Fengying Gong * and Linjie Wang *

Key Laboratory of Endocrinology of National Health Commission, Department of Endocrinology, Peking Union Medical College Hospital, Chinese Academy of Medical Sciences and Peking Union Medical College, Beijing 100730, China; jadef21@foxmail.com (H.X.); lvxiaorui0219@foxmail.com (X.L.); guoxiaonan199708@163.com (X.G.); hongbo.yang7@gmail.com (H.Y.); duanlianpumc@163.com (L.D.); shengxin2004@163.com (H.Z.); panhui20111111@163.com (H.P.)

* Correspondence: fygong@sina.com (F.G.); eileenwood@163.com (L.W.)

Abstract: Nuciferine (Nuci), the main aporphine alkaloid component in lotus leaf, was reported to reduce lipid accumulation in vitro. Herein we investigated whether Nuci prevents obesity in high fat diet (HFD)-fed mice and the underlying mechanism in liver/HepG2 hepatocytes and epididymal white adipose tissue (eWAT) /adipocytes. Male C57BL/6J mice were fed with HFD supplemented with Nuci (0.10%) for 12 weeks. We found that Nuci significantly reduced body weight and fat mass, improved glycolipid profiles, and enhanced energy expenditure in HFD-fed mice. Nuci also ameliorated hepatic steatosis and decreased the size of adipocytes. Furthermore, Nuci remarkably promoted the phosphorylation of AMPK, suppressed lipogenesis (SREBP1, FAS, ACC), promoted lipolysis (HSL, ATGL), and increased the expressions of adipokines (FGF21, ZAG) in liver and eWAT. Besides, fatty acid oxidation in liver and thermogenesis in eWAT were also activated by Nuci. Similar results were further observed at cellular level, and these beneficial effects of Nuci in cells were abolished by an effective AMPK inhibitor compound C. In conclusion, Nuci supplementation prevented HFD-induced obesity, attenuated hepatic steatosis, and reduced lipid accumulation in liver/hepatocytes and eWAT/adipocytes through regulating AMPK-mediated FAS/HSL pathway. Our findings provide novel insight into the clinical application of Nuci in treating obesity and related complications.

Keywords: Nuciferine (Nuci); AMP-activated protein kinase (AMPK); fatty acid synthase (FAS); hormone sensitive lipase (HSL); obesity; hepatic steatosis

Citation: Xu, H.; Lyu, X.; Guo, X.; Yang, H.; Duan, L.; Zhu, H.; Pan, H.; Gong, F.; Wang, L. Distinct AMPK-Mediated FAS/HSL Pathway Is Implicated in the Alleviating Effect of Nuciferine on Obesity and Hepatic Steatosis in HFD-Fed Mice. *Nutrients* **2022**, *14*, 1898. https://doi.org/10.3390/nu14091898

Academic Editor: Susanna Iossa

Received: 20 March 2022
Accepted: 27 April 2022
Published: 30 April 2022

1. Introduction

Obesity exerts much pressure on the public healthcare nationwide [1]. Characterized by excess fat accumulation in adipose and non-adipose tissues, obesity is closely associated with multiple metabolic diseases and contributes to all-cause mortality [2]. Other than surplus food intake, imbalance between lipogenesis, lipolysis, and energy expenditure also exacerbate the deterioration of obesity. Thus, suppressing lipogenesis, inducing lipolysis as well as promoting energy expenditure (EE) would be beneficial in preventing and managing obesity and related morbidities.

Liver and white adipose tissue (WAT) are classic target organs that regulate lipid metabolism. Nonalcoholic steatohepatitis and enlarged visceral WAT, especially epididymal WAT (eWAT) were reported to be notorious obesity comorbidities [3,4]. AMP-activated protein kinase (AMPK), a ubiquitously distributed serine/threonine protein kinase which is involved in a spectrum of metabolic processes plays a vital role in energy metabolism including lipid synthesis and catabolism in both liver and eWAT [5]. It was reported that AMPK inhibited de novo fat synthesis by regulating the expressions of transcription factor sterol-regulatory element binding protein 1 (SREBP1) and its downstream rate-limiting enzymes including fatty acid synthase (FAS) and acetyl-CoA carboxylase (ACC) [6]. Lipolysis was mainly catalyzed by hormone-sensitive lipase (HSL) and adipose triglyceride lipase (ATGL) [7]. Studies have also shown that the AMPK phosphorylated Ser406 of ATGL to stimulate TG hydrolase activity and activate lipolysis [8]. In hepatocytes, a central catabolism process, fatty acid oxidation, is delicately regulated by key transcription factor peroxisome proliferator-activated receptor α (PPARα) and its downstream enzyme carnitine palmitoyltransferase 1α (CPT1α) [9]. In adipocytes, there is another catabolism mechanism where stored fat can be decomposed to generate heat, and the best effector that characterized this non-shivering thermogenesis is uncoupling protein 1 (UCP1) [10]. Activating UCP1 in WAT to develop thermogenic capacity remain a promising approach for enhancing energy expenditure thus ameliorating obesity [11]. Besides adipokines that are secreted primarily in adipose tissue, fibroblast growth factor 21 (FGF21) [12,13] and zinc-α2-glycoprotein (ZAG) [14] were also reported to positively serve the whole-body metabolism targeting liver, eWAT, and other tissues. To date, both thermogenesis and the expressions of FGF21 and ZAG were also found to be associated with the AMPK pathway [15,16].

In decades, many anti-obesity drugs have been developed, yet their use was banned due to severe side effects [17]. In traditional Chinese medicine, there are various botanic products derived from natural plants used for treating obesity [18,19], and they have been increasingly studied by the modern medical sciences since less side effects were observed. Among these products, lotus leaf (Nelumbo nucifera leaf) embraces a long history of treating obesity and related complications. Nuciferine (Nuci), the main aporphine alkaloid component in lotus leaf, were found to possess beneficial effects on inflammation [20], cardiovascular diseases [21], and cancer [22]. Our previous study revealed that in vitro, Nuci is also capable of inhibiting proliferation and differentiation of preadipocytes [23], suggesting that it may exhibit an anti-obesity effect. In 2018, Zhang C, et al. reported that Nuci may act through a PPARα/PPARγ coactivator-1α pathway to ameliorate hepatic steatosis [24]. Another in vivo report found that Nuci decreased the lipid droplets and promoted the expression of glucose transporter type 4 (GLUT-4) in mature 3T3-L1 adipocytes with possible involvement of activation of AMPK [25]. However, full awareness of the underlying molecular mechanism remains lacking. Therefore, in this study, enlightened by previous reports, we aimed at exploring the alleviating effect of Nuci on obesity and related hepatic steatosis, and at investigating the role of AMPK in Nuci-induced decrease of lipid accumulation in both liver and eWAT in HFD-fed mice. Another highlight of this study is that to our knowledge, this is the first report of Nuci promoting energy expenditure in HFD-fed mice, further shedding light on another potential mechanism by which the anti-obesity effect of Nuci occurs.

2. Materials and Methods

2.1. Animal Experiments

Nuci (purity $\geq$ 98% by HPLC, CAS: 475-83-2) was purchased from Biopurify Phytochemicals Ltd. (Chengdu, China). The molecular weight of Nuci is 295.38. Six-week-old C57BL/6J male mice were purchased from HFK Bioscience Co., Ltd. (Beijing, China) and housed in a standard facility with 12 h dark/light cycle (3 mice/cage). After 7 days acclimation, the animals were randomly allocated to four groups. Mice in the control group were fed a normal diet (ND, 10% kcal fat, AIN93M, Trophic Animal Feed High-Tech Co., Ltd., Nantong, China; n = 12). Mice in the high-fat diet (HFD) group were fed a HFD (45% kcal fat, D12451, Trophic Animal Feed High-Tech Co., Ltd., Nantong, China; n = 12); mice in the HFD-Nuci group (HFD-Nuci) were fed a HFD supplemented with 0.10% Nuci (n = 12). The customized feeds were made by a commercial company from where the regular feeds were bought. According to the manufacturer of the experimental diets (Trophic Animal Feed High-Tech Co., Ltd., Nantong, China), the Nuci in powder form was first completely dissolved in ethanol and mixed with the diets at room temperature until it is uniformly distributed. Then the ethanol was evaporated by adequate air blowing evaporation, and the mixture was then made in the form of bar as regular diet for experimental application. Moreover, liraglutide, a glucagon like peptide-1 (GLP-1) receptor agonists approved for the treatment of obesity [26] was used as reference treatment. Mice in the HFD-liraglutide group (HFD-Lira) were fed an HFD with daily subcutaneous injection of liraglutide (15676, Novo Nordisk, Bagsværd, Denmark) at 200 μg/kg (n = 12). Mice in the ND, HFD, and HFD-Nuci groups received daily subcutaneous administration with equal volume of saline as controls compared to the HFD-Lira group. Mice in the ND, HFD, HFD-Lira groups were fed with standard ND/HFD, respectively, without Nuci supplementation as controls to the HFD-Nuci group. Body weight and food intake of each mouse were recorded twice a week. After treatment of 12 weeks, intraperitoneal glucose tolerance test (IPGTT) and intraperitoneal insulin tolerance test (IPITT) were conducted, then the 48-h energy expenditure was monitored by Promethion Metabolic Cage System (Sable Systems, North Las Vegas, NV, USA). At the end of the experiment, animals were fasted overnight (12 h) and sacrificed with CO_2 gas. Blood samples were collected for biochemical detections including liver function markers ALT and AST, lipid metabolism-related traits including total cholesterol (TC), triglycerides (TG), high density lipoprotein-cholesterol (HDL-C), low density lipoprotein-cholesterol (LDL-C), fasting blood glucose (FBG), and fasting serum insulin (FINS). Insulin resistance (HOMA-IR) indexes were calculated as HOMA-IR = FBG × FINS/22.5. Liver, brown adipose tissue (BAT), and white adipose tissue (WAT) including eWAT, inguinal WAT (iWAT), and perirenal WAT (pWAT) were collected, weighed, and immediately frozen in liquid nitrogen and stored at −80 °C for molecular experiments. A bulk of the liver and eWAT were fixed in 4% paraformaldehyde for histology observation; another portion of liver from each mouse was placed in optimum cutting temperature (OCT) medium (Sakura Finetek Co., Ltd., Torrance, CA, USA) and stored at −80 °C.

All experiments conducted on animals were approved by the Medical Ethics Committee of Peking Union Medical College (XHDW-2019-010, Beijing, China) and followed by the National Institutes of Health regulations regarding animal care and use (Beijing, China).

2.2. Indirect Calorimetry

Indirect calorimetry was conducted by Promethion Metabolic Cage System in accordance with the manufacturer's instructions. Mice were acclimated to the chambers for 24 h before the measurements began. Then the animals were kept under 12 h day/night cycle (from 8:00 a.m. to 8:00 p.m.) at 24 °C with feeding regimen described in Materials and Methods 2.1Day and night parameters of thermogenesis including O_2 consumption (VO_2), CO_2 consumption (VCO_2), energy expenditure (EE), and home-cage locomotor activity were measured in the following 48 h. VO_2, VCO_2, and EE were normalized against total body weight of each mouse. Respiratory exchange ratio (RER) was calculated as the ratio of VCO_2 versus VO_2.

2.3. Histology of Liver and eWAT

Livers and eWATs were fixed in 4% paraformaldehyde, then dehydrated and embedded in paraffin. About 3 μm thick slides of tissue were obtained and stained using hematoxylin and eosin (H&E) according to the standard protocol. Histological images were obtained using a light microscope with digital camera (Nikon eclipse Ti, Nikon Corporation, Tokyo, Japan) and a stereoscopic microscope (BX53, Olympus Optical Co., Ltd., Tokyo, Japan) equipped with a charged-coupled device (CCD) camera (DP80; Olympus) at 200× and 400× magnifications, respectively.

Furthermore, H&E stained slides of eWAT were further used to determine the diameters of each adipocyte in the field (400×) using Image J software (Olympus). Next, six fields of H&E stained slides of liver were chosen randomly for measurement per mouse. Liver characteristics of non-alcoholic fatty liver disease (NAFLD) were quantified by a NAFLD scoring system for rodent as described previously [27]. Briefly, the scoring system consists of steatosis score and inflammation score. For steatosis scores, there are three individual features including macrovesicular steatosis, microvesicular steatosis, and hypertrophy. Macrovesicular and microvesicular steatosis were scored based on the percentage of the total affected area: 0 (<5%), 1 (5–33%), 2 (34–66%), and 3 (>66%). The difference between macrovesicular and microvesicular steatosis was quantified by the position of the nucleus, i.e., the vacuoles displace the nucleus sideways (macrovesicular) or not (microvesicular). Hepatocellular hypertrophy was merely defined as cellular enlargement more than 1.5 times the normal hepatocyte diameter (distinct from ballooning). The level of hypertrophy was also scored based on the percentage of the affected area: 0 (<5%), 1 (5–33%), 2 (34–66%), and 3 (>66%). For inflammation scores, it was evaluated by counting the number of inflammatory foci per field. Five different fields were counted, and the average was subsequently scored into the following categories: 0 (<0.5 foci), 1 (0.5–1.0 foci), 2 (1.0–2.0 foci), 3 (>2.0 foci). The sum of the NAFLD scores ranged from 0~12, the greater the number, the more severe the condition of steatosis. These sections were read by the same pathologist in a blind manner.

For Oil-red O staining of frozen sections of liver, the frozen sections of liver were prepared in 8 μm thickness and stained with freshly diluted and filtered Oil Red O dye (Sigma, St. Louis, MO, USA). Hematoxylin staining was used as counterstaining. Samples were observed and photographed by a stereoscopic microscope (BX53, Olympus Optical Co, Ltd., Japan) equipped with a charged-coupled device (CCD) camera (DP80; Olympus) at magnification of 200×.

2.4. IPGTT and IPITT

IPITT and IPGTT were performed at the 10th week of intervention. In IPGTT, mice were fasted for 12 h and intraperitoneal administration of 50% glucose (2 g/kg) was conducted. In IPITT, mice were fasted for 6-h and intraperitoneal administration of insulin (0.4 IU/kg for the ND group, 0.5 IU/kg for HFD-fed groups) (Novolin R, Novo Nordisk, Denmark) was conducted. Blood glucose from caudal vein were measured at 0, 30, 60, 90, 120 min after the injections. Blood glucose curve were drawn and area under curve (AUC) of glucose was calculated accordingly.

2.5. Cell Culture of 3T3-L1 Preadipocytes and HepG2 Hepatocytes

The 3T3-L1 preadipocytes and HepG2 hepatocytes were cultured in accordance with previously used protocols in our laboratory [28]. In brief, 3T3-L1 cells were cultured in DMEM/F12 medium, mixed with 10% fetal bovine serum (FBS) and 1% Penicillin-Streptomycin (Hyclone, Logan, UT, USA) at 37 °C under 5% CO_2. The HepG2 hepatocytes were cultured in minimum essential medium with Earle's Balanced Salts (Hyclone, Logan, USA) supplemented with 1% penicillin-streptomycin and 0.1 mM non-essential amino acids (Solarbio Life Sciences, Beijing, China) at 37 °C under 5% CO_2.

2.6. Cell Isolation and Culture of Human Primary Preadipocytes and Murine Primary Mature Adipocytes

Human primary preadipocytes were obtained from the visceral WAT harvested during laparoscopic sleeve gastrectomy of a 21-year-old obese female (BMI = 35.5 kg/m^2). The comorbidities of the visceral WAT donor include insulin resistance, polycystic ovarian syndrome (PCOS), hypertension (grade 1, low risk), and chronic superficial gastritis and atopic dermatitis. The study was approved by the Medical Ethics Committee of Peking Union Medical College (No. JS-1093), and an informed consent was given to the patient before the surgery. Preadipocytes was isolated and cultured as previously described in our published paper [23]. Briefly, approximately 8 g of human visceral fat pad was harvested and digested by the same volume of type I collagenase (2 mg/mL) (Life Technologies, Van Allen Way, CA, USA). Then digestion was stopped by same volume of basal culture medium consisting of DMEM/F12 medium, 1% penicillin-streptomycin, 10% FBS, 17 μM pantothenic acid, 10 μg/mL transferrin, and 33 μM biotin (Sigma, St. Louis, MO, USA). The mixture was filtered and centrifuged at 600× *g* for 5 min. Sediment at the bottom of the tube was re-suspended and cultured in a cell culture flask (T25, Corning, NY, USA) at 37 °C under 5% CO_2.

The murine primary mature adipocytes were isolated from bilateral eWAT of 4 mice in each group by the "celling culture" method [29]. In brief, 0.6~2.0 g murine eWAT was digested by type I collagenase, and the digestion was terminated by the same volume of basal medium. Then the mixture was filtered and centrifuged. The floating mature adipocytes were re-suspended with culture medium and seeded in a T25 filled with basal medium. The flask was turned upside down and incubated at 37 °C under 5% CO_2 for 48 h. Then the culture medium was gently poured, and the flask was turned back and filled with 5 mL culture medium and continued incubating at 37 °C under 5% CO_2. Images of adipocytes were captured by a digital camera ((Nikon eclipse Ti, Nikon Corporation, Japan) at 100× magnification.

2.7. Cell Viability Assay

Effects of palmitic acid (PA) and Nuci of different concentrations on HepG2 hepatocytes, 3T3-L1 or human primary preadipocytes viability were determined by Cell Count Kit-8 (CCK-8, MedChem Express, HY-K0301, Monmouth Junction, NJ, USA). Cells were seeded in 96-well plates with a density of 1×10^3/well. After 24 h, cells were treated with 0, 2.5, 5, 10, 20 μM Nuci for 48 h, or with 0, 0.2, 0.4, 0.6, 0.8 mM PA for 12 h. Cell medium was then changed to fresh culture medium and mixed with 10 μL 2-(methoxy-4-nitrophenyl)-3-(4-nitrophenyl)-5-(2,4-disulfophenyl)2H-tetrazolium (WST-8) solution for each well and incubated for 4 h. Optical density (OD) values were measured at a wavelength of 450 nm by a spectrophotometer (Thermo, Waltham, MA, USA).

2.8. Cell Experiments of HepG2 Hepatocytes

HepG2 hepatocytes were cultured as described in Section 2.5. Nuci was dissolved in vehicle (dimethyl sulfoxide, DMSO) (Solarbio Life Sciences, China) for cell experiments. The maximum concentration of DMSO is 0.5%. After cells achieved 70~80% of confluence, medium was changed to fresh culture medium containing 0, 2.5, 5, 10, 20 μM Nuci. About 48 h later, cells were harvested and lysed for further molecular experiments.

2.9. Cell Experiments for AMPK Activities in Adipocytes and HepG2 Hepatocytes

3T3-L1 preadipocytes and primary human preadipocytes were induced differentiation as previously described [23]. Briefly, after 48 h of confluence, cells were cultured in differentiation medium supplemented with 10 μM dexamethasone, 10 μg/mL insulin, and 0.5 mM 3-isobutyl-1-methylxanthine (IBMX) (Gibco BRL, Grand Island, NY, USA) for 4 days. Then the medium was changed to fresh culture medium supplemented with 10 μg/mL insulin for another 2 days. Subsequently, the culture medium was refreshed every 2 days. Fully differentiated adipocytes were divided into five groups. The control group was treated with vehicle, the Nuci groups were treated with either 10 or 20 μM Nuci for 48 h, and the AMPK inhibitor Compound C (CC) (MedChem Express, USA, HY-13418A) groups were pretreated with 10 μM CC for 1 h, followed by treatment with either 10 or 20 μM Nuci for 48 h.

For HepG2 hepatocytes, 0.4 mM palmitic acid (PA) was used to induce lipid accumulation. Cells were divided into six groups. The control group was treated with vehicle, the PA group was treated with PA (0.4 mM) for 12 h. The Nuci groups were treated with either 10 or 20 μM Nuci for 48 h and together with PA for the last 12 h. The CC groups were pretreated with 10 μM CC for 1 h, followed by treatment with either 10 or 20 μM Nuci for 48 h and together with PA for the last 12 h.

2.10. Oil Red O Staining and TG Contents Determination in Adipocytes and HepG2 Hepatocytes

The experiment procedures were described in our published paper [30]. In brief, cells were stained with freshly diluted and filtered Oil Red O dye (Sigma, St. Louis, MO, USA) for 2 h. Representative photographs were captured by a digital camera (Nikon) at 100× magnification. Oil red O dye was extracted by isopropanol and the OD value was measured at 492 nm. TG contents were measured using a commercial kit (Comin Biotechnology, Suzhou, China). Briefly, cells were disrupted using an ultrasonic cell-crushing machine (Q700, Qsonica, Newtown, CT, USA). OD values were then measured by a spectrometer at 505 nm and the intracellular lipid contents were normalized against total protein concentration.

2.11. Real-Time Quantitative PCR

Real-time quantitative PCR (RT-qPCR) was performed in accordance with previous description [28]. In brief, the expression of housekeeping gene PPIA [31], lipid metabolism-related genes (FAS, ACC, SREBP1, HSL, ATGL), fatty acid oxidation-related genes (PPARα, CPT1α), adipokine genes (FGF21, ZAG), and thermogenesis gene (UCP1) were measured using a PCR system (ABI7500, Applied Biosystems, Bedford, MA, USA). Primer sequences were listed in Supplementary Table S1. The relative expressions of target genes were calculated by the $2^{-\triangle\triangle Ct}$ method [32].

2.12. Simple Western Analysis

Total protein content was extracted by RIPA lysis buffer containing Protease and Phosphatase Inhibitor Cocktail (MedChem Express, USA) and measured using a BCA Protein Assay Reagent kit (Beyotime Biotechnology, Shanghai, China). Protein expressions were detected by Simple Western using a Size Separation Master Kit with Split buffer (12–230 kD) following the instructions of the manufacturer. The reliability of this experimental approach has been verified by previous studies [33,34]. In brief, 1.5 μg protein of each sample was separated by capillary electrophoresis on Wes instrument (Protein Simple, San Jose, CA, USA) under default settings. Protein levels were probed with rabbit anti-AMPK (1:50, #2532, Cell Signaling Technology, Danvers, MA, USA), rabbit anti-phospho-AMPK (Thr172) (1:50, #2535, Cell Signaling Technology, USA), and rabbit anti-β-actin (13E5) (1:50, #4970, Cell Signaling Technology, USA). An Anti-Rabbit Detection Module was purchased from Protein Simple. Data were analyzed by Compass software (Protein Simple, San Jose, CA, USA) version 5.0.1.

2.13. Plasmid Transfection and Dual Luciferase Reporter Assay System

PGL3-hFAS (−622 ~ +3 bp)-Luc (hFAS625-Luc) plasmid which contains human FAS promoter from −622 ~ +3 bp [23] and PGL3-hHSL (−697 ~ +53 bp)-Luc (hHSL750-Luc) plasmid which contains human HSL promoter from −697 ~ +53 bp [30] were previously constructed in our laboratory. In brief, hFAS625-Luc, hHSL750-Luc plasmids and their internal control plasmid pRL-SV40 were transiently transfected into the cells, and 0~20 μM Nuci was added for 48 h. Both firefly and renilla luciferase activities were measured by a Dual-Luciferase Reporter Assay System Kit (Promega, Madison, WI, USA) in an automated optical immunoassay analyzer (Beijing Pilot Biotechnology Corporation, Beijing, China). Each concentration was repeated for 9~12 wells.

2.14. Statical Analyses

Cells experiments were carried out independently for at least three times. Results were expressed as mean ± standard error (SE). Statical analyses were performed by SPSS software for Windows (version 25.0) (SPSS Inc., Chicago, IL, USA). Chi square test, one-way and two-way ANOVA were used in data analyses. The Kruskal–Wallis test was used if ANOVA was not applicable. For post-hoc test, the Dunnett *t* test (two-side) or Dunnett t3 post-test was used for comparisons between three or more groups. $p < 0.05$ was considered as statistically significant in all analyses.

3. Results

3.1. Nuci Prevented Obesity in HFD-Fed Mice

The molecular structure of Nuci was displayed in Figure 1a. As shown in Figure 1b,c, compared with the ND group, the body weight and body weight gain of the HFD group were remarkedly increased from the fourth week till the last. Treatment of Nuci remarkably reduced the body weight of mice at the first week, with the maximal reduction effect being observed by the end of the experiment, where the body weight of mice decreased by 28.8% compared to the HFD group. Similarly, Lira remarkedly decreased body weight since the first week, with the maximal reduction effect being 23.0% less than the HFD group. Both Nuci and Lira effectively impeded the HFD-induced weight gain by 72.6% and 59.4% ($p < 0.05$). As shown in Figure 1d, HFD significantly elevated energy intake compared to the ND, yet Nuci and Lira had little effect on energy intake in HFD-fed mice.

As depicted in Figure 1e,f, HFD remarkedly increased the amount of fat mass (in cluding sWAT, eWAT and pWAT), while Nuci notably decreased fat mass and fat mass percentage by 73.8% and 58.7% compared with the HFD group ($p < 0.05$). Lira exhibited similar effect ($p < 0.05$). There was no effect of Nuci or Lira on the weight of BAT or liver (Figure 1g,h).

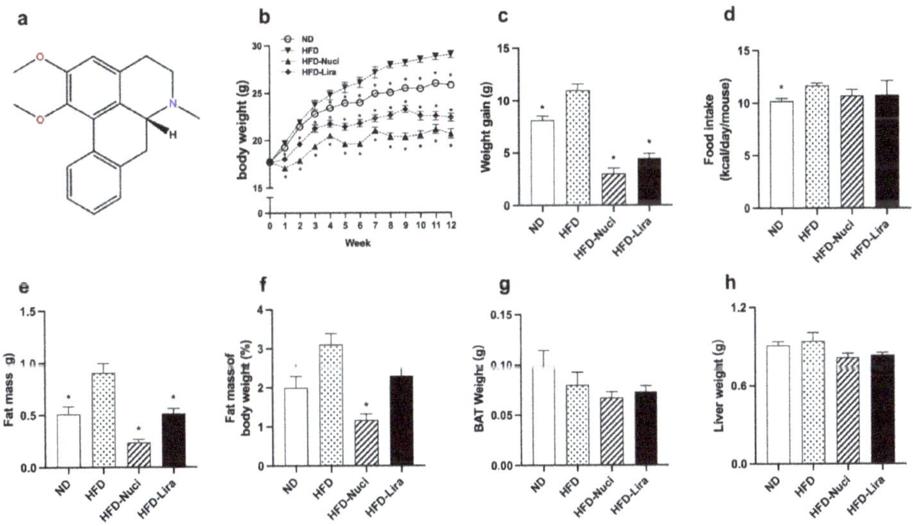

Figure 1. Nuci prevented obesity in HFD-fed mice. (**a**) The molecular structure of Nuci. Seven-weeks male C57BL/6J mice were fed with ND, HFD, HFD supplemented with Nuci and HFD with subcutaneous injection of Lira at 200 μg/kg/day for 12 weeks, body weight (**b**) and energy intake (**d**) were recorded, and body weight gain (**c**) was calculated by the end of the experiment. The amount of fat mass (**e**), including sWAT, eWAT, and pWAT, as well as the weight of BAT (**g**) and liver (**h**) were recorded. Fat mass of body weight (**f**) was calculated by dividing the sum of fat mass by body weight. Data were presented as mean ± SE, $n = 12$ in each group. * $p < 0.05$ vs. the HFD group.

3.2. Nuci Improved Dyslipidemia, Glucose Tolerance and Insulin Resistance in HFD-Fed Mice

Liver function remained similar among the groups (Figure 2a). As shown in Figure 2b, Nuci remarkedly decreased serum levels of TC, TG, and LDL-C by 10.6%, 55.9%, and 30.1%, respectively, when compared with the HFD group (all $p < 0.05$). As shown in Figure 2c, Nuci and Lira notably decreased FBG levels of HFD-fed mice ($p < 0.05$). IPGTT and IPITT were performed (Figure 2d and f, $p < 0.05$) and as depicted in Figure 2e,g, Nuci and Lira reduced AUC of IPGTT and IPITT compared with the HFD group, respectively ($p < 0.05$). Nuci also effectively improved insulin sensitivity as reflected by decreased FINS levels and

HOMA-IR. The hypoglycemic effect of Nuci was similar to Lira (Figure 2h,i, $p < 0.05$). Next, serum levels of FGF21 and ZAG were measured. As shown in Figure 2j,k, HFD led to a notable decrease in both adipokines compared to the ND group ($p < 0.05$), whereas the serum levels of which were remarkedly increased to the similar content as the ND group after treatment of Nuci or Lira ($p < 0.05$).

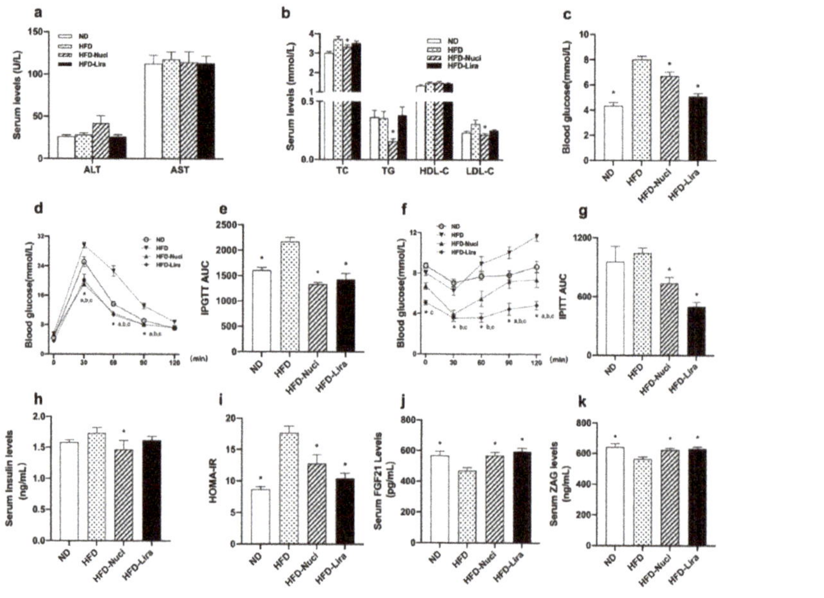

Figure 2. Nuci improved dyslipidemia, glucose tolerance, and insulin resistance in HFD-fed mice. Seven-weeks male C57BL/6J mice were fed with ND, HFD, HFD supplemented with Nuci and HFD with subcutaneous injection of Lira at 200 μg/kg/day for 12 weeks. Liver function markers including ALT and AST (**a**), lipid metabolism-related traits including TC, TG HDL-C, and LDL-C (**b**) and fasting blood glucose (**c**) were examined. The intraperitoneal glucose tolerance test (IPGTT, (**d**)) and the intraperitoneal insulin tolerance test (IPITT, (**f**)) were performed at the 10th week, and area under curves (AUCs) of IPGTT (**e**) and IPITT (**g**) were calculated. FINS (**h**) was determined, and HOMA-IR (**i**) was calculated accordingly. Serum levels of FGF21 (**j**) and ZAG (**k**) were determined by ELISA. Data were presented as mean ± SE, $n = 12$ in each group. * $p < 0.05$ vs. the HFD group. In Figure 2d and f, * a, the ND vs. the HFD group; * b the HFD-Nuci vs. the HFD group; * c the HFD-Lira vs. the HFD group.

3.3. Nuci Remarkedly Increased Energy Metabolism in HFD-Fed Mice

As shown in Figure 3a, the representative general pictures of mice showed that the HFD-fed mouse was larger than the ND-fed ones and the treatment of Nuci made them comparatively smaller. Next, the effect of Nuci on energy metabolism including VO_2, VCO_2, RER, and locomotor activity levels were measured in metabolic cages. As depicted in Figure 3b,c, Nuci remarkedly elevated VO_2 consumption, especially in the light cycle where the VO_2 of HFD-Nuci group increased to 1.33-fold of the HFD group ($p < 0.05$). Similar effect was observed in VCO_2 (Figure 3d,e). RER was not remarkedly altered by Nuci treatment (Figure 3f). Moreover, it was observed that Nuci significantly increased EE to 1.41-fold in the light cycle and 1.17-fold in the dark cycle in comparison with the HFD group, respectively (Figure 3g, $p < 0.05$). However, the locomotor activity reflected by beam breaks showed no statistically significant difference among the three groups (Figure 3h).

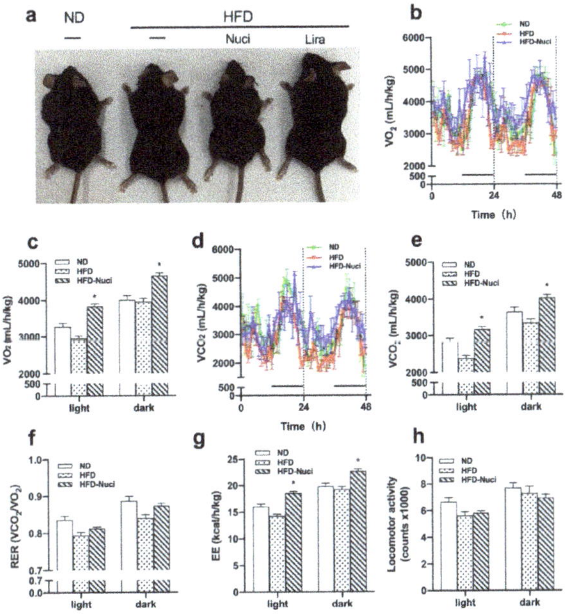

Figure 3. Nuci increased energy metabolism in HFD-fed mice. Seven-weeks male C57BL/6J mice were fed with ND, HFD, HFD supplemented with Nuci and HFD with subcutaneous injection of Lira at 200 µg/kg/day for 12 weeks. General pictures of mice in ND, HFD, HFD-Nuci, and HFD-Lira group were presented (**a**). After 11 weeks of treatment, mice in the ND, HFD, and HFD-Nuci group were placed in metabolic cages at 22 °C at 12 h light–dark cycle. VO_2 (**b**), VCO_2 (**d**) were detected and quantification of VO_2 (**c**), VCO_2 (**e**), RER (**f**), EE (**g**), and locomotor activity (**h**) during both the light and dark cycle were determined by Promethion. Data were presented as mean ± SE, $n = 4$ in each group. * $p < 0.05$ vs. the HFD group.

3.4. Nuci Ameliorated Hepatic Steatosis by Activating AMPK Phosphorylation in Liver of HFD-Fed Mice

As shown in Figure 4a,c, the histology of liver by H&E staining in HFD-fed mice presented obvious steatosis, indicated by distinct hepatocellular ballooning and parenchymal steatosis. The NAFLD scores also remarkedly increased in the HFD group. However, Nuci and Lira remarkedly decreased NAFLD scores when compared with the HFD group ($p < 0.05$). Moreover, as shown in Figure 4b, the Oil-red O staining also revealed that the liver of mice in the HFD group showed greater amount of stained lipid deposition compared to that in the ND group, indicating that HFD feeding induced significant hepatic steatosis. After treatment of Nuci and Lira, the hepatic lipid accumulation was ameliorated as evidenced by reduced Oil-red O-stained lipid droplets in livers of mice in the HFD-Nuci and HFD-Lira group. Next, as depicted in Figure 4d, HFD-fed mice presented substantially elevated expressions in lipogenesis-related genes including SREBP1, FAS, and ACC to 9.67-, 3.62-, and 6.64-fold of the ND group, respectively, and Nuci and Lira remarkably reduced the expressions of these genes ($p < 0.05$). Besides, they also remarkably increased the expressions of HSL and ATGL (Figure 4e, $p < 0.05$), as well as PPARα and CPT1α (Figure 4f, $p < 0.05$). Interestingly, the expressions of FGF21 and ZAG in liver were also prominently increased to 1.82- and 2.20-fold of HFD group by Nuci treatment, and the expression of ZAG also increased upon Lira treatment (Figure 4g, $p < 0.05$). Western blot revealed that HFD feeding remarkedly downregulated the phosphorylation of AMPK in liver, while Nuci treatment remarkedly increased the p-AMPK upregulation by 3.38-fold in liver when compared to the HFD group (Figure 4h,i, $p < 0.05$).

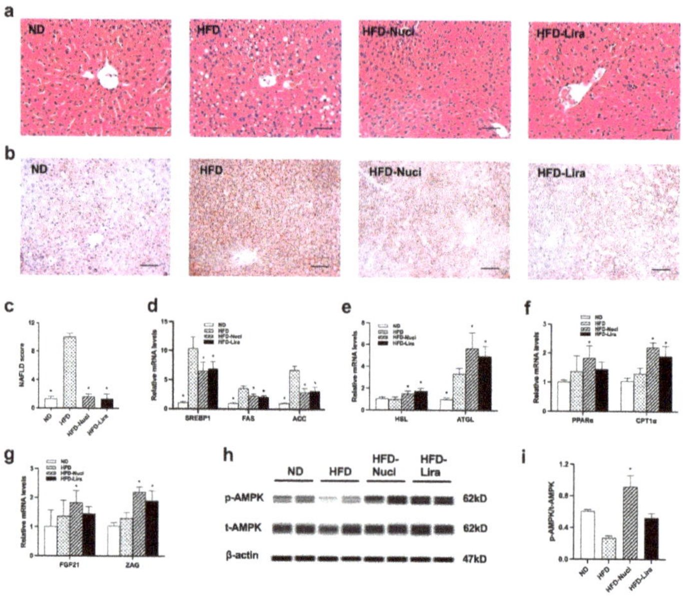

Figure 4. Nuci ameliorated hepatic steatosis by activating AMPK phosphorylation in liver of HFD-fed mice. Seven-weeks male C57BL/6J mice were fed with ND, HFD, HFD supplemented with Nuci and HFD with subcutaneous injection of Lira at 200 μg/kg/day for 12 weeks. H&E-staining (**a**) and Oil-red O staining (**b**) images of representative sections of mice liver in the ND, HFD, HFD-Nuci, and HFD-Lira group; photographs were taken at 200× magnification. NAFLD scores of the four groups according to a general NAFLD scoring system for rodent models (**c**). The mRNA expressions of lipogenesis-related genes SREBP1, FAS, and ACC (**d**), lipolysis-related genes HSL and ATGL (**e**), fatty acid oxidation-related genes PPARα and CPT1 (**f**), and adipokines FGF21 and ZAG (**g**) were determined by RT-qPCR. The protein levels of p-AMPK, t-AMPK and β-actin were analyzed by Simple western (**h**). the expression of p-AMPK was normalized against t-AMPK (**i**). Data were presented as mean ± SE, $n = 12$ in each group. * $p < 0.05$ vs. the HFD group.

3.5. Nuci Suppressed Lipid Accumulation by Reducing Lipogenesis and Promoting Lipolysis in HepG2 Hepatocytes

HepG2 hepatocytes were treated with 0, 10, 20 μM Nuci for 48 h after achieving 70–80% confluence. As shown in Figure 5a, the viability of HepG2 hepatocytes treated with 5~20 μM Nuci yielded between 97.3~101.1%, suggesting Nuci of applied concentrations did not elicit cell toxicity. Next, as shown in Figure 5b,c, 5~20 μM Nuci dose-dependently decreased the intracellular lipid contents; the maximal effect was observed at 20 μM where the lipid contents were reduced by 35.8% compared to the control group (0 μM) ($p < 0.05$). Similarly, TG contents were also remarkedly decreased (Figure 5d, $p < 0.05$). Moreover, similar to the results obtained in liver tissue, 20 μM Nuci remarkedly suppressed the expressions of SREBP1, FAS, and ACC by 27.7%, 27.0%, and 29.4%, respectively (Figure 5e, $p < 0.05$), while promoted the expressions of HSL, ATGL, PPARα, and CPT1α (Figure 5f,g, $p < 0.05$). Moreover, the expressions of FGF21 and ZAG were also promoted by Nuci with the maximal effect observed at 20 μM, where they were increased to 2.16- and 1.61-fold of the control group (0 μM), respectively (Figure 5h, $p < 0.05$). In order to further investigate the underlying mechanism by which Nuci stably regulates the expression of FAS and HSL, hFAS625-Luc and hHSL750-Luc plasmids were transiently transfected into HepG2 s cells and treated with Nuci for 48 h. It was found that the 20 μM Nuci notably suppressed the luciferase activities of FAS by 44.2%, while promoted HSL to 1.39-fold of the control group (0 μM) (Figure 5i,j, $p < 0.05$).

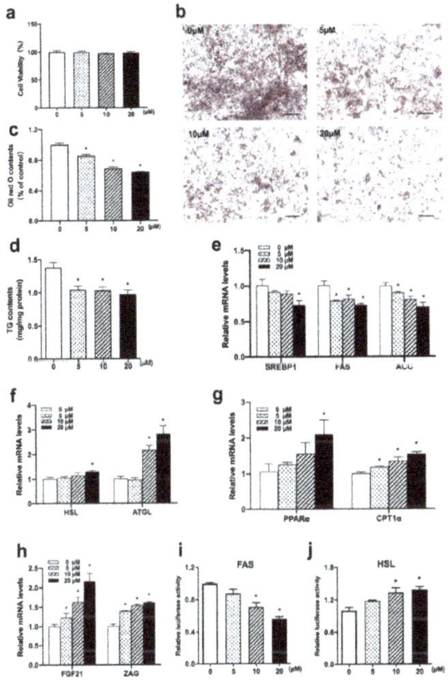

Figure 5. Nuci suppressed lipid accumulation by reducing lipogenesis and promoting lipolysis in HepG2 hepatocyte. Cell viability after treatment of 0~20 μM Nuci were determined by CCK-8 (**a**). Cells were stained by Oil red O and photographed at 100× magnification (**b**) and Oil red O dye was extracted and detected (**c**). Intracellular TG contents (**d**) were determined by a commercial kit following the manufacturer's instructions. The mRNA expressions of lipogenesis-related genes SREBP1, FAS, and ACC (**e**), lipolysis-related genes HSL and ATGL (**f**), fatty acid oxidation-related genes PPARα and CPT1 (**g**), and adipokines FGF21 and ZAG (**h**) were determined by RT-qPCR. Cells were transiently transfected with hFAS625-Luc and hHSL750-Luc plasmids and the luciferase activities of FAS (**i**) and HSL (**j**) were measured. The firefly luciferase activities were adjusted by the renilla luciferase. Data were presented with mean ± SE from three independent experiments, each concentration was repeated for 9~12 wells. * $p < 0.05$ vs. the control group (0 μM).

3.6. Nuci Prevented PA-Induced Cellular Steatosis by Activating AMPK Phosphorylation in HepG2 Hepatocytes

As remarkedly elevated AMPK phosphorylation was observed in liver of HFD-Nuci group, we further deployed PA-induced steatosis cell model to investigate whether AMPK was involved in alleviating the effect of Nuci on lipid accumulation in PA-induced HepG2 cells. As shown in Figure 6a–e, PA notably decreased AMPK phosphorylation while remarkedly increased intracellular lipid droplets and TG contents in HepG2 cells ($p < 0.05$). However, 10 and 20 μM Nuci restored the AMPK phosphorylation and ameliorated the PA-induced lipid accumulation. Next, a selective AMPK inhibitor CC was used to further confirm the role of AMPK. The result showed that CC remarkedly abolished the promotion of AMPK phosphorylation induced by Nuci (Figure 6b). Consequently, the prevention of lipid accumulation caused by Nuci was notably impeded in the presence of CC as demonstrated by the deepened Oil red O staining (Figure 6c), the increased lipid contents (Figure 6d), and intracellular TG contents (Figure 6e). Similarly, as shown in Figure 6f–k, the reduced expressions of SREBP1, FAS, and ACC and the promoted expression of HSL and FGF21 caused by 20 μM Nuci were also notably impeded after CC pretreatment ($p < 0.05$). All these findings suggested that Nuci suppressed lipogenesis-related genes and promoted lipolysis gene via activating AMPK signaling pathway in HepG2 hepatocytes.

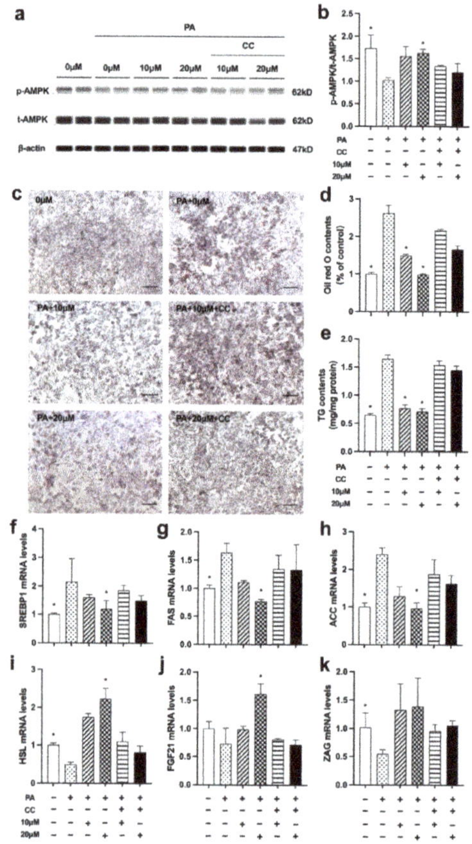

Figure 6. Nuci prevented PA-induced cellular steatosis by activating AMPK phosphorylation in HepG2 hepatocytes. HepG2 hepatocytes were cultured and treated according to the scheme described in Method 2.9. The protein levels of p-AMPK, t-AMPK, and β-actin were analyzed by Simple western (**a**), the expression of p-AMPK was normalized against t-AMPK (**b**). Oil red O staining was conducted and photographed at 100× magnification (**c**) and Oil red O dye was extracted and detected (**d**). Intracellular TG contents I were determined by a commercial kit following the manufacturer's instructions. The mRNA expressions of SREBP1 (**f**), FAS (**g**), ACC (**h**), HSL (**i**), ZAG (**j**), and FGF21 (**k**) were determined by RT-qPCR. Data were presented with mean ± SE from three independent experiments, * $p < 0.05$ vs. the PA group (PA + 0 μM).

3.7. Nuci Decreased the Weight of eWAT by Activating AMPK Phosphorylation in HFD-Fed Mice

eWATs in the ND, HFD, HFD-Nuci group were harvested and photographed (Figure 7b). As shown in Figure 7d, both Nuci and Lira remarkedly decreased the weight of eWAT in HFD-fed mice ($p < 0.05$). Moreover, HFD-fed mice presented remarkedly enlarged adipocytes as demonstrated by H&E staining, while Nuci and Lira inhibited the morbid expansion of adipocytes to the extent where it is much alike to the mice fed with ND (Figure 7a,e, $p < 0.05$). As shown in Figure 7c,f, Oil red O staining of primary mature adipocytes in eWATs revealed that the size and number of lipid droplets were remarkedly enlarged in the HFD group, whereas Nuci decreased the lipid contents by 43.2% compared to the HFD group ($p < 0.05$). Notably, HFD led to significant downregulation of UCP1, a vital marker for non-shivering thermogenesis, whereas Nuci completely blocked this phenomenon and strongly promoted the mRNA levels of UCP1 in eWAT to 14.0-fold of that in the HFD group (Figure 7g, $p < 0.05$). Likewise, it was shown that Nuci remarkably reduced the mRNA levels of SREBP1, FAS, and ACC by 63.7%, 82.6%, and 78.5%, respectively (Figure 7h, $p < 0.05$), and promoted the mRNA levels of HSL, ATGL (Figure 7i, $p < 0.05$), FGF21, and ZAG (Figure 7j, $p < 0.05$). Lira showed similar effects on the expressions of these genes. Further mechanical study demonstrated that the phosphorylation of AMPK was remarkably decreased in HFD group when compared to ND group as shown in Figure 7k,l, and the treatment of both Nuci and Lira remarkedly increased the p-AMPK/t-AMPK ratio to 2.55- and 2.48-fold of HFD group, respectively ($p < 0.05$).

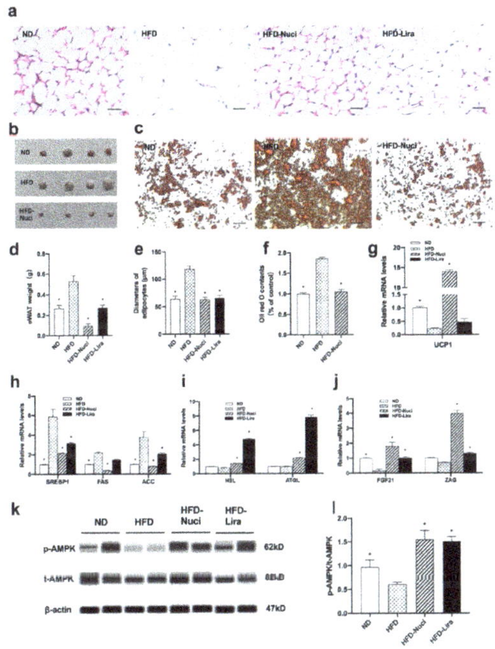

Figure 7. Nuci decreased the weight of eWAT by activating AMPK phosphorylation in HFD-fed mice. Seven-weeks male C57BL/6J mice were fed with ND, HFD, HFD supplemented with Nuci and HFD with subcutaneous injection of Lira at 200 μg/kg/day for 12 weeks. H&E-staining images of representative sections of mice eWAT in the ND, HFD, HFD-Nuci, and HFD-Lira group, photographs were taken at 400× magnification (**a**). Diameters of adipocytes were measured from six random fields using Image J software (**e**). The weight of eWAT was measured (**d**). Photographs of isolated mice eWATs in the ND, HFD, HFD-Nuci group (**b**). Representative images of the Oil red O stained mice primary mature adipocytes, photographs were taken at 100× magnification (**c**) and Oil red O dye was extracted and detected (**f**). The mRNA expressions of UCP1 (**g**) lipogenesis-related genes SREBP1, FAS, and ACC (**h**), lipolysis-related genes HSL and ATGL (**i**), adipokines FGF21 and ZAG (**j**) were determined by RT-qPCR. The protein levels of p-AMPK, t-AMPK, and β-actin were analyzed by Simple western (**k**), the expression of p-AMPK was normalized against t-AMPK (**l**). Data were presented as mean ± SE, $n = 12$ in each group. * $p < 0.05$ vs. the HFD group.

3.8. *Nuci Inhibited Lipid Accumulation by Activating AMPK Phosphorylation in Fully Differentiated 3T3-L1 Cells*

Since the remarkedly elevated AMPK phosphorylation was observed in eWAT of HFD-Nuci group, the causal relationship between the increased phosphorylation of AMPK and the fat loss remained unclear. In order to answer this question, we further explore the effect of Nuci on phosphorylation of AMPK and lipid accumulation in fully differentiated 3T3-L1 cells. Similar to Result 3.5, first, the viability of 3T3-L1 adipocytes was confirmed to be unaffected upon administration of 5~20 μM Nuci (93.5~104.6%) (Figure 8d). Next, as shown in Figure 8a,b, 10 and 20 μM Nuci directly and remarkedly increased the phosphorylation of AMPK, with this effect being abolished after pretreatment of CC. Subsequently, 10 and 20 μM Nuci directly decreased the lipid contents by 30.3% and 56.4%, respectively, as demonstrated by Oil red O staining, and this effect was also remarkedly abolished with CC (Figure 8c,e, $p < 0.05$). Similar results were observed in TG contents measurements (Figure 8f, $p < 0.05$). Furthermore, 10~20 μM Nuci directly and notably decreased the expressions of SREBP1, FAS, and ACC by 35.0~58.0% (Figure 8h–j, $p < 0.05$), while promoted the expressions of FGF21, ZAG as well as UCP1, and these effects being blocked by CC as well (Figure 8g,I,m, $p < 0.05$). Together, these findings suggested that Nuci directly and remarkedly reduced intracellular lipid accumulation through regulating AMPK-mediated FAS/HSL pathway and promoting FGF21, ZAG as well as UCP1 expression.

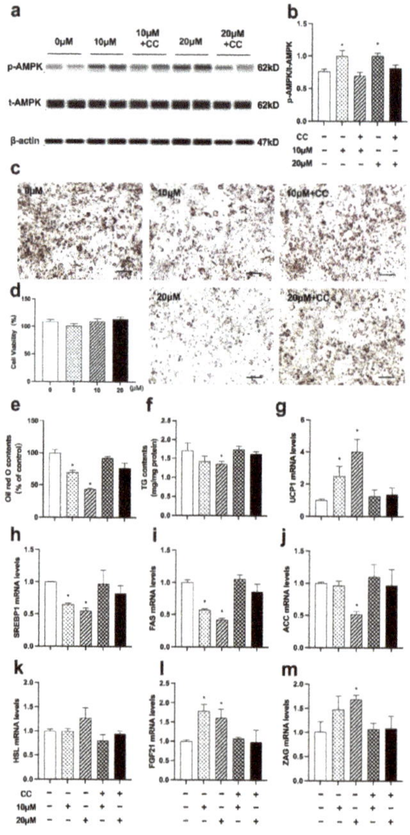

Figure 8. Nuci inhibited lipid accumulation by activating AMPK phosphorylation in fully differentiated 3T3-L1 adipocytes. 3T3-L1 preadipocytes were cultured and induced differentiation, then fully differentiated adipocytes were treated as described in the Method 2.9. The protein levels of p-AMPK, t-AMPK, and β-actin were analyzed by Simple western (**a**), the expression of p-AMPK was normalized against t-AMPK (**b**). Oil red O staining was conducted and photographed at 100× magnification (**c**) and Oil red O dye was extracted and detected (**e**). Cell viability after treatment of 0~20 μM Nuci were determined by CCK-8 (**d**). Intracellular TG contents (**f**) were determined by a commercial kit following the manufacturer's instructions. The mRNA expressions of UCP1 (**g**), SREBP1 (**h**), FAS (**i**), ACC (**j**), HSL (**k**), FGF21 (**l**), and ZAG (**m**) were determined by RT-qPCR. Data were presented with mean ± SE from three independent experiments, * $p < 0.05$ vs. the control group (0 μM).

3.9. Nuci Inhibited Lipid Accumulation by Activating AMPK Phosphorylation in Fully Differentiated Human Primary Adipocytes

Primary cells have the additional advantages of being unmodified, which is reputed to reflect in vivo conditions more closely than cell lines. Therefore, we harvested primary preadipocytes from human visceral WAT to further investigate the beneficial role of Nuci. As expected, 0~20 μM Nuci did not alter cell viability in human primary adipocytes (103.9~125.5%) (Figure 9d). As shown in Figure 9a,b, 20 μM Nuci remarkedly boosted AMPK phosphorylation by 1.31-fold in the control group (0 μM) ($p < 0.05$). In the meantime, intracellular lipid contents were also decreased by Nuci (Figure 9c,e,f, $p < 0.05$). Similar to Results 3.8, CC remarkedly decreased the phosphorylation of AMPK, led to the subsequent inhibition of lipid accumulation induced by Nuci rendered ineffective. Furthermore, as demonstrated in Figure 9g–m, 20 μM Nuci notably decreased the expressions of SREBP1 and FAS by 39.0% and 41.3%, while increased the mRNA levels of UCP1, HSL, and FGF21 to 1.92, 1.67, 2.26-fold of the control group, respectively ($p < 0.05$), with all these effects being reversed by CC.

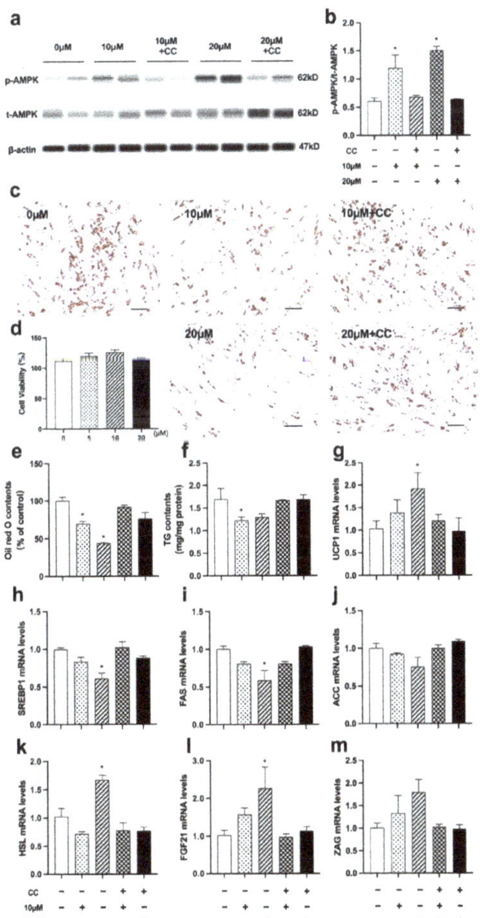

Figure 9. Nuci inhibited lipid accumulation by activating AMPK phosphorylation in fully differentiated human primary adipocytes. Human primary preadipocytes were cultured and induced

differentiation, then fully differentiated adipocytes were treated as described in the Method 2.9. The protein levels of p-AMPK, t-AMPK, and β-actin were analyzed by Simple western (**a**), the expression of p-AMPK was normalized against t-AMPK (**b**). Oil red O staining was conducted and photographed at 100× magnification (**c**) and Oil red O dye was extracted and detected (**e**). Cell viability after treatment of 0~20 μM Nuci was determined by CCK-8 (**d**). Intracellular TG contents (**f**) were determined by a commercial kit following the manufacturer's instructions. The mRNA expressions of UCP1 (**g**), SREBP1 (**h**), FAS (**i**), ACC (**j**), HSL (**k**), FGF21 (**l**), and ZAG (**m**) were determined by RT-qPCR. Data were presented with mean ± SE from three independent experiments, * $p < 0.05$ vs. the control group (0 μM).

4. Discussion

Recently, herbal medicines have attracted much attention due to their potential for easy availability, less side effects, and wide application. Nuci, as the main component of lotus leaf, have been reported with anti-obesity effect on murine models, yet the mechanism remains controversial. Thus, in this study, we demonstrated that Nuci remarkedly prevented HFD-induced weight gain and improved glycolipid metabolism. It was also worth noting that for the first time, we reported that Nuci increased energy expenditure in HFD-fed mice. Besides, Nuci also ameliorated hepatic steatosis and decreased weight and size of eWAT. In hepatocytes and adipocytes, Nuci was found to suppress lipogenesis, promote lipolysis by directly regulating a AMPK-mediated FAS/HSL pathway, and increased fatty acid oxidation, thermogenesis, and the expressions of adipokines including FGF21 and ZAG.

As indicated by glucose indicators including IPGTT, IPITT, and HOMA-IR, Nuci restored HFD-induced impairment of glucose tolerance and insulin sensitivity similar to that of the positive control drug liraglutide. In consistence with our results, several studies reported that Nuci notably prevented obesity and improved glucose metabolism in HFD-fed diabetic mice [24,35]. Furthermore, Nuci also improved plasma lipid profiles without affecting liver function, indicating that the anti-obesity effect of Nuci did not attribute to toxicity. This finding was supported by Guo F et al. who observed similar results in HFD-fed hamsters treated with Nuci [36]. One striking finding of our present study is that Nuci was first found to notably increase VO_2, VCO_2, and EE in HFD-fed mice without altering locomotor activity or energy intake. It is well established that the increase of EE can be further attributed to the increase of RER or basal metabolic rate [37]. Since there were no significant differences found in RER between the groups in the present study, it is more likely that Nuci prevents obesity in HFD-fed mice, at least partly, via enhancing EE, and the increase of EE was thought to be associated with the reinforced basal metabolic rate.

Excessive weight gain is often accompanied by hepatic steatosis and morbid expansion of adipose tissue. In the present study, we found that HFD-fed mice exhibit severe ballooning degeneration in liver, while Nuci ameliorated the development of HFD-induced hepatic steatosis. Consistently, Wang et al. also found that 8 weeks of Nuci supplementation achieved similar effects in HFD-fed rats [38]. Besides, in this study, Nuci also remarkedly reduced the amount of eWAT, deflated the volume of adipocytes, and decreased the intracellular lipid contents. There were studies that reported that Nuci decreased the weight and protected against histological degeneration of visceral WAT induced by high-fat diet in hamsters and mice [35,36], yet the mechanism was not thoroughly investigated. In the present study, Nuci supplementation remarkedly decreased the mRNA levels of lipogenesis-related transcription factors including SREBP1 and its target genes FAS and ACC, upregulated the expressions of key enzymes in lipolysis including HSL and ATGL in both liver and eWAT. Similarly, Guo F et al. revealed that 8 weeks of Nuci downregulated hepatic expressions of lipogenesis-related genes in HFD-fed hamsters [36]. Enlightened by these results, we further conducted in vitro experiments and observed comparable effects of Nuci in HepG2 hepatocytes, 3T3-L1 adipocytes, and human primary adipocytes, confirming that Nuci directly modulate lipid metabolism at cellular levels. Moreover, for the first time, we found that 0~20 μM Nuci remarkedly inhibited FAS and promoted HSL promoter activities in HepG2 cells, suggesting that Nuci may suppress lipogenesis and activate lipolysis by regulating the FAS/HSL pathway.

Fatty acid oxidation contributes to lipid catabolism as well, especially in the liver. Herein, we found that the expressions of a critical transcription factor in fatty acid oxidation, PPARα, as well as its downstream gene CPT1α in liver were boosted by Nuci. In consistence with our finding, Guo F et al. also revealed that in HFD-fed hamsters, Nuci had protective effects on hepatic lipid metabolism by promoting PPARα and CPT1α [36]. Zhang et al. established HFD/STZ-induced diabetic mice and observed similar phenomenon; further molecular experiments showed that the binding of activated PPARα to PPRE was reinforced by Nuci [24]. Moreover, in vitro experiments in the present study further confirmed that Nuci promoted the expression of PPARα and CPT1α in hepatocytes, indicating that Nuci alleviated fat accumulation by promoting fatty acid oxidation in liver and hepatocytes.

Previous studies have shed light on the involvement of adipokines in lipid mobilization in metabolic organs, among which FGF21 and ZAG possessed renowned regulatory effect on fat metabolism [14,39]. It has been reported that FGF21 suppressed hepatic lipogenesis in an autocrine and paracrine manner [40], and overexpressing ZAG promoted lipolysis and fatty acid oxidation in hepatocytes and boosted browning in adipocytes [41]. Our previous work demonstrated that Nuci promoted the expression of FGF21 and ZAG in adipocytes [23], and in this study, we found for the first time that in vivo, Nuci also promoted the expression of these two adipokines in both liver and eWAT, leading to higher levels of serum FGF21 and ZAG. Together, Nuci may reduce fat accumulation by enhancing the expression of FGF21 and ZAG, which further exerted their beneficial role on glycolipid metabolism in an endocrine manner. Moreover, browning of WAT is emerging as a potential therapeutic target for weight loss. When stimulated, WAT acquires a brown fat-like phenotype, leading to increased thermogenesis through dissipating energy as heat [42]. Notably, we first demonstrated that Nuci led to dramatic elevation in mRNA levels of the key non-shivering thermogenesis activator UCP1 in eWAT. Likewise, Nuci also promoted the mRNA levels of UCP1 in fully differentiated adipocytes derived from 3T3-L1 and human primary preadipocytes, suggesting Nuci may promote browning of WAT by directly increasing the expression of UCP1 in HFD-fed mice. It also provides an explanation that the enhanced EE induced by Nuci in HFD-fed mice might be resulting from stimulating eWAT browning, thus promoting thermogenesis and increasing basal metabolic rate.

The AMPK pathway is crucial for maintaining cellular homeostasis, especially for lipid metabolism. Activation of AMPK leads to weight loss and amelioration of NAFLD [43]. Ma C et al. demonstrated that Nuci activated AMPK phosphorylation and inhibited lipogenesis in insulin-resistant 3T3-L1 adipocytes [25], yet the in vivo situation remained unclear. In the present study, we first demonstrated that in HFD-fed mice, AMPK pathway was involved in the beneficial role of Nuci in preventing HFD-induced obesity and attenuating hepatic steatosis. First, HFD remarkedly decreased AMPK activity, while Nuci significantly restored AMPK activity to the extent similar to that of ND group in both liver and eWAT. Second, cell experiment performed in hepatocytes and adipocytes showed that Nuci directly and remarkedly increased the phosphorylation of AMPK, and this effect was abolished after pretreatment of CC. Finally, the Nuci-induced suppression of lipid accumulation, which was derived from the decreased lipogenesis (SREBP1, FAS and ACC), increased lipolysis (HSL), as well as fatty acid oxidation (PPARα and CPT1α) or thermogenesis (UCP1), were all notably impeded after inhibition of AMPK. A previous study showed that AMPK is an upstream kinase of SREBP-1c which directly inhibits its nuclear translocation [44], hereby suppressing downstream lipogenesis. Furthermore, the activation of AMPK also increases lipolysis and fatty acid oxidation [15], and upregulated the expression of adipokine FGF21 levels [45]. Consistently, in this study, the elevation of FGF21 and ZAG mRNA levels induced by Nuci in hepatocytes and adipocytes was also abolished by CC, indicating that AMPK may be involved in the effect of Nuci on regulating the expression of both adipokines. All these findings suggest that Nuci improves whole body glucose and lipid metabolism and ameliorates hepatic steatosis through directly activating AMPK signaling pathway, which subsequently leads to the inhibition of lipogenesis, promotion of lipolysis and fatty acid oxidation, as well as upregulation of the expressions of FGF21 and ZAG in liver/hepatocytes and eWAT/adipocytes. Still, how Nuci acts with the blockage of AMPK pathway in HFD-induced obesity mice was not confirmed, hence further in vivo studies centering on Nuci and AMPK pathway are needed.

Moreover, it should be noted that different from our study, there are research that reported Nuci may also exert its anti-obesity effect via regulation of gut microbiota. Gut microbiota is essential for maintaining normal gastrointestinal and immune functions and efficient digestion of nutrients [46], and increasingly abundant evidence has shown that the gut microbiota affects whole-body nutrient acquisition, energy regulation, and fat storage [47]. Pharmacokinetics studies of Nuci showed that the absolute bioavailability is relatively low (1.9 $\pm$ 0.8%), yet mild dosage of Nuci was reported to show robust anti-obesity effect. Therefore, it is reasonable to speculate that apart from the ability of regulating the expressions of lipid-metabolism genes, the metabolic benefits of Nuci may also be related to the regulation of gut microbiota. Wang et al. [38] found that Nuci significantly reduced the body weight of HFD-fed rats after 8 weeks of intervention, 16S rRNA sequencing revealed that it decreased the ratio of Firmicutes/Bacteroidetes, the relative abundance of the LPS-producing Desulfovibrio, and other bacteria involved in lipid metabolism, whereas it increased the relative abundance of short-chain fatty acids (SCFA)-producing bacteria, enhanced intestinal integrity, leading to lower blood endotoxemia to reduce inflammation in HFD-fed rats. Consistent with this result, Yu et al. [48] also found that Nuci reduced body weight in HFD-fed mice. By applying metagenomic sequencing, it was found that Nuci significantly changed the construction of gut microbiota, enriched the abundance of a potential probiotic Akkermansia muciniphila. However, the anti-obesity effect of Nuci was abolished after pretreatment of antibiotics, suggesting that gut microbiota plays an important role in the metabolic benefits of Nuci. Similarly, Tang et al. [35] found that Nuci corrected intestinal dysbacteriosis in gestational diabetic mice, enriched probiotic abundances of Akkermansia, Lactobacillus and Bifidobacterium, while reduced conditional pathogen abundances including Escherichia-Shigella and Staphylococcus. Additionally, the study of Shi et al. [47] found that Nuci reduced body weight gain and fat accumulation in HFD-fed mice by improving intestinal permeability and autophagy [49].

5. Conclusions

To summarize, the present study illustrated that Nuci prevented obesity, improved glycolipid metabolism, attenuated hepatic steatosis and enhanced energy expenditure in HFD-fed mice. These beneficial role of Nuci was closely associated with the AMPK-mediated inhibition of lipogenesis and promotion of lipolysis, together with upregulation of adipokines, fatty acid oxidation, and thermogenesis in liver/hepatocytes and eWAT/adipocytes. Overall, our results show for the first time that AMPK-mediated FAS/HSL pathway was implicated in the beneficial effect of Nuci in reducing lipid accumulation, providing experimental evidence for clinical application of Nuci in treating obesity and related metabolic dysfunctions.

Supplementary Materials: The following are available online at https://www.mdpi.com/article/10.3390/nu14091898/s1. Table S1: Primers used for RT-qPCR analysis.

Author Contributions: H.X. designed and performed the experiments, drafted the manuscript; X.L. and X.G. contributed to the animal experiments and the collection of data; H.Y. and L.D. contributed to the analyses and screening of data. H.Z. and H.P. contributed to the design of the study and the review of manuscript. F.G. and L.W. supervised the study and contributed to the interpretation of data and revision of the manuscript. All authors have read and agreed to the published version of the manuscript.

Funding: This research was funded by grants from the Beijing Natural Science Foundation (Nos. 7222137, 7182130), the China Diabetes Young Scientific Talent Research Project (No. 2020-N-01-10), the Innovation fund for postgraduate students of Peking Union Medical College (No. 2019-1002-26), the National Natural Science Foundation of China (Nos. 81370898, 30771026), CAMS Innovation Fund for Medical Sciences (CIFMS) (No. 2021-1-I2M-002), the National Key Program of Clinical Science (No. WBYZ2011–873) and the PUMCH Foundation (No. 2013–020).

Institutional Review Board Statement: The study was conducted according to the guidelines of the Declaration of Helsinki, and approved by the Medical Ethics Committee of Peking Union Medical College (XHDW-2019-010, Beijing, China) and followed by the National Institutes of Health regulations regarding animal care and use (Beijing, China).

Informed Consent Statement: Informed consent was obtained from all subjects involved in the study.

Data Availability Statement: The corresponding authors are responsible for data availability on reasonable request.

Conflicts of Interest: The authors declare no conflict of interest.

References

1. Seidell, J.C.; Halberstadt, J. The global burden of obesity and the challenges of prevention. *Ann. Nutr. Metab.* **2015**, *66* (Suppl. S2), 7–12. [CrossRef] [PubMed]
2. Apovian, C.M. Obesity: Definition, Comorbidities, Causes, and Burden. *Am. J. Manag. Care* **2016**, *22*, S176–S185. [PubMed]
3. Gustafson, B.; Smith, U. Regulation of white adipogenesis and its relation to ectopic fat accumulation and cardiovascular risk. *Atherosclerosis* **2015**, *241*, 27–35. [CrossRef] [PubMed]
4. Suzuki, A.; Diehl, A.M. Nonalcoholic Steatohepatitis. *Annu. Rev. Med.* **2017**, *68*, 85–98. [CrossRef] [PubMed]
5. Day, E.A.; Ford, R.J.; Steinberg, G.R. AMPK as a Therapeutic Target for Treating Metabolic Diseases. *Trends Endocrinol. Metab.* **2017**, *28*, 545–560. [CrossRef] [PubMed]
6. Liu, G.; Kuang, S.; Cao, R.; Wang, J.; Peng, Q.; Sun, C. Sorafenib kills liver cancer cells by disrupting SCD1-mediated synthesis of monounsaturated fatty acids via the ATP-AMPK-mTOR-SREBP1 signaling pathway. *FASEB J.* **2019**, *33*, 10089–10103. [CrossRef] [PubMed]
7. Bolsoni-Lopes, A.; Alonso-Vale, M.I. Lipolysis and lipases in white adipose tissue—An update. *Arch. Endocrinol. Metab.* **2015**, *59*, 335–342. [CrossRef]
8. Wang, Q.; Liu, S.; Zhai, A.; Zhang, B.; Tian, G. AMPK-Mediated Regulation of Lipid Metabolism by Phosphorylation. *Biol. Pharm. Bull.* **2018**, *41*, 985–993. [CrossRef]
9. Zhang, Q.; Kong, X.; Yuan, H.; Guan, H.; Li, Y.; Niu, Y. Mangiferin Improved Palmitate-Induced-Insulin Resistance by Promoting Free Fatty Acid Metabolism in HepG2 and C2C12 Cells via PPARα: Mangiferin Improved Insulin Resistance. *J. Diabetes Res.* **2019**, *2019*, 2052675. [CrossRef]
10. Lo, K.A.; Sun, L. Turning WAT into BAT: A review on regulators controlling the browning of white adipocytes. *Biosci. Rep.* **2013**, *33*, 711–719. [CrossRef]
11. Langin, D. Recruitment of brown fat and conversion of white into brown adipocytes: Strategies to fight the metabolic complications of obesity? *Biochim. Biophys. Acta* **2010**, *1801*, 372–376. [CrossRef] [PubMed]
12. Fisher, F.M.; Kleiner, S.; Douris, N.; Fox, E.C.; Mepani, R.J.; Verdeguer, F.; Wu, J.; Kharitonenkov, A.; Flier, J.S.; Maratos-Flier, E.; et al. FGF21 regulates PGC-1α and browning of white adipose tissues in adaptive thermogenesis. *Genes. Dev.* **2012**, *26*, 271–281. [CrossRef] [PubMed]
13. Tillman, E.J.; Rolph, T. FGF21: An Emerging Therapeutic Target for Non-Alcoholic Steatohepatitis and Related Metabolic Diseases. *Front. Endocrinol.* **2020**, *11*, 601290. [CrossRef] [PubMed]
14. Wei, X.; Liu, X.; Tan, C.; Mo, L.; Wang, H.; Peng, X.; Deng, F.; Chen, L. Expression and Function of Zinc-α2-Glycoprotein. *Neurosci. Bull.* **2019**, *35*, 540–550. [CrossRef]
15. Smith, B.K.; Marcinko, K.; Desjardins, E.M.; Lally, J.S.; Ford, R.J.; Steinberg, G.R. Treatment of nonalcoholic fatty liver disease: Role of AMPK. *Am. J. Physiol. Endocrinol. Metab.* **2016**, *311*, E730–E740. [CrossRef]
16. Mottillo, E.P.; Desjardins, E.M.; Crane, J.D.; Smith, B.K.; Green, A.E.; Ducommun, S.; Henriksen, T.I.; Rebalka, I.A.; Razi, A.; Sakamoto, K.; et al. Lack of Adipocyte AMPK Exacerbates Insulin Resistance and Hepatic Steatosis through Brown and Beige Adipose Tissue Function. *Cell. Metab.* **2016**, *24*, 118–129. [CrossRef]
17. Derosa, G.; Maffioli, P. Anti-obesity drugs: A review about their effects and their safety. *Expert. Opin. Drug. Saf.* **2012**, *11*, 459–471. [CrossRef]
18. Chen, S.; Liu, X.; Peng, C.; Tan, C.; Sun, H.; Liu, H.; Zhang, Y.; Wu, P.; Cui, C.; Liu, C.; et al. The phytochemical hyperforin triggers thermogenesis in adipose tissue via a Dlat-AMPK signaling axis to curb obesity. *Cell. Metab.* **2021**, *33*, 565–580.e7. [CrossRef]
19. Sun, Y.; Xia, M.; Yan, H.; Han, Y.; Zhang, F.; Hu, Z.; Cui, A.; Ma, F.; Liu, Z.; Gong, Q.; et al. Berberine attenuates hepatic steatosis and enhances energy expenditure in mice by inducing autophagy and fibroblast growth factor 21. *Br. J. Pharmacol.* **2018**, *175*, 374–387. [CrossRef]
20. Zhang, L.; Gao, J.; Tang, P.; Chong, L.; Liu, Y.; Liu, P.; Zhang, X.; Chen, L.; Hou, C. Nuciferine inhibits LPS-induced inflammatory response in BV2 cells by activating PPAR-γ. *Int. Immunopharmacol.* **2018**, *63*, 9–13. [CrossRef]

21. Harishkumar, R.; Christopher, J.G.; Ravindran, R.; Selvaraj, C.I. Nuciferine Attenuates Doxorubicin-Induced Cardiotoxicity: An In Vitro and In Vivo Study. *Cardiovasc. Toxicol.* **2021**, *21*, 947–963. [CrossRef] [PubMed]
22. Li, Z.; Chen, Y.; An, T.; Liu, P.; Zhu, J.; Yang, H.; Zhang, W.; Dong, T.; Jiang, J.; Zhang, Y.; et al. Nuciferine inhibits the progression of glioblastoma by suppressing the SOX2-AKT/STAT3-Slug signaling pathway. *J. Exp. Clin. Cancer Res.* **2019**, *38*, 139. [CrossRef] [PubMed]
23. Xu, H.; Wang, L.; Yan, K.; Zhu, H.; Pan, H.; Yang, H.; Liu, M.; Gong, F. Nuciferine Inhibited the Differentiation and Lipid Accumulation of 3T3-L1 Preadipocytes by Regulating the Expression of Lipogenic Genes and Adipokines. *Front. Pharmacol.* **2021**, *12*, 632236. [CrossRef] [PubMed]
24. Zhang, C.; Deng, J.; Liu, D.; Tuo, X.; Xiao, L.; Lai, B.; Yao, Q.; Liu, J.; Yang, H.; Wang, N. Nuciferine ameliorates hepatic steatosis in high-fat diet/streptozocin-induced diabetic mice through a PPARα/PPARγ coactivator-1α pathway. *Br. J. Pharmacol.* **2018**, *175*, 4218–4228. [CrossRef]
25. Ma, C.; Li, G.; He, Y.; Xu, B.; Mi, X.; Wang, H.; Wang, Z. Pronuciferine and nuciferine inhibit lipogenesis in 3T3-L1 adipocytes by activating the AMPK signaling pathway. *Life Sci.* **2015**, *136*, 120–125. [CrossRef]
26. Liu, C.; Zou, Y.; Qian, H. GLP-1R agonists for the treatment of obesity: A patent review (2015-present). *Expert. Opin. Ther. Pat.* **2020**, *30*, 781–794. [CrossRef]
27. Liang, W.; Menke, A.L.; Driessen, A.; Koek, G.H.; Lindeman, J.H.; Stoop, R.; Havekes, L.M.; Kleemann, R.; Van Den Hoek, A.M. Establishment of a general NAFLD scoring system for rodent models and comparison to human liver pathology. *PLoS ONE* **2014**, *9*, e115922. [CrossRef]
28. Yan, K.; Wang, X.; Pan, H.; Wang, L.; Yang, H.; Liu, M.; Zhu, H.; Gong, F. Safflower Yellow and Its Main Component HSYA Alleviate Diet-Induced Obesity in Mice: Possible Involvement of the Increased Antioxidant Enzymes in Liver and Adipose Tissue. *Front. Pharmacol.* **2020**, *11*, 482. [CrossRef]
29. Sugihara, H.; Fau-Yonemitsu, N.; Yonemitsu, N.; Fau-Miyabara, S.; Miyabara, S.; Fau-Yun, K.; Yun, K. Primary cultures of unilocular fat cells: Characteristics of growth in vitro and changes in differentiation properties. *Differentiation* **1986**, *31*, 42–49. [CrossRef]
30. Zhu, H.J.; Wang, L.J.; Wang, X.Q.; Pan, H.; Li, N.S.; Yang, H.B.; Jin, M.; Zang, B.X.; Gong, F.Y. Hormone-sensitive lipase is involved in the action of hydroxysafflor yellow A (HYSA) inhibiting adipogenesis of 3T3-L1cells. *Fitoterapia* **2014**, *93*, 182–188. [CrossRef]
31. Fan, X.; Yao, H.; Liu, X.; Shi, Q.; Lv, L.; Li, P.; Wang, R.; Tang, T.; Qi, K. High-Fat Diet Alters the Expression of Reference Genes in Male Mice. *Front. Nutr.* **2020**, *7*, 589771. [CrossRef]
32. Livak, K.J.; Schmittgen, T.D. Analysis of relative gene expression data using real-time quantitative PCR and the 2(-Delta Delta C(T)) Method. *Methods* **2001**, *25*, 402–408. [CrossRef] [PubMed]
33. Mccoll, E.R.; Piquette-Miller, M. Viral model of maternal immune activation alters placental AMPK and mTORC1 signaling in rats. *Placenta* **2021**, *112*, 36–44. [CrossRef] [PubMed]
34. Zhang, J.; Sun, H.; Jiang, K.; Song, X.; Wang, X.; Yang, Y.; Liu, H.; Ji, Q.; Yu, X.; Liu, Y.; et al. Cudraxanthone L inhibits gastric cancer by regulating the MAPK signalling and promoting FAS-mediated pathway. *Biomed. Pharmacother.* **2021**, *141*, 111876. [CrossRef] [PubMed]
35. Tang, Z.; Luo, T.; Huang, P.; Luo, M.; Zhu, J.; Wang, X.; Lin, Q.; He, Z.; Gao, P.; Liu, S. Nuciferine administration in C57BL/6J mice with gestational diabetes mellitus induced by a high-fat diet: The improvement of glycolipid disorders and intestinal dysbacteriosis. *Food Funct.* **2021**, *12*, 11174–11189. [CrossRef] [PubMed]
36. Guo, F.; Yang, X.; Li, X.; Feng, R.; Guan, C.; Wang, Y.; Li, Y.; Sun, C. Nuciferine prevents hepatic steatosis and injury induced by a high-fat diet in hamsters. *PLoS ONE* **2013**, *8*, e63770. [CrossRef]
37. Ricquier, D. Respiration uncoupling and metabolism in the control of energy expenditure. *Proc. Nutr. Soc.* **2005**, *64*, 47–52. [CrossRef]
38. Wang, Y.; Yao, W.; Li, B.; Qian, S.; Wei, B.; Gong, S.; Wang, J.; Liu, M.; Wei, M. Nuciferine modulates the gut microbiota and prevents obesity in high-fat diet-fed rats. *Exp. Mol. Med.* **2020**, *52*, 1959–1975. [CrossRef]
39. Tucker, B.; Li, H.; Long, X.; Rye, K.A.; Ong, K.L. Fibroblast growth factor 21 in non-alcoholic fatty liver disease. *Metabolism* **2019**, *101*, 153994. [CrossRef]
40. Zhang, Y.; Lei, T.; Huang, J.F.; Wang, S.B.; Zhou, L.L.; Yang, Z.Q.; Chen, X.D. The link between fibroblast growth factor 21 and sterol regulatory element binding protein 1c during lipogenesis in hepatocytes. *Mol. Cell. Endocrinol.* **2011**, *342*, 41–47. [CrossRef]
41. Xiao, X.H.; Qi, X.Y.; Wang, Y.D.; Ran, L.; Yang, J.; Zhang, H.L.; Xu, C.X.; Wen, G.B.; Liu, J.H. Zinc alpha2 glycoprotein promotes browning in adipocytes. *Biochem. Biophys. Res. Commun.* **2018**, *496*, 287–293. [CrossRef] [PubMed]
42. Bartelt, A.; Heeren, J. Adipose tissue browning and metabolic health. *Nat. Rev. Endocrinol.* **2014**, *10*, 24–36. [CrossRef] [PubMed]
43. Garcia, D.; Hellberg, K.; Chaix, A.; Wallace, M.; Herzig, S.; Badur, M.G.; Lin, T.; Shokhirev, M.N.; Pinto, A.F.M.; Ross, D.S.; et al. Genetic Liver-Specific AMPK Activation Protects against Diet-Induced Obesity and NAFLD. *Cell Rep.* **2019**, *26*, 192–208.e6. [CrossRef] [PubMed]
44. Li, Y.; Xu, S.; Mihaylova, M.M.; Zheng, B.; Hou, X.; Jiang, B.; Park, O.; Luo, Z.; Lefai, E.; Shyy, J.Y.; et al. AMPK phosphorylates and inhibits SREBP activity to attenuate hepatic steatosis and atherosclerosis in diet-induced insulin-resistant mice. *Cell Metab.* **2011**, *13*, 376–388. [CrossRef]

45. Hua, X.; Sun, D.Y.; Zhang, W.J.; Fu, J.T.; Tong, J.; Sun, S.J.; Zeng, F.Y.; Ouyang, S.X.; Zhang, G.Y.; Wang, S.N.; et al. P7C3-A20 alleviates fatty liver by shaping gut microbiota and inducing FGF21/FGF1, via the AMP-activated protein kinase/CREB regulated transcription coactivator 2 pathway. *Br. J. Pharmacol.* **2021**, *178*, 2111–2130. [CrossRef]
46. Bäckhed, F.; Ley, R.E.; Sonnenburg, J.L.; Peterson, D.A.; Gordon, J.I. Host-bacterial mutualism in the human intestine. *Science* **2005**, *307*, 1915–1920. [CrossRef]
47. Rosenbaum, M.; Knight, R.; Leibel, R.L. The gut microbiota in human energy homeostasis and obesity. *Trends Endocrinol. Metab.* **2015**, *26*, 493–501. [CrossRef]
48. Yu, Y.; Lu, J.; Sun, L.; Lyu, X.; Chang, X.Y.; Mi, X.; Hu, M.G.; Wu, C.; Chen, X. Akkermansia muciniphila: A potential novel mechanism of nuciferine to improve hyperlipidemia. *Biomed. Pharmacother.* **2021**, *133*, 111014. [CrossRef]
49. Shi, Z.; Fang, Z.Y.; Gao, X.X.; Yu, H.; Zhu, Y.W.; Ouyang, H.L.; Song, Y.X.; Du, X.L.; Wang, Z.; Li, X.W.; et al. Nuciferine improves high-fat diet-induced obesity via reducing intestinal permeability by increasing autophagy and remodeling the gut microbiota. *Food Funct.* **2021**, *12*, 5850–5861. [CrossRef]

nutrients

Article

Liquiritigenin Inhibits Lipid Accumulation in 3T3-L1 Cells via mTOR-Mediated Regulation of the Autophagy Mechanism

Hong Qin [1], Ziyu Song [1], Chunyu Zhao [1], Jinxin Yang [1], Fan Xia [1], Lewen Wang [1], Anwar Ali [2] and Wenya Zheng [1,*]

1 Department of Nutrition Science and Food Hygiene, Xiangya School of Public Health, Central South University, 110 Xiangya Road, Changsha 410078, China; qinhong@csu.edu.cn (H.Q.); song0305ziyu@126.com (Z.S.); zcy67807189@163.com (C.Z.); yangjinxin20212021@163.com (J.Y.); xx126948@163.com (F.X.); wlwcsu@163.com (L.W.)
2 Department of Epidemiology and Health Statistics, Xiangya School of Public Health, Central South University, 110 Xiangya Road, Changsha 410078, China; 206908003@csu.edu.cn
* Correspondence: wenyazheng@csu.edu.cn; Tel.: +86-0731-84805454

Abstract: Liquiritigenin (LQG) is a natural flavonoid from the herb Glycyrrhiza uralensis Fisch that exhibits multiple biological activities. However, its specific role in antiobesity and its related underlying molecular mechanisms remain unknown. The primary purpose of this study is to explore the effects and regulatory mechanisms of LQG on lipid accumulation in 3T3-L1 adipocytes. The results show that LQG significantly reduced triglyceride levels and downregulated the expression of transcription factors such as CCAAT/enhancer-binding protein α (C/EBPα) and peroxisome proliferator-activated receptor γ (PPARγ) in 3T3-L1 adipocytes. Additionally, the expression of sterol-regulatory element-binding protein 1c (SREBP1c), acetyl-CoA carboxylase 1 (ACC1), and fatty acid synthase (FASN) involved in lipogenesis was reduced by treatment with LQG. The protein expression levels of light chain 3B (LC3B), autophagy-related protein 7 (ATG7) and p62 were also modulated by LQG, leading to the suppression of autophagy. Further, LQG activated the phosphorylation of the mammalian target of rapamycin (mTOR), the inhibition of which was followed by the restored expression of autophagy-related proteins. Pretreatment with an mTOR inhibitor also reverted the expression of several genes or proteins involved in lipid synthesis. These results suggest that LQG inhibited lipid accumulation via mTOR-mediated autophagy in 3T3-L1 white adipocytes, indicating the role of LQG as a potential natural bioactive component for use in dietary supplements for preventing obesity.

Keywords: liquiritigenin; obesity; lipid accumulation; mTOR; autophagy

Citation: Qin, H.; Song, Z.; Zhao, C.; Yang, J.; Xia, F.; Wang, L.; Ali, A.; Zheng, W. Liquiritigenin Inhibits Lipid Accumulation in 3T3-L1 Cells via mTOR-Mediated Regulation of the Autophagy Mechanism. *Nutrients* **2022**, *14*, 1287. https://doi.org/10.3390/nu14061287

Academic Editors: Anna Gramza-Michałowska and Paola Nieri

Received: 17 January 2022
Accepted: 16 March 2022
Published: 18 March 2022

1. Introduction

Obesity causes great public health concerns because of its association with conditions such as hepatic steatosis, dyslipidemia, and cardiovascular diseases [1]. To date, six medications, such as phentermine and orlistat, have been approved for and are commonly used for the long-term treatment of obesity [2]. However, they can cause adverse effects such as headache, insomnia, constipation, and steatorrhea [3]. Several exogenous factors, such as diet and lifestyle, can affect energy expenditure and are therefore also associated with the risk of developing obesity [4]. Therefore, it is important to explore safe and effective natural dietary phytochemicals as candidates for obesity prevention and to understand their underlying mechanisms of action.

Obesity is characterized by the expansion of white adipose tissue mass and morphological changes in the adipocytes due to excess triglyceride accumulation [5,6]. In the past several decades, studies have shown that lipid accumulation in adipose tissues is influenced by a variety of factors, including a complicated network of adipocyte-specific genes, lipogenic enzymes, transcription factors, and signaling intermediates from numerous pathways [7]. In vivo and in vitro studies have shown that peroxisome proliferation-activated receptor γ (PPARγ) is positively associated with fat-cell size in mice [8] and with obesity in humans [9]. Additionally, CCAAT/enhancer-binding protein α (C/EBPα) is well characterized as a critical transcription factor for the upstream regulator of PPARγ [10]. Moreover, sterol-regulatory element-binding protein 1c (SREBP1c) is well recognized as a key transcription factor in regulating lipogenesis and preferentially activates the transcription of genes required for fatty acid and triglyceride synthesis.

Macroautophagy is a highly regulated process whereby cytosolic proteins or whole organelles are firstly engulfed inside double-membrane autophagosomes and then translocated to lysosomes for fusion and degradation [11]. This process is customarily referred to as autophagy since it is the most well-known autophagic pathway. It was demonstrated that autophagy regulates nutrition supply and cell death and that it also plays a role in cell differentiation and development [12–14]. Recent research showed that, due to the removal of some intracellular components that are crucially required for cellular remodeling functions, autophagy is pivotal in the development of adipocytes [12]. This suggests that suppressing autophagy in white adipose tissue contributes to the blocking of fat depots and the amelioration of obesity [15]. It was reported that the adipose tissue of obese individuals exhibits enhanced autophagic activity [16]. Thus, interventions that influence autophagy might be a novel strategy for obesity prevention.

Liquiritigenin (LQG) is a natural flavonoid substance isolated from the herb Glycyrrhiza uralensis Fisch (also called licorice root) and has been widely used as a sweetener or antioxidant in various food items such as confectioneries, pastries, and beverages [17]. Licorice candy is popular in Europe, and teas and soups made with licorice are popular in various regions of Asia. It was shown that LQG exhibits multiple activities such as anti-inflammation, anti-hyperglycemia, and anti-neurotoxicity [18–20]. Studies show that Glycyrrhiza uralensis has antiobesity potential through the induction of adipocyte browning [21]. One study proved the presence of LQG in the blood samples of mice after dietary licorice-root intake [22]. However, the effects on antiobesity of LQG, the main aglycone component in Glycyrrhiza uralensis, remain unclear, and the underlying mechanisms of LQG in preventing adipose depot expansion need to be further investigated. Thus, we studied the effects of LQG on reducing lipid accumulation in adipocytes and its underlying molecular mechanisms of action. The finding of this research may provide scientific support for the use of LQG as a functional compound in dietary supplements for the prevention of obesity.

2. Materials and Methods

2.1. Chemicals and Reagents

LQG (99.49%) and rapamycin (RAP, 99.94%, an mTOR inhibitor) were purchased from Med Chem Express (South Brunswick, NJ, USA). Dulbecco's Modified Eagle's Medium (DMEM), penicillin–streptomycin solution, Trizol, RIPA lysis buffer, protease inhibitor, 1% phenylmethylsulfonyl fluoride (PMSF), bicinchoninic acid (BCA) protein assay kits, and thiazolyl blue tetrazolium bromide (MTT) assay kits were purchased from Ding Guo Changsheng Biotechnology Co., Ltd. (Beijing, China). Newborn calf serum (CS) was purchased from Gibco of Thermo Fisher Scientific (Waltham, MA, USA). Jiancheng Bioengineering Institute (Nanjing, China) provided the triglyceride (TG) test kits, and NCM Biotech (Suzhou, China) provided the protein loading buffer. Goat anti-Rabbit IgG (H&L)-HRP was purchased from Bioworld Technology (Bloomington, MN, USA). Antibodies against the mammalian target of rapamycin (mTOR), fatty acid synthase (FASN), PPARγ, and β-actin were purchased from Abclonal (Woburn, MA, USA). Phospho-mTOR-ser2448 (p-mTOR), light chain 3B (LC3B), recombinant autophagy-related protein 7 (ATG7), and p62 were purchased from ZEN-BIOSCIENCE (Chengdu, China).

2.2. T3-L1 Preadipocyte Culture and Differentiation

Preadipocytes 3T3-L1 were cultured in Dulbecco's Modified Eagle's Medium (DMEM) containing 10% CS and 1% penicillin–streptomycin solution at 37 °C under 5% CO_2 atmosphere. In confluent cells, differentiation was induced as previously described [23]. The cell medium with CS was replaced with fetal bovine serum (FBS) in the first differentiation induction medium (day 0), consisting of 8 μg/mL insulin, 1 μM dexamethasone, and 0.5 mM IBMX in DMEM. After 48 h, cells were switched to the second differentiation medium, consisting of 8 μg/mL insulin only for another 48 h. Then, cells were switched to a maintenance medium of DMEM and 10% FBS to form mature adipocytes until day 8. During the differentiation process, cells were treated with or without LQG. In the experiments with the mTOR inhibitor, 3T3-L1 white adipocytes were pretreated with 20 nM RAP for 2 h before LQG treatment.

2.3. Cell Viability Assay

Preadipocytes were seeded in a 96-well plate at a density of 5.0×10^3 cells/well and incubated until mature adipocytes formed, as described above. To determine the intervention concentrations, cells were treated with variable concentrations of LQG (0.01, 0.1, 1, 10, 25 and 50 μM) for 72 h. After the incubation period for treating cells with LQG, 20 μL MTT solution was added to each well and the cells were incubated for another 4 h. The media was then removed and 150 μL of dimethyl sulfoxide (DMSO) was added for 10 min, with shaking to fully dissolve the crystallization. Absorbance was measured at 570 nm by a microplate spectrophotometer (Thermo Scientific, Waltham, MA, USA).

2.4. Lipid Content Assays

The accumulation of intracellular lipids was measured and quantified using a TG assay kit and normalized to total intracellular protein using a BCA kit, following the manufacturer's protocol. Additionally, cells were stained with Oil Red O. The cells treated with LQG were fixed with 4% paraformaldehyde for 10 min, and then stained with Oil Red O working solution for 30 min. The cells were washed with distilled water after the staining solution was removed, and the stained lipid droplets were examined under an inverted microscope (Invitrogen EVOS M7000, Waltham, MA, USA).

2.5. Western Blotting Analysis

The adipocytes treated with LQG and/or RAP were harvested with RIPA lysis buffer containing a protease inhibitor and 1% PMSF. The protein samples (25 μg) were separated by 8% or 10% sodium dodecyl sulfate polyacrylamide gel electrophoresis (SDS−PAGE), and then the gels were run at a constant voltage of 80 V and 120 V. Then, using a voltage of 80 V, proteins were transferred to polyvinylidene fluoride (PVDF) membranes. The membranes were incubated overnight at 4 °C with the following primary antibodies: mTOR (1:1000), p-mTOR-Ser2448 (1:1000), PPARγ (1:1000), FASN (1:750), LC3B (1:1000), ATG7 (1:1000), p62 (1:1000), and β-actin (1:300,000). Afterwards, bands were incubated for 1 h with a peroxidase-conjugated secondary antibody. Electrochemiluminescence (ECL) was used to visualize immunoreactive bands, and a chemiluminescence imager (Tanon-5500, Shanghai, China) was used to detect and measure them. Bands were quantified via densitometry by Image J 1.52a.

2.6. Quantitative Real-Time Reverse Transcription Polymerase Chain Reaction (RT-PCR)

Total RNA from the adipocytes treated with LQG and/or RAP was isolated utilizing a Trizol reagent and 1 μg RNA was converted to cDNA using HiScriptIIQ RT SuperMix (Vazyme, Nanjing, China). Gene expression was quantified using Hieff UNICON® qPCR SYBR Green Master Mix (Yeasen Biotech, Shanghai, China) and a LightCycler 480 II (Roche, Basel, Switzerland). RT-PCR reactions were run in triplicate for each sample, and relative gene expressions were calculated using the $2^{-\Delta\Delta CT}$ method after values were normalized to β-actin. The oligonucleotide primers (Sangon Biotech, Shanghai, China) used for amplification are shown in Table 1.

Table 1. Primer Sequences.

Name	Primer Sequence (5′ to 3′)
pparγ	forward: TCTCCCCACATCCTTTCT reverse: CTGCCGTTGTCTGTCACT
c/ebpα	forward: CTGGAAAGAAGGCCACCTC reverse: AAGAGAAGGAAGCGGTCCA
srebp1c	forward: GCAACACAGCAACCAGAA reverse: GAAAGGTGAGCCAGCATC
acc1	forward: AAAACAGGGAGGAAGCAA reverse: TCACCCCGAATAGACAGC
fasn	forward: GCCCAAGGGAAGCACATT reverse: CGAAGCCACCCAGACCAC
β-actin	forward: TTGCGTTACACCCTTTCT reverse: ACCTTCACCGTTCCAGTT

2.7. Statistical Analysis

Data were expressed as the means ± standard errors of the mean (SEM) and statistical analyses were performed with SPSS 18.0 software (Chicago, IL, USA). The comparisons among different groups were calculated by one-way analysis of variance (ANOVA) and further analyzed by independent *t*-test. When $p \leq 0.05$, results were considered statistically significant.

3. Results

3.1. Effects of LQG on Lipid Accumulation in 3T3-L1 White Adipocytes

Firstly, the cytotoxic effects of LQG on 3T3-L1 adipocytes were determined using the MTT assay at different doses. As shown in Figure 1A, LQG at all tested concentrations did not affect cell viability. Based on the results, 10, 25 and 50 μM of LQG were used to further analyze lipid accumulation in the adipocytes. The intracellular TG levels were notably decreased when 50 μM of LQG was present in 3T3-L1 cells; however, the concentration of LQG at 25 μM only induced a tendency to reduce TG levels compared with the control group (Figure 1B). Moreover, the inhibitory effects on the accumulation of lipids by LQG were investigated by Oil Red O staining. Compared with the control group, apparent decreases in the lipid droplets in 3T3-L1 cells were observed with LQG treatments of 25 and 50 μM (Figure 1C). These results indicate that LQG treatment effectively reduced the level of lipid accumulation in 3T3-L1 cells; therefore, 25 μM and 50 μM of LQG were chosen as the experimental doses in later tests.

3.2. Effects of LQG on Modulating Molecular Regulators of Lipid Synthesis in 3T3-L1 Adipocytes

To explore the molecular and cellular mechanisms underlying the lipid-inhibitory actions of LQG, the expressions of c/ebpα and pparγ, the two important genes related to adipocyte development and lipid accumulation in fat cells, were examined using RT-PCR. As shown in Figure 2A,B, 50 μM of LQG significantly inhibited the mRNA expression of c/ebpα and pparγ, while LQG at 25 μM did not show obvious inhibitory effects. In addition, the protein expression level of PPARγ was also found to be downregulated by 50 μM of LQG (Figure 2F). To further determine the effects of LQG on lipid synthesis, we examined the changes in mRNA expression or protein expression in lipogenic regulators in 3T3-L1 cells. LQG at 50 μM significantly reduced the mRNA expression of srebp1c, fasn, and acetyl-CoA carboxylase 1 (acc1) (Figure 2C–E), which are genes closely associated with lipogenesis. A 25 μM dose of LQG showed a tendency to decrease the mRNA expression of srebp1c and fasn; however, there were no significant differences after statistical analysis. Additionally, FASN, a multifunctional enzyme that catalyzes the de novo biosynthesis of long-chain, saturated fatty acids starting with acetyl-CoA and malonyl-CoA, was downregulated by LQG at 25 and 50 μM (Figure 2G).

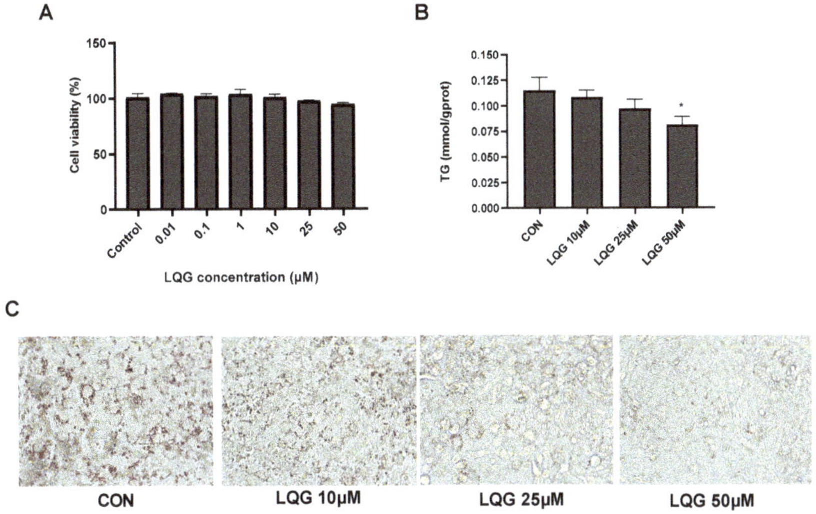

Figure 1. Effects of liquiritigenin (LQG) on cell viability and lipid accumulation in 3T3-L1 adipocytes. (**A**) 3T3-L1 cells were treated with different concentrations of LQG (0, 0.01, 0.1, 1, 10, 25 and 50 μM) and cytotoxicity of the cells was expressed as optical density percentage. (**B**) LQG induced suppression of intracellular triglyceride (TG) levels in 3T3-L1 cells. TG levels of LQG at 10, 25 and 50 μM were assayed and are expressed in bar charts. (**C**) Mature 3T3-L1 adipocytes treated with 10, 25 and 50 μM LQG were stained with Oil Red O to observe lipid droplets at 400× magnification. Data are presented as means ± SEMs and analyzed with one-way ANOVA (n = 3). * marks significant differences compared with control group ($p \leq 0.05$).

3.3. Effects of LQG on Autophagy-Related Protein Expression in 3T3-L1 Adipocytes

Since autophagy is a significant factor involved in adipocyte development, the expression of autophagy-related proteins after LQG treatment was examined. Endogenous LC3 is converted to LC3-I and then LC3-II during autophagosome formation. As shown in Figure 3A, the ratio of LC3BII/LC3BI was reduced after treatment with LQG. Similarly, the protein expression of ATG7 and p62 (Figure 3B,C), the substrates of specific autophagy, was decreased and increased, respectively, by LQG treatment.

mTOR is a key regulator of cell metabolism as well as autophagy. To determine how the expression of autophagy-related proteins was affected by LQG, the effect of LQG on the protein expression of mTOR was examined using Western blotting analysis. The activation of the phosphorylation of mTOR, a major negative modulator of autophagy, was notably increased in the presence of 50 μM of LQG (Figure 3D).

3.4. LQG Regulated Autophagy via mTOR in 3T3-L1 Adipocytes

To investigate the correlation between mTOR and the regulation of autophagy induced by LQG, we examined whether autophagy-related proteins are modulated by RAP, which is an mTOR inhibitor. In Figure 4A, our results show that 50 μM of LQG induced phosphorylation of mTOR, whereas RAP inhibited the expression levels of p-mTOR; on the other hand, the results demonstrate that pretreatment of RAP with LQG could abrogate the activated phosphorylation of mTOR by LQG. Subsequently, expression levels of autophagy-related proteins were determined. As shown in Figure 4B,C, compared with the treatment with LQG and RAP together, treatment with LQG alone decreased the protein levels of LC3BII/LC3BI and ATG7. Meanwhile, the upregulation of p62 induced by LQG alone could be restored by the pretreatment of RAP (Figure 4D). These results suggest that the signal molecule of mTOR was involved in LQG-reduced autophagy during the process of 3T3-L1 adipocyte differentiation.

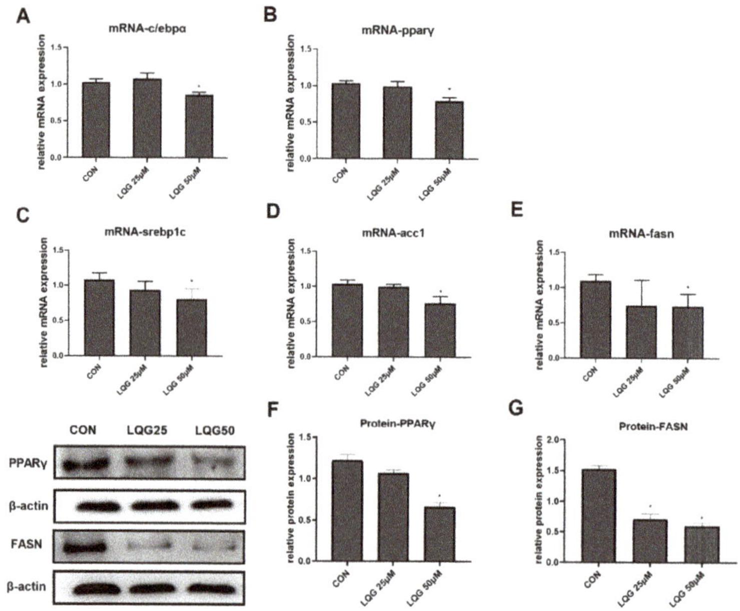

Figure 2. Effects of LQG on molecular regulators of lipid synthesis in 3T3-L1 adipocytes. The cells were treated with 25 and 50 μM LQG. The mRNA-expression levels of (**A**) c/ebpα, (**B**) pparγ, (**C**) srebp1c, (**D**) acc1, and (**E**) fasn were measured by RT-PCR, and the protein levels of (**F**) PPARγ and (**G**) FASN were analyzed by Western blotting. Protein expression was quantified by densitometry, and the relative intensities are expressed in the bar charts. Values are presented as means ± SEMs (n = 3) and analyzed with one-way ANOVA. * shows a statistically significant difference compared with control group ($p \leq 0.05$).

3.5. The Role of mTOR in Lipogenic Regulatory Factors in LQG-Treated Adipocytes

To investigate whether the effects of LQG on the process of lipid accumulation were mediated by mTOR activation, the expression of genes or proteins correlated with lipogenesis were detected by presenting LQG with RAP. As shown in Figure 5A,B, the inhibitory effects of LQG on c/ebpα and pparγ mRNA expression were abrogated by the pretreatment of RAP with LQG, which was similar to the impacts induced by RAP alone. Further, the protein levels of PPARγ (Figure 5F) were similarly modulated as their corresponding mRNA levels. The protein expression of PPARγ modulated by LQG alone was significantly downregulated compared with either the control group or the LQG+RAP group. To further elucidate whether the molecular mechanisms of the repression of intracellular lipid synthesis by LQG were correlated with mTOR, we investigated the effects of LQG on the expression of lipogenic regulatory factors by using an mTOR inhibitor. As shown in Figure 5C,D, the mRNA expression of lipogenic genes srebp1c and fasn was significantly downregulated by LQG; interestingly, RAP alone or together with LQG showed inhibitory effects on srebp1c and fasn mRNA expression as well. The acc1 mRNA-expression level was also reduced by LQG, while these effects could not be observed when LQG was presented with RAP together, or when the RAP treatment was used alone (Figure 5E), which was similar to the results for c/ebpα and pparγ mRNA expression. In order to confirm whether the effects of LQG on lipogenesis were modulated by mTOR, the protein expression levels of FASN were determined with a pretreatment of RAP with LQG. As shown in Figure 5G, the FASN protein-expression level was inhibited by LQG, and the pretreatment of RAP with LQG as well as RAP alone decreased FASN protein-expression levels compared with the control group.

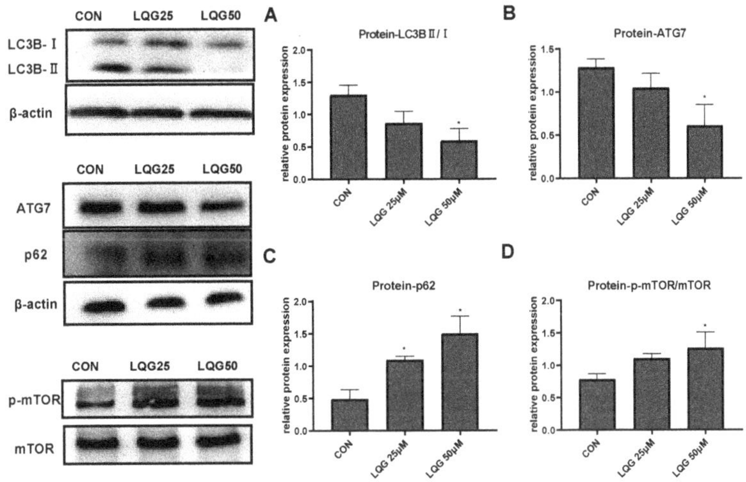

Figure 3. Effects of LQG on autophagy-related protein expression in 3T3-L1 adipocytes. The cells were treated with 25 and 50 μM LQG. The protein expression levels of (**A**) LC3B, (**B**) ATG7, (**C**) p62, and (**D**) phosphorylated-mTOR (p-mTOR) were analyzed by Western blotting. The results are means ± SEMs of three independent experiments and analyzed with one-way ANOVA. * marks significant differences compared with control group ($p \leq 0.05$).

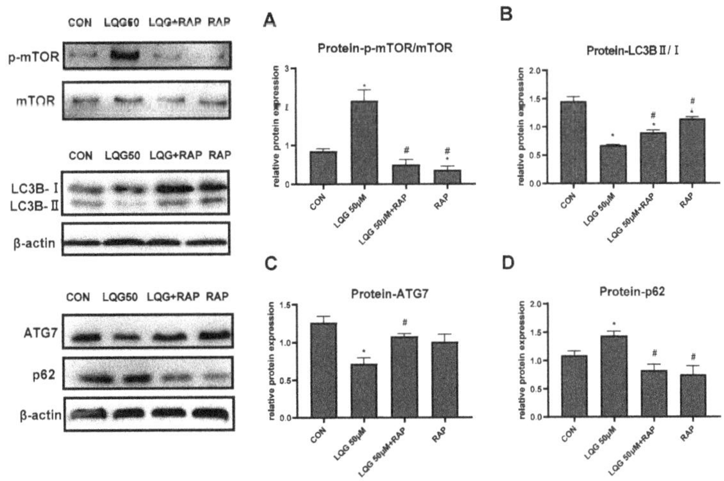

Figure 4. Effects of rapamycin (RAP) on modulating autophagy-related factors in LQG-treated 3T3-L1 cells. The cells were divided into four groups: control group, 50 μM LQG group, LQG + RAP group, and RAP-alone group. Protein expression of (**A**) p-mTOR and mTOR, (**B**) LC3BII/LC3BI, (**C**) ATG7, and (**D**) p62 were quantified by densitometry, and the relative intensities are expressed in the bar charts. Western blotting bands represent detection of the proteins from three independent tests. Data are presented as means ± SEMs and analyzed with one-way ANOVA. * vs. control group ($p \leq 0.05$); # vs. 50 μM LQG-alone treatment group ($p \leq 0.05$).

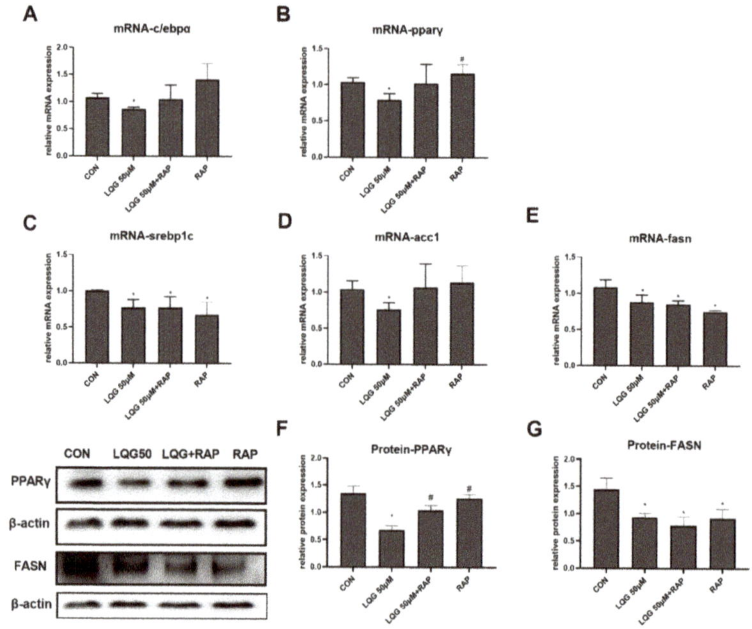

Figure 5. Effects of RAP on modulating lipogenic factors in LQG-treated 3T3-L1 cells. The cells were divided into four groups: control group, 50 μM LQG group, LQG + RAP group, and RAP-alone group. The mRNA expression levels of (**A**) c/ebpα, (**B**) pparγ, (**C**) srebp1c, (**D**) acc1, and (**E**) fasn were measured by RT-PCR and the protein levels of (**G**) PPARγ and (**F**) FASN were analyzed by Western blotting. Protein expression was quantified by densitometry and the relative intensities are expressed in the bar charts. Western blotting bands represent the detection of the proteins from three independent tests. Data are presented as means ± SEMs and analyzed with one-way ANOVA. * vs. control group ($p \leq 0.05$); # vs. 50 μM LQG-alone treatment group ($p \leq 0.05$).

4. Discussion

Obesity is a chronic disease that arouses great public health concern globally. Increasing attention is paid to the functions of phytochemicals, especially flavonoids, as alternative strategies in obesity prevention and amelioration. In the current study, we demonstrated that LQG administration was able to inhibit lipid accumulation in 3T3-L1 white adipocytes through modulating the genes or proteins involved in lipogenesis. Our data also showed for the first time that the function of LQG on suppressing the adipocyte accumulation of lipids occurred via the activation of mTOR, which led to a reduction of autophagy.

Obesity triggers the development of various metabolic diseases [1,24] and is defined as an over accumulation of fat; thus, adipose tissue mass in obese individuals is expanded by a large number of lipids [25,26]. In a fed state, the principal function of white adipose tissue is lipid storage [26,27]. Our results show that 50 μM of LQG significantly reduced lipid and fat accumulation in 3T3-L1 adipocytes. There are a limited number of studies reporting LQG's antiobesity effects. One study demonstrated that Glycyrrhiza uralensis exhibited antiobesity effects and that several substances, including the LQG in licorice extract, were able to induce UCP1 expression in 3T3-L1 adipocytes, indicating a potential role of LQ in adipocyte browning [21]. LQG is an important constituent of the flavones in Glycyrrhiza uralensis, which has shown biological properties including anti-inflammation, antihyperglycemia, antineurotoxicity, and cytoprotection [18–20,28]. Additionally, LQG was reported as an aglycone form in vivo [28] and could be detected in the blood samples of mice administrated licorice-root extract [22]. This current study shows that LQG is a bioactive compound in Glycyrrhiza uralensisto, which is in line with previous studies.

Specifically, based on the results of this study, we proved that LQG exhibited effects on reducing lipid accumulation in adipose cells.

Our observation of reduced TG accumulation in LQG-treated 3T3-L1 adipocytes indicates that LQG might affect the regulatory mechanisms for controlling lipogenesis during the process of adipose tissue expansion. Lipogenesis includes the process of triglyceride synthesis leading to increased fat mass, while a reduction in lipogenesis protects against the development of obesity [29,30]. In our study, we demonstrated that LQG at 50 μM significantly downregulated the mRNA expression levels of c/ebpα and pparγ, and the remarkable changes in pparγ expression in mRNA levels were consistent with the changes in its levels of encoded proteins. C/EBPα and PPARγ are two important transcription factors that have been implicated in 3T3-L1 adipocyte development, including cell differentiation and hypertrophy [5,8]. PPARγ could cooperate with C/EBP to cause mutual activation of other lipogenic genes or proteins, leading to triglyceride synthesis or lipid accumulation in cells [31]. Moreover, it is well known that de novo lipid synthesis could be decreased by targeting the fatty-acid synthesis signals of SREBP1c and its downstream molecules such as ACC and FASN. Many studies demonstrated that SREBP1c can directly bind to the acc1 promoter and the regulatory factor of the fasn gene enhancer [32], which is supported by our results. Furthermore, one study by Payne et al. [33] demonstrated that the transcription factor for C/EBPα positively regulated SREBP1c expression by using cells in which C/EBPα was inhibited with shRNA and ChIP assays. Our results clearly indicate that LQG exerted the ability to inhibit t lipogenesis by decreasing the expression levels of C/EBPα, PPARγ, SREBP1c, ACC1, and FASN. Many other natural flavonoids, such as genistein and fisetin, were also shown to suppress lipid accumulation in 3T3-L1 white adipocytes by modulating those lipogenic molecules [7,34], which is inconsistent with the results of our study. It is interesting that the protein expression of FASN was also significantly modulated at a 25 μM dose of LQG, which was not observed for the mRNA expression of srebp1c or acc1. One explanation for this is that FASN may not be regulated by only SREBP1c; even though SREBP–FASN is an important axis in the regulation of lipid metabolism, studies have proved that there could be other upstream molecular signals that modulate FASN. For example, some studies have demonstrated that the downregulation of FASN in adipocytes is partially mediated by the PKA and MAPK signaling pathways [35,36]. Therefore, even at lower doses, LQG may have a significant effect on the modulation of FASN. Additionally, the protein expression may not always be the same as the mRNA expression of its gene or relative genes, since there are complex processes of protein translation and protein post-translational modification. In general, the results that LQG suppressed de novo triglyceride synthesis suggest that LQG might be a beneficial flavonoid compound for the prevention of obesity.

Recently, increasing evidence has been put forward for the role of autophagy in the process of the accumulation of lipids in fat cells [37]. A study by Singh et al. [15] found that the inhibition of autophagy limited TG accumulation in 3T3-L1 adipocytes, indicating that the lipid accumulation in cell development had been blocked. That study also demonstrated that the knockdown of autophagy-related gene atg7 or atg5 in 3T3-L1 preadipocytes notably inhibited lipid accumulation and decreased the protein levels of regulatory factors in adipocyte development [15]. Similar effects were seen when lysosome function was inhibited pharmacologically [15]. Additionally, an adipocyte-specific mouse knockout of atg7 or atg5 generated lean mice with decreased white adipose mass and enhanced insulin sensitivity [38,39]. Although the mechanism of how autophagy influences lipid accumulation in cell development is not yet clear, a study that investigated autophagy-related protein expression at different time points proved that there was an increase in autophagy on day 6 after 3T3-L1 adipocyte differentiation, as evidenced by the significantly increased expression of LC3BII/LC3BI and ATG12 [40]. LC3B is a well-known autophagic marker during the whole process of autophagy, and ATG7 is an essential protein for the induction of autophagosomes. In the present study, we found that LQG was able to inhibit autophagic activity, since it was observed that the autophagy-related protein expression levels of LC3BII/LC3BI and ATG7 were significantly repressed. In addition, p62 was significantly increased after LQG treatment. It is commonly recognized that p62 is responsible for the cargo selection and transport of protein aggregates, and it is degraded together with its cargo in lysosomes. The loss of atg genes may block the fusion of autophagosomes with lysosomes, leading to an increase in p62 protein expression [41]. Thus, the level of p62 is usually inversely correlated with autophagy. In this study, the expression of p62 was significantly increased after LQG treatment. These findings add to growing evidence that administration of LQG may inhibit autophagy.

The phosphorylation expression of mTOR, an autophagy inhibitor, was significantly activated by LQG treatment. Many published studies have demonstrated that mTOR is a key regulator of autophagy [40,42]. Zhang et al. [43] showed that LQG protects liver tissue by enhancing mTOR-mediated autophagy in arsenic-trioxide-induced liver injury. The inconsistent results for LQG modulating mTOR expression and autophagy might be due to different tissues being under-investigated such that the role of autophagy in adipocyte development would differ from that in hepatic cell damage [44]. Additionally, the mechanisms correlating with autophagy in different periods of disease development could also differ. One study showed that arsenite exposure reduced the differentiation of murine brown adipocytes via autophagy inhibition in brown adipose tissue [45]. However, several other studies demonstrated that the suppression of autophagy reduced lipid accumulation or induced brown-like adipocyte formation in adipocytes and adipose tissue [15,38–40], which was inconsistent with our results. The current study shows that the inhibition of mTOR by RAP significantly decreased the p-mTOR expression activated by LQG. On the one hand, this supports the assumption that mTOR was stimulated in the LQG modulation of 3T3-L1 adipocyte development, and on the other hand, it shows that RAP principally regulated mTOR when treated together with LQG. A study investigating resveratrol also proved that polyphenols reduced lipid accumulation in 3T3-L1 cells by activating mTORC1 and its downstream site p70S6 [46]. Another polyphenol, raspberry ketone, was confirmed to considerably stimulate the expression of p-mTOR and inhibit the expression of SIRT1 on day 6 of 3T3-L1 adipocyte differentiation, resulting in a reduction in lipid accumulation [40].

mTOR also plays a central role in regulating fat and lipid homeostasis. The inhibition of LQG in the gene or protein expression of C/EBPα, PPARγ and ACC1 was abrogated by pretreatment with RAP, demonstrating that mTOR was involved in the lipid synthesis process affected by LQG. Considering the role of mTOR in the autophagy induced by LQG, this suggests that LQG inhibited lipogenesis in 3T3-L1 white adipocytes through mTOR-mediated autophagy. Some studies have shown that the inhibition of mTORC1 signaling genetically impairs lipogenesis by regulating the SREBP transcriptional network via S6K1 or Lipin1 [47]. However, the regulating network of mTOR in lipid homeostasis is very complex and shows crosstalk with a variety of molecules, including different substrates of mTOR. For example, Lipin1, which could be inhibited by mTORC1, was shown on the one hand to negatively regulate SREBP1c [48] and was found, on the other hand, to act as a key factor in adipocyte maturation and maintenance by stimulating PPARγ [49]. Thus, mTOR activation might lead to the suppression of PPARγ, which supports the results of the mTOR and PPARγ expression patterns in our current study. Moreover, even though mTOR is often recognized as a stimulator of SREBP in research regarding lipid metabolism [50,51], it is notable that mTORC1 signaling is essential, but not sufficient, to activate SREBP1c. For example, C/EBPα was also able to regulate expression of SREBP1c [33], indicating a parallel regulation of C/EBPα, PPARγ, and SREBP1c by the autophagic pathway via mTOR, which was in line with the results of LQG in our current study. Interestingly, the current study only shows that the co-treatment of LQG and RAP diminished the effects induced by LQG, while RAP alone did not result in significant opposite effects. Some studies have shown that RAP inhibited C/EBPα or PPARγ protein expression [52], which could not be observed in our data. The inconsistent results may be due to different concentrations and distinct stages of cell development in which RAP was added to treat the cells. In contrast, RAP treatment alone significantly downregulated the protein and mRNA expressions of SREBP1c and

FASN in our results, showing similar impacts to those induced by LQG. This indicates that RAP, an mTOR inhibitor, regulates the expression of SREBP1c and FASN through different signaling pathways from LQG, which may not involve the modulation of autophagy. It was reported that RAP was able to inhibit lipogenesis via the SREBP1c–FASN axis [53,54], which was also observed in our results. Although the mechanisms of LQG and RAP in modulating 3T3-L1 adipocyte lipogenesis might differ, the results from pretreatment of RAP with LQG demonstrate that LQG modulated mTOR activity. Moreover, the activation of mTOR was correlated with reductions in the autophagy and lipid-synthesis processes. This also suggests the important role of mTOR in the molecular mechanisms of LQG in modulating adipocyte lipid accumulation.

5. Conclusions

LQG exhibited a significant reduction in lipid accumulation in 3T3-L1 white adipocytes in our current study. The positive effect was associated with the mTOR-mediated autophagy process (Figure 6). Together, our findings make an essential contribution to understanding the biological activities of LQG. This present study provides scientific evidence for supporting the utilization of LQG as a bioactive flavonoid to prevent obesity by inhibiting the accumulation of lipids in fat cells. This in vitro study provides evidence supporting further research regarding the properties of LQG with respect to preventing adipose tissue expansion in in vivo animal studies or even in human clinical studies. LQG can be expected to be developed as a new functional compound for application in dietary supplements. Moreover, the study supports the crucial role of autophagy in 3T3-L1-cell lipid accumulation. In addition to the bioactive phytochemicals from herbs, the molecular signals correlated with autophagy regulation might be novel and efficient targets for preventing obesity in the future.

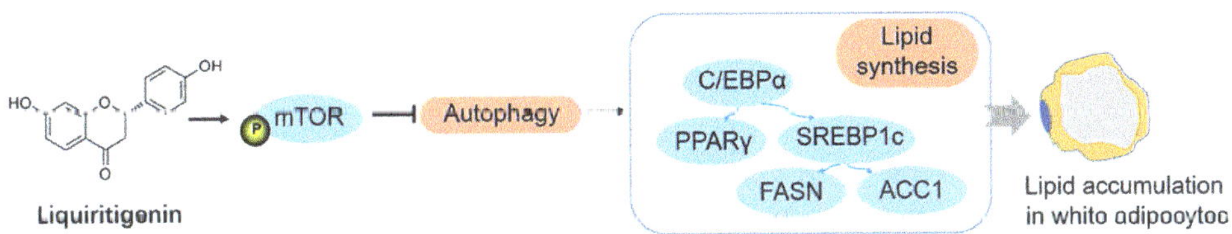

Figure 6. LQG reduced lipid accumulation in 3T3-L1 white adipocytes via mTOR-mediated autophagy mechanism.

Author Contributions: Conceptualization, W.Z.; methodology, Z.S., C.Z., J.Y., F.X., L.W. and W.Z.; software, Z.S. and W.Z.; validation, H.Q., Z.S., C.Z., J.Y., F.X., L.W., A.A. and W.Z.; formal analysis, W.Z.; investigation, H.Q., Z.S. and W.Z.; resources, H.Q. and W.Z.; data curation, W.Z.; writing—original draft preparation, W.Z.; writing—review and editing, H.Q., A.A. and W.Z.; visualization, H.Q. and W.Z.; supervision, W.Z.; project administration, W.Z.; funding acquisition, H.Q. and W.Z. All authors have read and agreed to the published version of the manuscript.

Funding: This research was funded by the National Natural Science Foundation of China, grant number 82073556; the Natural Science Foundation of Hunan Province, grant number 2020JJ5795; the Hunan Province Health Commission Project, grant number 202112021806; and the College Students' Innovation and Entrepreneurship Project, grant number S2021105330392.

Conflicts of Interest: The authors declare no conflict of interest.

References

1. Kopelman, P.G. Obesity as a medical problem. *Nature* **2000**, *404*, 635–643. [CrossRef] [PubMed]
2. Saunders, K.H.; Umashanker, D.; Igel, L.I.; Kumar, R.B.; Aronne, L.J. Obesity Pharmacotherapy. *Med. Clin. N. Am.* **2018**, *102*, 135–148. [CrossRef]
3. Derosa, G.; Maffioli, P. Anti-obesity drugs: A review about their effects and their safety. *Expert Opin. Drug Saf.* **2012**, *11*, 459–471. [CrossRef] [PubMed]

4. Hall, K.D.; Guo, J. Obesity Energetics: Body Weight Regulation and the Effects of Diet Composition. *Gastroenterology* **2017**, *152*, 1718–1727.e3. [CrossRef]
5. Rosen, E.D.; MacDougald, O.A. Adipocyte differentiation from the inside out. *Nat. Rev. Mol. Cell Biol.* **2006**, *7*, 885–896. [CrossRef]
6. Hammarstedt, A.; Gogg, S.; Hedjazifar, S.; Nerstedt, A.; Smith, U. Impaired Adipogenesis and Dysfunctional Adipose Tissue in Human Hypertrophic Obesity. *Physiol. Rev.* **2018**, *98*, 1911–1941. [CrossRef] [PubMed]
7. Choi, Y.R.; Shim, J.; Kim, M.J. Genistin: A Novel Potent Anti-Adipogenic and Anti-Lipogenic Agent. *Molecules* **2020**, *25*, 2042. [CrossRef]
8. Kubota, N.; Terauchi, Y.; Miki, H.; Tamemoto, H.; Yamauchi, T.; Komeda, K.; Satoh, S.; Nakano, R.; Ishii, C.; Sugiyama, T.; et al. PPAR gamma mediates high-fat diet-induced adipocyte hypertrophy and insulin resistance. *Mol. Cell* **1999**, *4*, 597–609. [CrossRef]
9. Sewter, C.; Blows, F.; Considine, R.; Vidal-Puig, A.; O'Rahilly, S. Differential effects of adiposity on peroxisomal proliferator-activated receptor gamma1 and gamma2 messenger ribonucleic acid expression in human adipocytes. *J. Clin. Endocrinol. Metab.* **2002**, *87*, 4203–4207. [CrossRef]
10. Mota de Sá, P.; Richard, A.J.; Hang, H.; Stephens, J.M. Transcriptional Regulation of Adipogenesis. *Compr. Physiol.* **2017**, *7*, 635–674.
11. Mizushima, N.; Levine, B.; Cuervo, A.M.; Klionsky, D.J. Autophagy fights disease through cellular self-digestion. *Nature* **2008**, *451*, 1069–1075. [CrossRef] [PubMed]
12. Cecconi, F.; Levine, B. The role of autophagy in mammalian development: Cell makeover rather than cell death. *Dev. Cell* **2008**, *15*, 344–357. [CrossRef]
13. Goldman, S.J.; Zhang, Y.; Jin, S. Autophagic degradation of mitochondria in white adipose tissue differentiation. *Antioxid. Redox Signal.* **2011**, *14*, 1971–1978. [CrossRef]
14. Dong, H.; Czaja, M.J. Regulation of lipid droplets by autophagy. *Trends Endocrinol. Metab. TEM* **2011**, *22*, 234–240. [CrossRef]
15. Singh, R.; Xiang, Y.; Wang, Y.; Baikati, K.; Cuervo, A.M.; Luu, Y.K.; Tang, Y.; Pessin, J.E.; Schwartz, G.J.; Czaja, M.J. Autophagy regulates adipose mass and differentiation in mice. *J. Clin. Investig.* **2009**, *119*, 3329–3339. [CrossRef] [PubMed]
16. Kovsan, J.; Blüher, M.; Tarnovscki, T.; Klöting, N.; Kirshtein, B.; Madar, L.; Shai, I.; Golan, R.; Harman-Boehm, I.; Schön, M.R.; et al. Altered autophagy in human adipose tissues in obesity. *J. Clin. Endocrinol. Metab.* **2011**, *96*, E268–E277. [CrossRef] [PubMed]
17. Kao, T.C.; Wu, C.H.; Yen, G.C. Bioactivity and potential health benefits of licorice. *J. Agric. Food Chem.* **2014**, *62*, 542–553. [CrossRef]
18. Kim, Y.W.; Zhao, R.J.; Park, S.J.; Lee, J.R.; Cho, I.J.; Yang, C.H.; Kim, S.G.; Kim, S.C. Anti-inflammatory effects of liquiritigenin as a consequence of the inhibition of NF-kappaB-dependent iNOS and proinflammatory cytokines production. *Br. J. Pharmacol.* **2008**, *154*, 165–173. [CrossRef]
19. Kim, G.-H.; Ju, J.-Y.; Chung, K.-S.; Cheon, S.-Y.; Gil, T.-Y.; Cominguez, D.C.; Cha, Y.-Y.; Lee, J.-H.; Roh, S.-S.; An, H.-J. Rice Hull Extract (RHE) Suppresses Adiposity in High-Fat Diet-Induced Obese Mice and Inhibits Differentiation of 3T3-L1 Preadipocytes. *Nutrients* **2019**, *11*, 1162. [CrossRef]
20. Yuan, X.; Wang, Z.; Zhang, L.; Sui, R.; Khan, S. Exploring the inhibitory effects of liquiritigenin against tau fibrillation and related neurotoxicity as a model of preventive care in Alzheimer's disease. *Int. J. Biol. Macromol.* **2021**, *183*, 1184–1190. [CrossRef]
21. Lee, H.E.; Yang, G.; Han, S.H.; Lee, J.H.; An, T.J.; Jang, J.K.; Lee, J.Y. Anti-obesity potential of Glycyrrhiza uralensis and licochalcone A through induction of adipocyte browning. *Biochem. Biophys. Res. Commun.* **2018**, *503*, 2117–2123. [CrossRef] [PubMed]
22. Madak-Erdogan, Z.; Gong, P.; Zhao, Y.C.; Xu, L.; Wrobel, K.U.; Hartman, J.A.; Wang, M.; Cam, A.; Iwaniec, U.T.; Turner, R.T.; et al. Dietary licorice root supplementation reduces diet-induced weight gain, lipid deposition, and hepatic steatosis in ovariectomized mice without stimulating reproductive tissues and mammary gland. *Mol. Nutr. Food Res.* **2016**, *60*, 369–380. [CrossRef]
23. Fan, L.; Xu, H.; Yang, R.; Zang, Y.; Chen, J.; Qin, H. Combination of Capsaicin and Capsiate Induces Browning in 3T3-L1 White Adipocytes via Activation of the Peroxisome Proliferator-Activated Receptor $\gamma/\beta(3)$-Adrenergic Receptor Signaling Pathways. *J. Agric. Food Chem.* **2019**, *67*, 6232–6240. [CrossRef] [PubMed]
24. Eckel, R.H.; Grundy, S.M.; Zimmet, P.Z. The metabolic syndrome. *Lancet* **2005**, *365*, 1415–1428. [CrossRef]
25. Gong, X.M.; Li, Y.F.; Luo, J.; Wang, J.Q.; Wei, J.; Wang, J.Q.; Xiao, T.; Xie, C.; Hong, J.; Ning, G.; et al. Gpnmb secreted from liver promotes lipogenesis in white adipose tissue and aggravates obesity and insulin resistance. *Nat. Metab.* **2019**, *1*, 570–583. [CrossRef] [PubMed]
26. Song, Z.; Xiaoli, A.M.; Yang, F. Regulation and Metabolic Significance of De Novo Lipogenesis in Adipose Tissues. *Nutrients* **2018**, *10*, 1383. [CrossRef]
27. Heinonen, S.; Jokinen, R.; Rissanen, A.; Pietiläinen, K.H. White adipose tissue mitochondrial metabolism in health and in obesity. *Obes. Rev. Off. J. Int. Assoc. Study Obes.* **2020**, *21*, e12958. [CrossRef]
28. Kim, Y.W.; Ki, S.H.; Lee, J.R.; Lee, S.J.; Kim, C.W.; Kim, S.C.; Kim, S.G. Liquiritigenin, an aglycone of liquiritin in Glycyrrhizae radix, prevents acute liver injuries in rats induced by acetaminophen with or without buthionine sulfoximine. *Chem.-Biol. Interact.* **2006**, *161*, 125–138. [CrossRef]
29. Strable, M.S.; Ntambi, J.M. Genetic control of de novo lipogenesis: Role in diet-induced obesity. *Crit. Rev. Biochem. Mol. Biol.* **2010**, *45*, 199–214. [CrossRef]
30. Ameer, F.; Scandiuzzi, L.; Hasnain, S.; Kalbacher, H.; Zaidi, N. De novo lipogenesis in health and disease. *Metabolism* **2014**, *63*, 895–902. [CrossRef]

31. Tontonoz, P.; Spiegelman, B.M. Fat and beyond: The diverse biology of PPARgamma. *Annu. Rev. Biochem.* **2008**, *77*, 289–312. [CrossRef] [PubMed]
32. Magaña, M.M.; Lin, S.S.; Dooley, K.A.; Osborne, T.F. Sterol regulation of acetyl coenzyme A carboxylase promoter requires two interdependent binding sites for sterol regulatory element binding proteins. *J. Lipid Res.* **1997**, *38*, 1630–1638. [CrossRef]
33. Payne, V.A.; Au, W.S.; Lowe, C.E.; Rahman, S.M.; Friedman, J.E.; O'Rahilly, S.; Rochford, J.J. C/EBP transcription factors regulate SREBP1c gene expression during adipogenesis. *Biochem. J.* **2009**, *425*, 215–223. [CrossRef] [PubMed]
34. Watanabe, M.; Hisatake, M.; Fujimori, K. Fisetin Suppresses Lipid Accumulation in Mouse Adipocytic 3T3-L1 Cells by Repressing GLUT4-Mediated Glucose Uptake through Inhibition of mTOR-C/EBPα Signaling. *J. Agric. Food Chem.* **2015**, *63*, 4979–4987. [CrossRef] [PubMed]
35. Chen, J.; Zhao, H.; Ma, X.; Zhang, Y.; Lu, S.; Wang, Y.; Zong, C.; Qin, D.; Wang, Y.; Yingfeng Yang, Y.; et al. GLP-1/GLP-1R Signaling in Regulation of Adipocyte Differentiation and Lipogenesis. *Cell. Physiol. Biochem. Int. J. Exp. Cell. Physiol. Biochem. Pharmacol.* **2017**, *42*, 1165–1176. [CrossRef] [PubMed]
36. Chen, J.; Ren, J.; Jing, Q.; Lu, S.; Zhang, Y.; Liu, Y.; Yu, C.; Gao, P.; Zong, C.; Li, X.; et al. TSH/TSHR Signaling Suppresses Fatty Acid Synthase (FASN) Expression in Adipocytes. *J. Cell Physiol.* **2015**, *230*, 2233–2239. [CrossRef] [PubMed]
37. Levine, B.; Yuan, J. Autophagy in cell death: An innocent convict? *J. Clin. Investig.* **2005**, *115*, 2679–2688. [CrossRef] [PubMed]
38. Zhang, Y.; Goldman, S.; Baerga, R.; Zhao, Y.; Komatsu, M.; Jin, S. Adipose-specific deletion of autophagy-related gene 7 (atg7) in mice reveals a role in adipogenesis. *Proc. Natl. Acad. Sci. USA* **2009**, *106*, 19860–19865. [CrossRef] [PubMed]
39. Baerga, R.; Zhang, Y.; Chen, P.H.; Goldman, S.; Jin, S. Targeted deletion of autophagy-related 5 (atg5) impairs adipogenesis in a cellular model and in mice. *Autophagy* **2009**, *5*, 1118–1130. [CrossRef]
40. Leu, S.Y.; Chen, Y.C.; Tsai, Y.C.; Hung, Y.W.; Hsu, C.H.; Lee, Y.M.; Cheng, P.Y. Raspberry Ketone Reduced Lipid Accumulation in 3T3-L1 Cells and Ovariectomy-Induced Obesity in Wistar Rats by Regulating Autophagy Mechanisms. *J. Agric. Food Chem.* **2017**, *65*, 10907–10914. [CrossRef]
41. Bartlett, B.J.; Isakson, P.; Lewerenz, J.; Sanchez, H.; Kotzebue, R.W.; Cumming, R.C.; Harris, G.L.; Nezis, I.P.; Schubert, D.R.; Simonsen, A.; et al. p62, Ref(2)P and ubiquitinated proteins are conserved markers of neuronal aging, aggregate formation and progressive autophagic defects. *Autophagy* **2011**, *7*, 572–583. [CrossRef] [PubMed]
42. Kim, Y.C.; Guan, K.L. mTOR: A pharmacologic target for autophagy regulation. *J. Clin. Investig.* **2015**, *125*, 25–32. [CrossRef] [PubMed]
43. Zhang, M.; Xue, Y.; Zheng, B.; Li, L.; Chu, X.; Zhao, Y.; Wu, Y.; Zhang, J.; Han, X.; Wu, Z.; et al. Liquiritigenin protects against arsenic trioxide-induced liver injury by inhibiting oxidative stress and enhancing mTOR-mediated autophagy. *Biomed. Pharmacother.* **2021**, *143*, 112167. [CrossRef] [PubMed]
44. Singh, R. Autophagy and regulation of lipid metabolism. *Results Probl. Cell Differ.* **2010**, *52*, 35–46. [PubMed]
45. Bae, J.; Jang, Y.; Kim, H., Mahato, K.; Schaecher, C.; Kim, I.M.; Kim, E.; Ro, S.H. Arsenite exposure suppresses adipogenesis, mitochondrial biogenesis and thermogenesis via autophagy inhibition in brown adipose tissue. *Sci. Rep.* **2019**, *9*, 14464. [CrossRef]
46. Liu, Z.; Liao, W.; Yin, X.; Zheng, X.; Li, Q.; Zhang, H.; Zheng, L.; Feng, X. Resveratrol-induced brown fat-like phenotype in 3T3-L1 adipocytes partly via mTOR pathway. *Food Nutr. Res.* **2020**, *64*, 3656. [CrossRef]
47. Lamming, D.W.; Sabatini, D.M. A Central role for mTOR in lipid homeostasis. *Cell Metab.* **2013**, *18*, 465–469. [CrossRef]
48. Peterson, T.R.; Sengupta, S.S.; Harris, T.E.; Carmack, A.E.; Kang, S.A.; Balderas, E.; Guertin, D.A.; Madden, K.L.; Carpenter, A.E.; Finck, B.N.; et al. mTOR complex 1 regulates lipin 1 localization to control the SREBP pathway. *Cell* **2011**, *146*, 408–420. [CrossRef]
49. Zhang, P.; Takeuchi, K.; Csaki, L.S.; Reue, K. Lipin-1 phosphatidic phosphatase activity modulates phosphatidate levels to promote peroxisome proliferator-activated receptor γ (PPARγ) gene expression during adipogenesis. *J. Biol. Chem.* **2012**, *287*, 3485–3494. [CrossRef]
50. Bakan, I.; Laplante, M. Connecting mTORC1 signaling to SREBP-1 activation. *Curr. Opin. Lipidol.* **2012**, *23*, 226–234. [CrossRef]
51. Qin, H.; Song, Z.; Shaukat, H.; Zheng, W. Genistein Regulates Lipid Metabolism via Estrogen Receptor β and Its Downstream Signal Akt/mTOR in HepG2 Cells. *Nutrients* **2021**, *13*, 4015. [CrossRef] [PubMed]
52. Gagnon, A.; Lau, S.; Sorisky, A. Rapamycin-sensitive phase of 3T3-L1 preadipocyte differentiation after clonal expansion. *J. Cell Physiol.* **2001**, *189*, 14–22. [CrossRef] [PubMed]
53. Li, S.; Brown, M.S.; Goldstein, J.L. Bifurcation of insulin signaling pathway in rat liver: mTORC1 required for stimulation of lipogenesis, but not inhibition of gluconeogenesis. *Proc. Natl. Acad. Sci. USA* **2010**, *107*, 3441–3446. [CrossRef] [PubMed]
54. Yecies, J.L.; Zhang, H.H.; Menon, S.; Liu, S.; Yecies, D.; Lipovsky, A.I.; Gorgun, C.; Kwiatkowski, D.J.; Hotamisligil, G.S.; Lee, C.H.; et al. Akt stimulates hepatic SREBP1c and lipogenesis through parallel mTORC1-dependent and independent pathways. *Cell Metab.* **2011**, *14*, 21–32. [CrossRef] [PubMed]

nutrients

MDPI

Article

Anti-Obesity Effects of Polymethoxyflavone-Rich Fraction from Jinkyool (*Citrus sunki* Hort. ex Tanaka) Leaf on Obese Mice Induced by High-Fat Diet

Yeong-Jun Jin [1,†], Mi-Gyeong Jang [2,†], Jae-Won Kim [2], Songyee Baek [2], Hee-Chul Ko [3], Sung-Pyo Hur [4] and Se-Jae Kim [1,2,*]

1 Department of Biology, Jeju National University, Jeju 63243, Korea; y2j-1004@hanmail.net
2 Biotech Regional Innovation Center, Jeju Nation University, Jeju 63423, Korea; mkjang@jejunu.ac.kr (M.-G.J.); kjw8839@jejunu.ac.kr (J.-W.K.); summerbee@jejunu.ac.kr (S.B.)
3 Jeju Institute of Korean Medicine, Jeju 63309, Korea; fly1007@jikom.or.kr
4 Jeju International Marine Science Research & Logistics Center, Korea Institute of Ocean Science & Technology, Gujwa, Jeju 63349, Korea; hursp@kiost.ac.kr
* Correspondence: sjkim@jejunu.ac.kr; Tel.: +82-64-754-3529
† These authors contributed equally to this work.

Abstract: Polymethoxyflavones (PMFs) are flavonoids exclusively found in certain citrus fruits and have been reported to be beneficial to human health. Most studies have been conducted with PMFs isolated from citrus peels, while there is no study on PMFs isolated from leaves. In this study, we prepared a PMF-rich fraction (PRF) from the leaves of *Citrus sunki* Hort ex. Tanaka (Jinkyool) and investigated whether the PRF could improve metabolic decline in obese mice induced by a high-fat diet (HFD) for 5 weeks. The HFD-induced obese mice were assigned into HFD, OR (HFD + orlistat at 15.6 mg/kg of body weight/day), and PRF (HFD + 50, 100, and 200 mg/kg of body weight/day) groups. Orlistat and PRF were orally administered for 5 weeks. At the end of the experiment, the serum biochemical parameters, histology, and gene expression profiles in the tissues of each group were analyzed. The body weight gain of the obese mice was significantly reduced after orlistat and PRF administration for 5 weeks. PRF effectively improved HFD-induced insulin resistance and dyslipidemia. Histological analysis in the liver demonstrated that PRF decreased adipocyte size and potentially improved the liver function, as it inhibited the incidence of fatty liver. PRF activated AMP-activated protein kinase (AMPK), acetyl-CoA carboxylase (ACC), and hormone-sensitive lipase (HSL) in HFD-induced obese mice. Moreover, liver transcriptome analysis revealed that PRF administration enriched genes mainly related to fatty-acid metabolism and immune responses. Overall, these results suggest that the PRF exerted an anti-obesity effect via the modulation of lipid metabolism.

Keywords: *Citrus sunki*; polymethoxyflavones; anti-obesity; high-fat diet; C57BL/6; fatty-acid oxidation; lipolysis

Citation: Jin, Y.-J.; Jang, M.-G.; Kim, J.-W.; Baek, S.; Ko, H.-C.; Hur, S.-P.; Kim, S.-J. Anti-Obesity Effects of Polymethoxyflavone-Rich Fraction from Jinkyool (*Citrus sunki* Hort. ex Tanaka) Leaf on Obese Mice Induced by High-Fat Diet. *Nutrients* **2022**, *14*, 865. https://doi.org/10.3390/nu14040865

Academic Editors: Lindsay Brown and Maria Luz Fernandez

Received: 19 January 2022
Accepted: 15 February 2022
Published: 18 February 2022

1. Introduction

Obesity is a complex, multifactorial disorder that is primarily caused by an imbalance between energy uptake and expenditure. The prevalence of obesity is increasing worldwide and its associated metabolic disorders are a major public health concern [1]. Moreover, obesity contributes to chronic diseases such as type 2 diabetes, nonalcoholic fatty liver disease, hypertension, heart disease, osteoarthritis, and obstructive sleep apnea, and increases the risk of several cancers and other diseases [2–5]. In addition, people living with obesity often experience mental health issues alongside different degrees of functional limitations, i.e., obesity-related disability, and they suffer from social bias, prejudice, and discrimination [6].

Currently, multiple therapeutic methods with which to treat obesity are available, including strengthening exercises, diet control, behavioral changes, surgical treatment, and pharmacotherapy. However, these therapeutic options have the following problems: their time requirements and economic costs are high, many have side effects, and the weight is easily regained after stopping the drugs [7–9]. Epidemiological studies have revealed a positive correlation between a high intake of flavonoid-rich foods and the incidence of metabolic syndrome [10,11]. Therefore, recently, bioactive compounds from plant extracts, including citrus flavonoids, have attracted more attention for curing obesity. Citrus flavonoids derived from citrus plants are polyphenolic compounds, and they are divided into three major subgroups: flavanones, flavone glycosides, and polymethoxyflavones (PMFs). PMFs are flavones with two or more methoxyl groups, which exhibit various physiological and pharmacological bioactivities. In particular, PMFs are of great interest due to their beneficial properties, such as anti-obesity, anticancer, and anti-inflammatory activities, both in vitro and in vivo [12–17].

Citrus sunki Hort ex. Tanaka (commonly known as Jinkyool) is a native citrus plant cultivated in Jeju-do, Korea, and its peel extract is known to contain a large amount of PMFs with various biological activities [18–20]. PMFs have been found in the seeds, leaves, juice, stems, and peels of citrus. However, most studies have been conducted on the flavedo and peels of citrus fruits, and there is no study on PMFs isolated from citrus leaves. In this study, we prepared a PMF-rich fraction (PRF) from *C. sunki* leaf and evaluated its efficacy as an anti-obesity ingredient for obese mice whose obesity was induced by a high-fat diet (HFD).

2. Materials and Methods

2.1. Plant Material and Preparation of Polymethoxyflavone-Rich Fraction (PRF)

The Jinkyool fruits and leaves used in this study were collected in Jeju-do, Korea, in February 2021. The collected fruits and leaves were washed immediately, dried at 60 °C for 24 h, and then powdered. The PMF-rich fraction (PRF) was made according to the method of Ko et al. [20]. Briefly, the pulverized powder was extracted with hot water at 100 °C for 5 h. The hot water extract was mixed with hexane in a 1:1 ratio; the hexane layer was recovered, concentrated, lyophilized, and stored at −20 °C until it was used in experiments. The contents of three major PMFs (nobiletin, tangeretin, and sinensetin) in the hexane fraction were quantified using a high-performance liquid chromatography (HPLC) system comprising two pumps, an autosampler, a column oven, and a PDA detector (Waters 2695 Alliance system). The instrument control and data acquisition were accomplished using the Empower software (version 2.0; Waters) and a Sunfire C18 column (4.6 × 250 mm, 5 μm; Waters), monitored at 330 nm. The fraction from whole fruits contained 22.03 mg/g of PMFs (nobiletin, 8.07 mg/g, and tangeretin, 13.96 mg/g), while that of the leaves contained 67.25 mg/g of PMFs (sinensetin, 1.10 mg/g; nobiletin, 15.81 mg/g; and tangeretin, 50.34 mg/g) (Figure 1). Thus, the hexane fraction from the leaves was designated as the PMF-rich fraction (PRF) and used in the subsequent experiment.

2.2. Animals and Experimental Design

Four-week-old male C57BL/6J mice were obtained from Cronex Inc. (Gyeonggi-do, Korea). The animals were acclimated for 1 week by housing in appropriate cages under a 12 h/12 h light/dark cycle at room temperature (23 ± 2 °C) and 55% ± 5% relative humidity. After the acclimation period, the animals were fed a normal diet consisting of 14% fat, 21% protein, and 65% carbohydrate (total energy 4.04 kcal/g) (ND, LabDiel 5L79, Orient Bio. Inc. Seongnam-si, Gyeonggi-do, Korea) or a high-fat-diet (HFD), consisting of 60% fat, 20% protein, and 20% carbohydrate (total energy 5.24 kcal/g) (D12492, Research Diets, Inc. New Brunswick, NJ, USA) to induce obesity, for 5 weeks [21]. Then, the HFD-induced obese mice were assigned into HFD, OR (HFD + orlistat at 15.6 mg/kg of body weight/day), and PRF (HFD + 50, 100, and 200 mg/kg of body weight/day) groups. Orlistat and PRF were orally administered at a given dose daily for 5 weeks. The concentration of orlistat was determined by referring to previous reports [22,23]; the drugs were diluted in 0.5% carboxymethyl cellulose (CMC) and administered once daily.

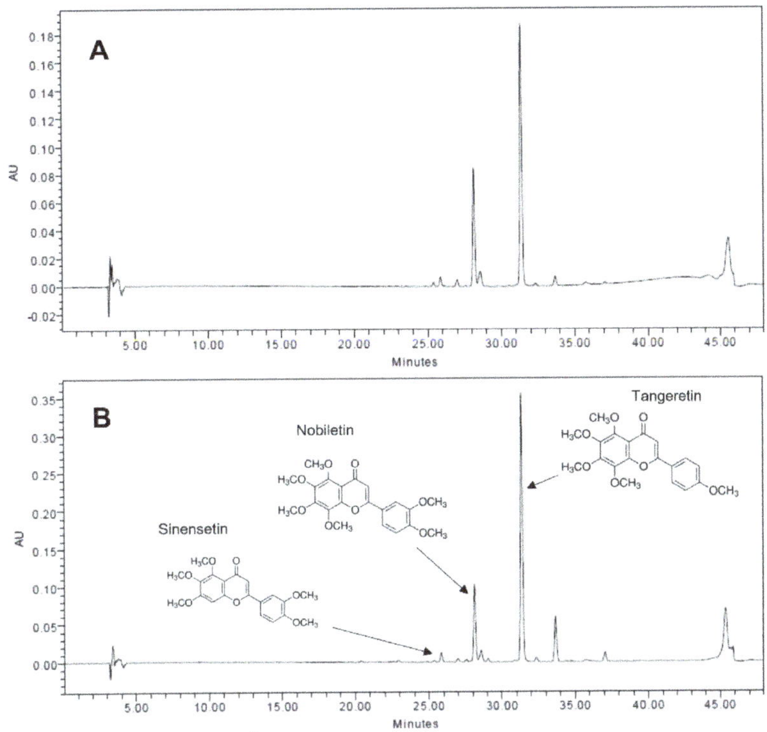

Figure 1. HPLC chromatograms of PMF-rich fraction (PRF) from fruits (**A**) and leaves of Jinkyool (*Citrus sunki* Hort ex. Tanaka) (**B**). Tangeretin, nobiletin, and sinensetin were the three major PMFs contained in PRF.

2.3. Oral Glucose Tolerance Test (OGTT)

The OGTT was performed at the final experimental day. The experimental animals fasted for at least 8 h, and then glucose (2 g/kg of body weight) was administered orally. After glucose administration, rat tail vein blood was taken at 30, 60, 90, and 120 min to measure the blood glucose level. The HOMA-IR index was assessed by multiplying the fasting serum insulin and fasting serum glucose, and then dividing the result by a constant number [24]: HOMA-IR = (fasting glucose (mg/dL) × fasting insulin (mU/L))/405.

2.4. Collection of Serum and Tissue Samples

Blood was taken from the hearts of mice that had fasted for 8 h and been anaesthetized with isoflurane from Vspharm (Gyeonggi, South Korea). The blood was allowed to stand at room temperature for 5 h or more to coagulate, and the coagulated blood was centrifuged at 3000× *g* for 20 min to separate the serum. The separated serum was stored at −70 °C until blood analysis. The liver, adipose tissue, spleen, and kidneys were extracted, weighed, and stored in liquid nitrogen for analysis.

2.5. Biochemical Parameter Assays

The serum triglyceride (TG), total cholesterol (TC), high-density lipoprotein cholesterol (HDL-C), and low-density lipoprotein cholesterol (LDL-C) levels were determined using commercial kits (DoGenBio, Seoul, South Korea). The serum glutamic oxaloacetic transaminase (GOT) and glutamic pyruvic transaminase (GPT) levels were determined using commercial kits from Asan Pharm (Gyeonggi, South Korea), according to the manufacturer's protocol. The insulin levels were determined using rat insulin ELISA kits (Mercodia, Uppsala, Sweden).

2.6. Histological Analysis

The adipose and liver tissues were extracted and fixed in 4% paraformaldehyde for 48 h. The tissues were dehydrated with an ethanol series, embedded in paraffin wax, and sectioned into 7 μm-thick slices. After deparaffinization and rehydration, the tissue sections were stained with hematoxylin and eosin (H&E). Then, the stained sections were dehydrated and sealed with Canada balsam (Junsei, Tokyo, Japan), and observed under a microscope (BX-51, Olympus, Tokyo, Japan).

2.7. Western Blot Analysis

Western blot analysis was performed according to a previous report [25]. The adipose tissue was homogenized in cold lysis buffer (1 × RIPA buffer, 1 mM phenylmethylsulfonyl fluoride, 1 mM Na_3VO_4, 1 mM NaF, 1 μg/mL aprotinin, 1 μg/mL pepstatin, and 1 μg/mL leupeptin) and gathered by centrifugation at 14,000 rpm for 20 mim. The lysate protein concentrations were quantified using a protein assay kit (Bio-Rad, Hercules, CA, USA). The proteins were separated using electrophoresis on sodium dodecyl sulfate–polyacrylamide gels and then moved to polyvinylidene fluoride membranes. The membranes were blocked with 5% (*w*/*v*) skim milk and 0.1% (*v*/*v*) Tween-20 in Tris-buffered saline (TBST). The membranes were incubated with primary antibodies (dilution 1:2000–10,000) against β-actin (Santa Cruz, CA, USA), phospho-AMP-activated protein kinase (pAMPK) and AMPK (Cell Signaling technology, Beverly, MA, USA), phosphor-acetyl-CoA carboxylase (pACC) and ACC (Cell Signaling Technology), and phosphor-hormone-sensitive lipase (pHSL) (Cell Signaling Technology) overnight at 4 °C. The membranes were cleaned with 0.1% TBST and then reacted at room temperature for 1 h with a peroxidase-conjugated secondary antibody (dilution 1:5000–10,000) (Vector Laboratories, Burlingame, CA, USA). The membranes were cleaned with 0.1% TBST; then, the proteins were identified using the Westar ETA C 2.0 substrate (Cyanagen, Bologna, Italy).

2.8. RNA-Seq Library Preparation and Sequencing

Liver tissue samples were homogenized to extract the RNA, and the total RNA was prepared using TRIzol reagent (Invitrogen, MA, USA). The RNA quality was evaluated using an Agilent 2100 Bioanalyzer (Agilent Technologies, Amstelveen, The Netherlands, USA), and RNA quantification was executed by an ND-2000 Spectrophotometer (Thermo Inc., DE, USA). Libraries were made from total RNA using the NEBNext Ultra II Directional RNA-Seq Kit (NEW ENGLAND BioLabs, Inc., Hitchin, UK). The mRNA was isolated using the Poly(A) RNA Selection Kit (LEXOGEN, Inc., Vienna, Austria), and used for the cDNA synthesis and shearing according to the manufacturer's instructions. Indexing was performed using the Illumina indices 1–12. The enrichment step was performed using PCR. Subsequently, the libraries were verified using the TapeStation HS D1000 Screen Tape (Agilent Technologies, Amstelveen, The Netherlands) to assess the mean fragment size. Quantification was conducted using the library quantification kit with a StepOne Real-Time PCR System (Life Technologies, Inc., Carlsbad, CA, USA). High-throughput sequencing was achieved as paired-end 100 sequencing using a NovaSeq 6000 (Illumina, Inc., San Diego, CA, USA).

2.9. Data Analysis

Quality control of the raw sequencing data was achieved using FastQC [26]. Adapters and low-quality reads (<Q20) were withdrawn using FASTX_Trimmer [27] and BBMap [28]. Then, the trimmed reads were mapped to the reference genome using TopHat [29]. The RC (read count) data were processed according to the FPKM + geometric normalization method using EdgeR within R [30]. FPKM (fragments per kb per million reads) values were

estimated using Cufflinks [31]. Data mining and graphic visualization were accomplished using ExDEGA (Ebiogen Inc., Seoul, Korea).

2.10. Statistical Analysis

All the statistical analyses were performed using SPSS version 21.0 for Windows (SPSS; Chicago, IL, USA). All the data are presented as means ± standard deviations (SDs). Differences between groups were examined by Duncan multiple test in one-way analysis of variance (ANOVA). A p-value < 0.05 was considered to indicate statistical significance.

3. Results

3.1. Effects of PRF on the Body Weight (BW) in HFD-Induced Obese Mice

It was confirmed that obesity was induced in mice fed a high-fat diet for 5 weeks by the considerable increase in body weight compared to the normal-diet group (the initial weights are shown in Table 1). The HFD-induced obese mice were further divided into five groups and administered orlistat or PRF orally for 5 weeks. The OR and PRF groups exhibited a significant decrease in body weight compared to the high-fat diet group (final weights in Table 1). In particular, PRF decreased the weight gain in a dose-dependent manner. During the feeding period, the food intake was higher in the ND group than in the HFD group, but it did not differ significantly among the HFD, OR, and three PRF groups. The weights of the liver, epididymal fat, kidney, and spleen were higher in the HFD group compared with the ND group. However, the weights of the liver were significantly decreased in the orlistat and PRF groups compared with the HFD group. In particular, the weights of white adipose tissue (epididymal fat) and the spleen were effectively decreased in the PRF group, indicating that the PRF modulated lipids and the immune response. There were no significant differences in kidney weights among the HFD, OR, PRF 50, PRF 100, and PRF 200 groups. Similarly to orlistat, PRF substantially reduced the body weight gain in HFD-induced obese mice; however, we did not detect any sign of PRF-induced toxicity.

Table 1. Effects of PRF on body weight food intake and organ weights in HFD-induced obese mice.

	ND	HFD	OR	PRF 50	PRF 100	PRF 200
Initial weight (g)	27.64 ± 1.26 ^{a}	34.60 ± 6.20 ^{b}	32.53 ± 3.78 ^{b}	33.01 ± 4.61 ^{b}	32.70 ± 3.11 ^{b}	32.84 ± 2.83 ^{b}
Final weight (g)	28.73 ± 0.30 ^{a}	43.26 ± 5.68 ^{c}	37.36 ± 3.98 ^{b}	37.54 ± 5.12 ^{b}	37.07 ± 3.56 ^{b}	36.71 ± 4.66 ^{b}
Weight gain (g)	1.26 ± 0.30 ^{a}	8.66 ± 1.20 ^{c}	4.83 ± 1.93 ^{b}	4.53 ± 2.10 ^{b}	4.37 ± 2.17 ^{b}	3.87 ± 2.22 ^{b}
Food intake (g/mouse/day)	4.08 ± 0.51 ^{a}	2,76 ± 0.18 ^{b}	3.05 ± 0.34 ^{b}	2.89 ± 0.50 ^{b}	2.61 ± 0.23 ^{b}	2.74 ± 0.56 ^{b}
Liver (g)	1.37 ± 0.08 ^{bc}	1.49 ± 0.41 ^{c}	1.09 ± 0.18 ^{a}	1.36 ± 0.15 ^{abc}	1.19 ± 0.17 ^{ab}	1.25 ± 0.08 ^{ab}
Epididymal fat (g)	0.67 ± 0.17 ^{a}	2.55 ± 0.60 ^{c}	2.11 ± 0.20 ^{bc}	1.70 ± 0.41 ^{b}	1.93 ± 0.41 ^{b}	1.74 ± 0.77 ^{b}
Kidney (g)	0.30 ± 0.03 ^{a}	0.36 ± 0.02 ^{b}	0.31 ± 0.04 ^{a}	0.35 ± 0.03 ^{b}	0.34 ± 0.02 ^{b}	0.34 ± 0.03 ^{b}
Spleen (g)	0.08 ± 0.01 ^{a}	0.13 ± 0.02 ^{b}	0.07 ± 0.02 ^{a}	0.09 ± 0.01 ^{a}	0.08 ± 0.02 ^{a}	0.09 ± 0.03 ^{a}

HFD-induced obese mice were orally administered the drug under an HFD for a further 5 weeks. ND, normal diet; HFD, high-fat diet; OR, HFD + 15.6 mg of orlistat; PRF 50, PRF 100, and PRF 200, HFD + 50, 100, and 200 mg of PRF, respectively. The results are expressed as the means ± SDs (n = 7). The values with different letters (a–c) are significantly different (p < 0.05) from each other according to one-way ANOVA and Duncan's multiple-range test.

3.2. Effects of PRF on Serum Lipid Profiles in HFD-Induced Obese Mice

At the end of the 10-week experimental period, we compared the serum lipid profiles of each group. The serum TG and TC levels were higher in the HFD group than in the ND group, but there was no difference in HDL levels between the two groups. The LDL level was rather high in ND group (Figure 2). Orlistat or PRF administration for 5 weeks to HFD-induced obese mice significantly improved the serum TG and HDL levels (Figure 2A,C). However, there was no difference in TC or LDL levels in the HFD, OR, PRF 100, and PRF 200 groups, in contrast to the PRF 50 group.

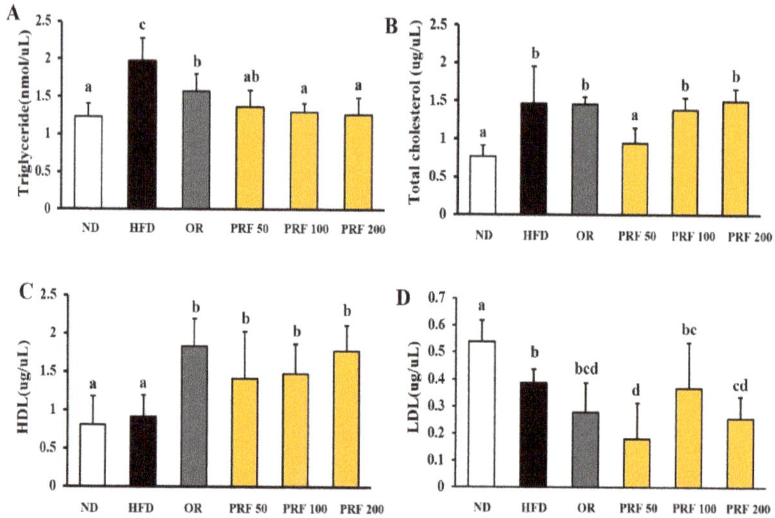

Figure 2. Effect of PRF on serum lipid profiles in HFD-induced obese mice. HFD-induced obese mice were orally administered the drug under an HFD for a further 5 weeks. (**A**) Triglyceride levels, (**B**) total cholesterol levels, (**C**) HDL-cholesterol levels, (**D**) LDL-cholesterol levels. ND, normal diet; HFD, high-fat diet; OR, HFD + 15.6 mg of orlistat; PRF 50, PRF 100, and PRF 200, HFD + 50, 100, and 200 mg of PRF, respectively. The results are expressed as the means ± SDs ($n = 7$). The values with different letters (a–d) are significantly different ($p < 0.05$) from each other according to one-way ANOVA and Duncan's multiple-range test.

3.3. Effects of PRF on the Insulin Resistance in HFD-Induced Obese Mice

The fasting blood glucose levels were higher in the HFD group than the ND group, indicating that the HFD induced hyperglycemia in this animal model. The OR and three PRF groups showed attenuation of HFD-induced hyperglycemia (Figure 3A). The oral glucose tolerance test showed that HFD-induced glucose tolerance was significantly improved by OR and PRF administration (Figure 3B). Moreover, the serum insulin level in HFD-induced obese mice was dramatically reduced by OR and PRF (Figure 3C). PRF effectively reduced the homeostasis model assessment insulin resistance (HOMA-IR) index in a dose-dependent manner (Figure 3D).

3.4. Effects of PRF on the Hypertrophy of Adipocytes in HFD-Induced Obese Mice

After 5 weeks of oral administration, the morphometric analysis of epididymal adipose tissue showed a difference among the experimental groups. The sizes of the epididymal adipocytes in each group are shown in Figure 4. The sizes of the adipocytes were appreciably larger in the HFD group than in the ND group. The sizes of visceral adipocytes were considerably decreased in the OR and PRF groups compared with the HFD group. This indicates that weight gain is a result of the accumulation of triglycerides in fat cells in white adipose tissue (WAT) and the expansion of the adipose tissue. In line with the action of orlistat (an anti-obesity drug), PRF ameliorated the hypertrophy of WAT adipocytes in HFD-induced obese mice.

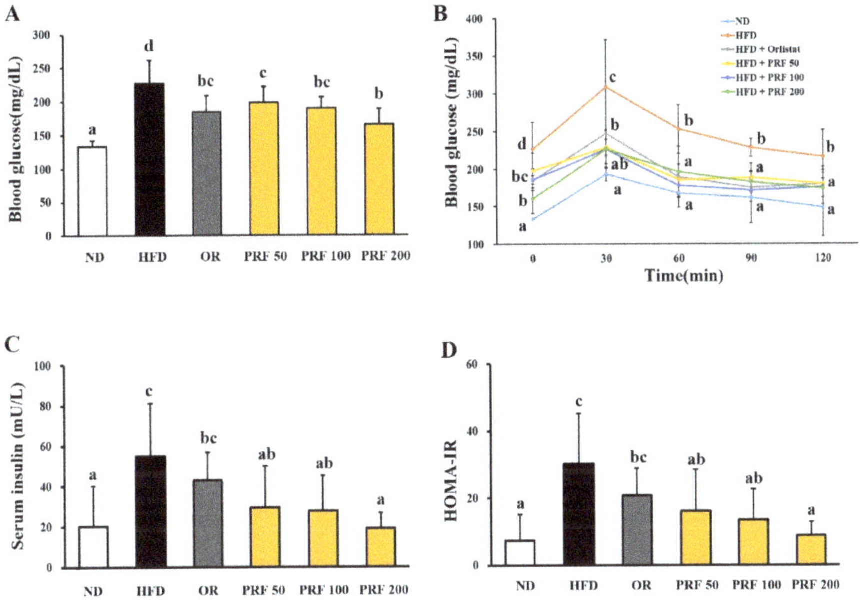

Figure 3. Effect of PRF on glucose homeostasis in HFD-induced obese mice. The mice were fasted for 8 h before measuring blood glucose. (**A**) Fasting blood glucose levels; (**B**) oral glucose tolerance test (OGTT); (**C**) serum insulin levels; (**D**) HOMA-IR index. ND, normal diet; HFD, high-fat diet; OR, HFD + 15.6 mg of orlistat; PRF 50, PRF 100, and PRF 200, HFD + 50, 100, and 200 mg of PRF, respectively. The results are expressed as the means ± SDs (n = 7). The values with different letters (a–d) are significantly different ($p < 0.05$) from each other according to one-way ANOVA and Duncan's multiple-range test.

3.5. Effects of PRF on Fatty-Acid Oxidation and Lipolysis in HFD-Induced Obese Mice

To explore the underlying molecular mechanisms of the anti-obesity effects of PRF administration, we investigated the expression of epididymal adipose proteins among the groups. PRF administration dramatically enhanced the phosphorylation of AMP-activated protein kinase (AMPK), acetyl-CoA carboxylase (ACC), and hormone-sensitive lipase (HSL) in HFD-induced obese mice (Figure 5). These results suggest that PRF exerted its anti-obesity effects by promoting fatty-acid oxidation and lipid lysis in HFD-induced obese mice.

3.6. Effects of PRF on Liver Function in HFD-Induced Obese Mice

The serum glutamic oxaloacetic transaminase (GOT) and glutamic pyruvic transaminase (GPT) levels, which are delicate clinical indicators of liver damage, were elevated in the HFD group compared with the ND group. However, OR and PRF administration effectively reduced the GOT/GPT levels in HFD-induced obese mice (Figure 6A,B), indicating their preventive effects on the liver damage induced by the HFD. Consistent with this result, the histological analysis revealed that the HFD group progressed to a fatty liver by accumulating larger lipid droplets. However, OR and PRF administration significantly ameliorated the HFD-induced fatty liver (Figure 6C–H). Since AMPK activation is linked with metabolic improvement, we compared the phosphorylation of AMPK and ACC among the experimental groups. As shown in Figure 6I, orlistat and PRF administration activated AMPK and ACC, indicating that AMPK played a vital role in mediating the beneficial effect of PRF in HFD-induced obese mice.

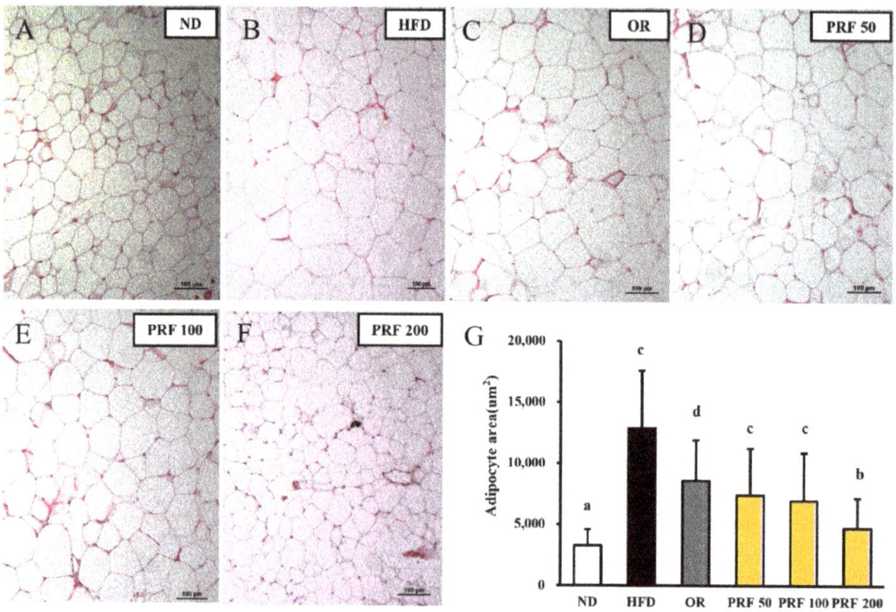

Figure 4. Effect of PRF on the size of epididymal adipocytes in HFD-induced obese mice. The sections of the epididymal fat tissues of ND (**A**), HFD (**B**), OR (**C**), PRF 50 (**D**), PRF 100 (**E**), and PRF 200 (**F**) were stained with H&E (100×). Scale bar = 100 μm. (**G**) The epididymal adipocyte sizes of each group. ND, normal diet; HFD, high-fat diet; OR, HFD + 15.6 mg of orlistat; PRF 50, PRF 100, and PRF 200, HFD + 50, 100, and 200 mg of PRF, respectively. The results are expressed as the means ± SDs (n = 7). The values with different letters (a–d) are significantly different ($p < 0.05$) from each other according to one-way ANOVA and Duncan's multiple-range test.

3.7. Effects of PRF on the Liver Transcriptome in HFD-Induced Obese Mice

To explore how PRF administration affected whole-gene expression in HFD-induced obese mice, we further analyzed the liver transcriptome profiles using RNA-Seq in the HFD and PRF 50 groups (n = 3). To identify the differentially expressed genes (DEGs) between the HFD and PRF 50 groups, the transcriptomic profiles were compared. Among 23,183 genes, 117 DEGs were identified (Table S1), comprising 61 upregulated genes and 56 downregulated genes in the PRF vs. HFD groups, respectively ($p < 0.05$; fold change >2). To elucidate the signatures of the DEGs, Gene Ontology (GO) and Kyoto Encyclopedia of Genes and Genomes (KEGG) pathways were analyzed. PRF administration mainly affected functions in the biological process category of GO terms, such as fatty-acid metabolic process, acyl-CoA metabolic process, lipid metabolic process, fatty-acid transport, and triglyceride metabolic process (Figure 7A). KEGG pathway analysis revealed that PRF administration altered the expression of genes implicated in pathways such as PPAR signaling, retinol metabolism, fatty-acid degradation, metabolic pathways, linoleic-acid metabolism, and fatty-acid metabolism (Figure 7B). Transcriptome analyses indicated that PRF could exert its anti-obesity effect by regulating the genes associated with the fatty-acid metabolic process, lipid metabolic process, and retinol metabolism.

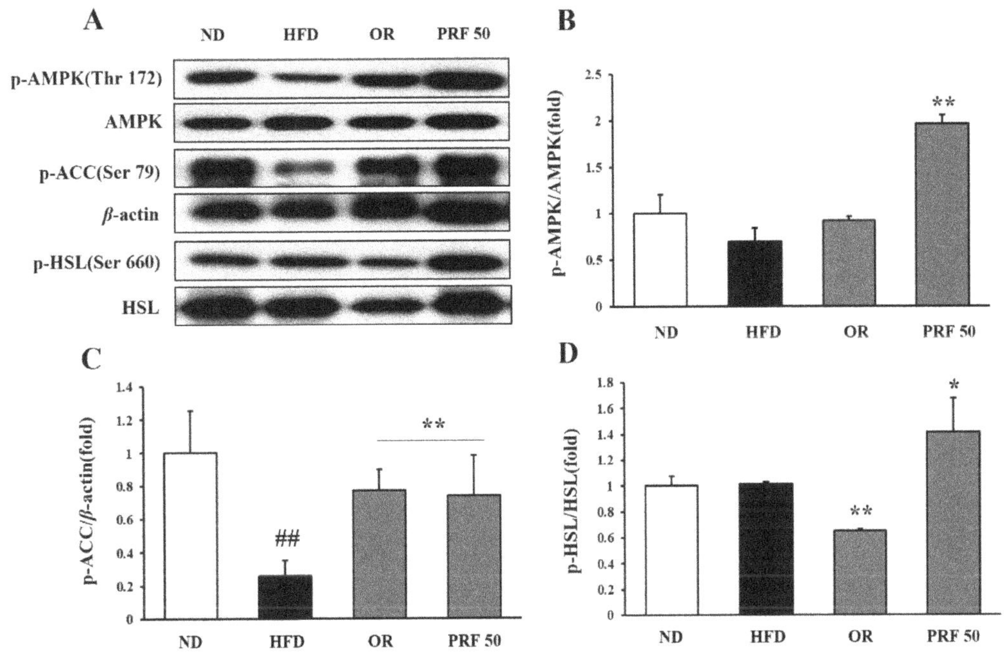

Figure 5. Effect of PRF on the regulation of lipid-metabolism-related protein expression in epididymal adipose tissue in HFD-induced obese mice. The expression of p-AMPK, AMPK, p-ACC, p-HSL, HSL, and β-actin was analyzed by Western blot analysis (**A**), and the intensity of each band was measured using ImageJ. The relative intensity of p-AMPK/AMPK (**B**), p-ACC/β-actin (**C**), and p-HSL/HSL (**D**). The results are expressed as the means ± SDs ($n = 3$; ## $p < 0.01$ compared to ND; * $p < 0.05$, ** $p < 0.01$ compared to HFD). ND, normal diet; HFD, high-fat diet; OR, HFD + 15.6 mg of orlistat; PRF 50, PRF 100, and PRF 200, HFD + 50, 100, and 200 mg of PRF, respectively.

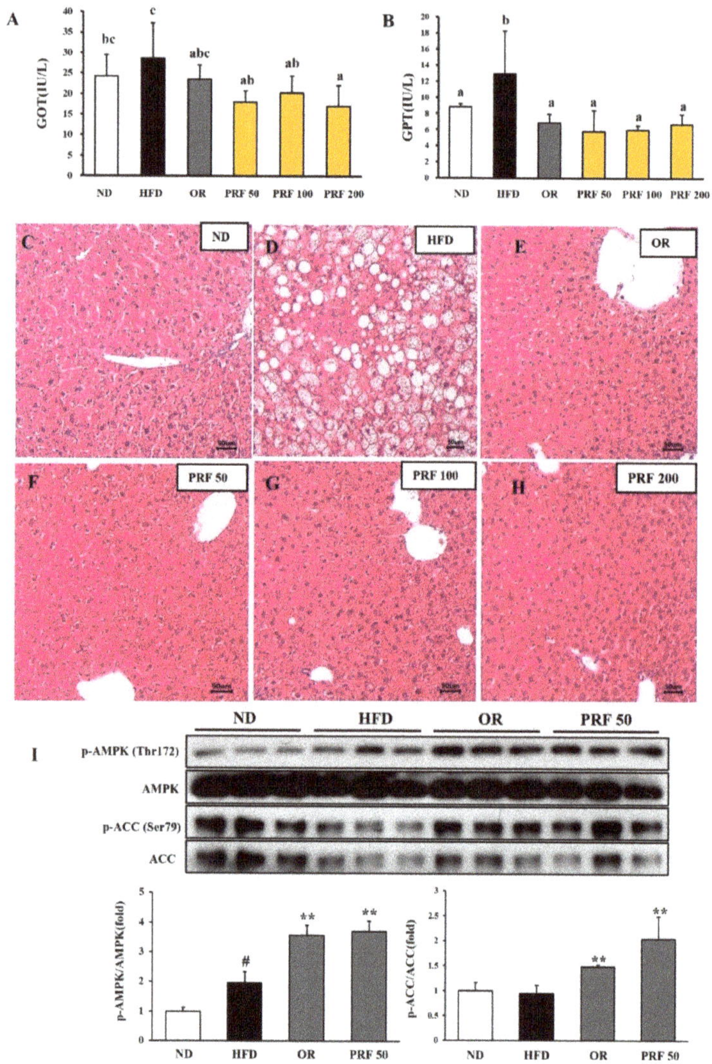

Figure 6. Effect of PRF on liver function in HFD-induced obese mice. (**A**) GOT levels; (**B**) GPT levels. The results are expressed as the means ± SDs ($n = 7$). The values with different letters (a–c) are significantly different ($p < 0.05$) from each other according to one-way ANOVA and Duncan's multiple-range test. The liver sections for ND (**C**), HFD (**D**), OR (**E**), PRF 50 (**F**), PRF 100 (**G**), and PRF 200 (**H**) stained with H&E (200×). Scale bar = 50 μm. (**I**) Western blot of AMPK and ACC. The results are expressed as the means ± SDs ($n = 3$; $^{\#}$ $p < 0.05$ compared to ND; ** $p < 0.01$ compared to HFD). ND, normal diet; HFD, high-fat diet; OR, HFD + 15.6 mg of orlistat; PRF 50, PRF 100, and PRF 200, HFD + 50, 100, and 200 mg of PRF, respectively.

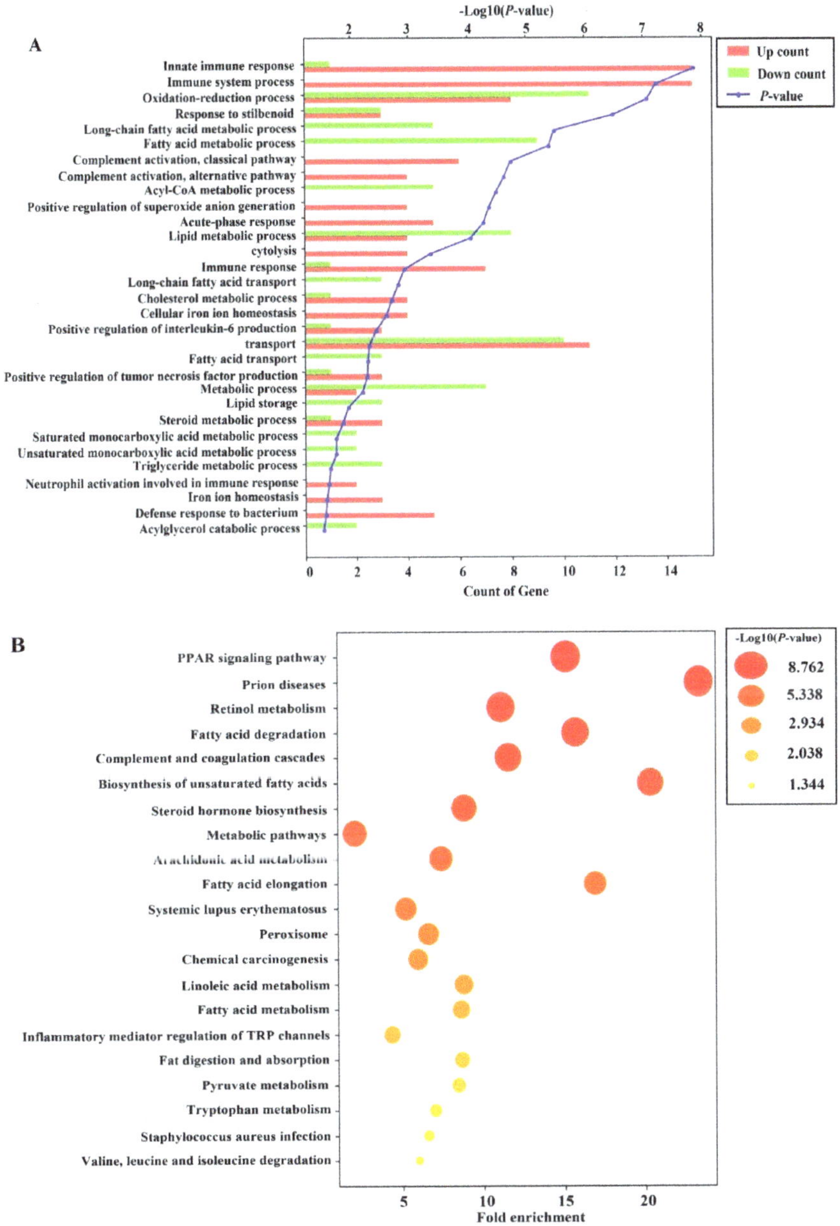

Figure 7. Effects of PRF administration on the liver transcriptome in HFD-induced obese mice. (**A**) Gene Ontology (GO) analysis using differentially expressed genes (DEGs) between the PRF 50 and HFD groups. Top 30 terms within the biological process category of GO function analysis. (**B**) Kyoto Encyclopedia of Genes and Genomes (KEGG) pathway analysis using DEGs between the PRF 50 and HFD groups. Top 20 terms among KEGG pathways. A *p*-value < 0.05 was considered to reflect statistical significance.

4. Discussion

Obesity-associated metabolic dysfunction is occurring as one of the most health-threatening diseases in the world [32]. Hence, there remains an insistent need for the development of efficient and safe therapeutic agents. Several reports have indicated that citrus PMFs could be appealing candidates due to their potent defensive effects against metabolic syndrome, as well as their safety in rodent experimental models [33,34]. Citrus PMFs, particularly nobiletin, tangeretin, and sinensetin, are abundant in the *Citrus* genus and exhibit various biological properties [15]. Recently, it was suggested that the beneficial effects of PMFs, which are groups of phenolic phytochemicals with poor bioavailability, are mostly exerted through the regulation of the gut microbiota [14,33]. Most of the research on PMFs has been conducted using fruits, and there is no study on PMFs isolated from citrus leaves. Here, we established an optimized and standardized procedure for preparing a PMF-rich fraction (PRF) from Jinkyool (*C. sunki*) leaves. The PRF was qualitatively and quantitatively characterized, and the compounds contained were clearly elucidated [20]. The PRF from leaves contained tangeretin (50.34 mg/g), nobiletin (15.81 mg/g), and sinensetin (1.10 mg/g) as the major constituents. The PRF from the leaves contained a three-times-higher content of PMFs compared to whole fruits that were harvested at the same time as the leaves. Jinkyool fruit is harvested once a year, but its leaves can be harvested year-round, and the leaves at all the harvesting times have a constant PMF content. Thus, we evaluated its potential as an anti-obesity dietary ingredient. There is no information on the average intake of PRF from citrus leaves. However, the animal studies using PMFs from citrus peels have shown beneficial effects at 30–120 mg/kg [14,15,35–37]. Based on this information, we implemented an administration dose of PRF at 50, 100, and 200 mg/kg of body weight.

Over 5 weeks of administration in HFD-induced obese mice, three doses of PRF exhibited an equal or even better obesity protective effect than a current anti-obesity drug, orlistat, whose mechanism involves inhibiting pancreatic lipase. We did not observe any obvious side effects of a high dose of PRF in HFD-induced obese mice. We noted no significant difference in the total food intake of mice in each group during the experimental period ($p > 0.05$), suggesting that PRF-induced weight loss involved a mechanism that was independent of hyperphagia as compared to the HFD group.

Adipose tissue can increase in size through hypertrophy or hyperplasia. The hyperplasia of adipose tissue is believed healthy and adaptive. However, the hypertrophy of adipocytes is related to an increase in the hypoxia experienced by these cells because of their massive expansion. Hypoxic adipose tissue increases the expression of pro-fibrotic genes and leads to tissue fibrosis. Hypoxic adipocytes occasionally experience necrosis, leading to infiltration by immune cells and tissue inflammation [38]. These factors work together to decrease adipose tissue function and consistently raise blood glucose and lipid levels. This might contribute to an earlier onset of metabolic disease and trigger toxic lipid deposition in other tissues [39–41]. PRF administration dose-dependently decreased the sizes of the epididymal adipocytes in a superior manner to orlistat in mice with HFD-induced obesity. This indicates that PRF might reduce HFD-induced weight gain by lowering the accumulation of triglycerides and the expansion of the adipose tissue. PRF effectively improved dyslipidemia and decreased the fasting blood glucose and insulin levels, resulting in improved glucose tolerance and insulin resistance in HFD-induced obese mice.

To explore the mechanism underlying the anti-obesity effect of PRF administration in HFD-induced obese mice, we analyzed the expression of lipid-catabolism-related proteins among the groups. PRF administration dramatically increased the phosphorylation of AMP-activated protein kinase (AMPK), acetyl-CoA carboxylase (ACC), and hormone-sensitive lipase (HSL) in the white adipose tissue of HFD-induced obese mice. These results suggest that PRF exerted its anti-obesity effects through the activation of AMPK. AMPK is a main regulator of cell and whole-body energy metabolism and is a crucial modulator of cell survival or death in response to pathological stress [42,43]. The activation of AMPK in adipocytes results in a decrease in triglyceride synthesis, lipogenesis, and inflammatory cytokine secretion, along with an increase in fatty-acid oxidation and GLUT4 translocation to cell membranes [42,43]. We found that the expression of HSL was significantly enhanced in the orlistat and PRF administration groups compared with the HFD group. HSL is a rate-limiting enzyme in the breakdown of triglycerides that are mainly involved in the hydrolysis of diglycerides to monoglycerides [44,45]. Thus, these results suggest that PRF exerted its anti-obesity effect by promoting both fatty-acid oxidation via the AMPK/ACC pathway and lipolysis via the PKA/HSL pathway in the WAT of HFD-induced obese mice. However, further studies on these mechanisms are needed.

PRF significantly decreased the serum GOT/GPT level and liver tissue weight in HFD-induced obese mice; thus, PRF might protect against the liver damage induced by an HFD. Consistent with this observation, the liver tissue of the HFD group had a typical fatty-liver shape. However, the number of lipid droplets was significantly reduced in the liver tissue of the groups administered PRF and orlistat. PRF administration increased the phosphorylation of AMPK/ACC in HFD-induced obese mice. Thus, it is suggested that PRF prevented fatty liver by both inhibiting de novo lipogenesis and promoting fatty-acid oxidation via AMPK activation. It is well known that AMPK plays a major role in maintaining the energy balance, including controlling lipid metabolism in the liver [46,47]. Indeed, our results suggest that AMPK activation could play a vital role in mediating the beneficial effects of PRF. Inflammatory factors, such as lipopolysaccharide (LPS) and tumor necrosis factor-α (TNF-α), reduce AMPK activity, whereas AMPK activation counteracts NF-κB signaling, which is a multifaceted protein complex that generates inflammatory factors in the fatty liver [48–51].

Thus, to further understand the mechanism underlying the anti-obesity effect of PRF in liver tissues, we analyzed differences in the liver transcriptome profiles between the HFD and PRF 50 groups by RNA-Seq. PRF administration mainly enriched GO terms associated with biological processes, such as the fatty-acid metabolic process, acyl-CoA metabolic process, lipid metabolic process, fatty-acid transport, and triglyceride metabolic process. The KEGG pathways affected by PRF administration in HFD-induced obese mice involved PPAR signaling, retinol metabolism, fatty-acid degradation, metabolic pathways, linoleic-acid metabolism, and fatty-acid metabolism. For example, the PPAR signaling pathway was shown to play a crucial role in energy homeostasis, glucose metabolism, and lipid metabolism [52]. PRF-enriched genes that are involved in the PPAR pathway include *CYP4A12A, FABP2, FABP5, ACSL1, EHHADH, FABP7, CYP4A10, ACAA1B, ACAA1B, CYP4A14, CD36*, and *CYP7A1*. The expression patterns of the fatty-acid uptake factors, such as CD36, FABPs, and CYP7A1, were differentially recovered to levels of the ND group by PRF administration in HFD-induced obese mice. Since these genes are involved in lipid deposition in the liver tissue [53–58], it is proposed that PRF prevents fatty liver by both suppressing lipogenesis and increasing fatty-acid oxidation in the liver. In this study, we did not validate the RNA-Seq data with qPCR; however, the changes in the transcriptome profiles were consistent with the difference in protein expression patterns between the PRF 50 and HFD groups. Taking into account all these results, we hypothesize that PRF might alleviate obesity-associated metabolic derangement by regulating relevant genes.

In conclusion, PRF suppressed body weight gain, white adipose tissue weight, hypertrophy of adipocytes, and fatty liver by regulating lipid-metabolism-related genes. PRF effectively improved dyslipidemia and insulin resistance in HFD-induced obese mice. These results demonstrate that the PRF might be a favorable therapeutic agent for metabolic disorders in terms of its efficacy and safety.

Supplementary Materials: The following supporting information can be downloaded at https://www.mdpi.com/article/10.3390/nu14040865/s1: Table S1. List of differentially expressed genes in HFD versus PRF 50 group.

Author Contributions: Conceptualization, investigation, and writing—original draft preparation, Y.-J.J. and M.-G.J.; methodology and visualization, H.-C.K. and J.-W.K.; investigation, S.B. and S.-P.H.;

writing—review and editing, supervision, project administration, and funding acquisition, S.-J.K. All authors have read and agreed to the published version of the manuscript.

Funding: This study was supported by the Basic Science Research Program through the National Research Foundation of Korea (NRF), with the backing of the Ministry of Education (NRF-2020R1I1A3A04037169 and NRF-2021R1I1A1A01055732), and by the Open Labs Program in partnership with Local Industrial Associates through the Commercialization Promotion Agency for R&D Outcomes of the Ministry of Science and Technology Information and Communication (NTIS-1711139489).

Institutional Review Board Statement: The study was conducted in accordance with the Declaration of Helsinki and approved by the Institutional Animal Care and Use Committee of Jeju National University (No. 2021-0012).

Informed Consent Statement: Not applicable.

Data Availability Statement: The datasets generated for this study can be found in GEO, Accession NO. GSE193785.

Conflicts of Interest: The authors declare no conflict of interest.

References

1. Klaauw, A.A.; Farooqi, I.S. The Hunger genes: Pathway to obesity. *Cell* **2015**, *26*, 119–132. [CrossRef]
2. Ng, M.; Fleming, T.; Robinson, M.; Thomson, B.; Graetz, N.; Margono, C.; Mullany, E.; Biryukov, S.; Abbafati, C.; Abera, S.F.; et al. Global, regional, and national prevalence of overweight and obesity in children and adults during 1980–2013: A systematic analysis for the Global Burden of Disease Study 2013. *Lancet* **2014**, *384*, 766–781. [CrossRef]
3. Blueher, M. Obesity: Global epidemiology and pathogenesis. *Nat. Rev. Endocrinol.* **2019**, *15*, 288–298. [CrossRef]
4. Flegal, K.M.; Kit, B.K.; Orpana, H.; Graubard, B.I. Association of all-cause mortality with overweight and obesity using standard body mass index categories: A systematic review and meta-analysis. *JAMA-J. Am. Med. Assoc.* **2013**, *309*, 71–82. [CrossRef] [PubMed]
5. Gadde, K.M.; Martin, C.K.; Verthoud, H.R.; Heymsfield, S.B. Obesity: Pathophysiology and management. *J. Am. Coll. Cardiol.* **2018**, *71*, 69–84. [CrossRef] [PubMed]
6. Sirtori, A.; Brunani, A.; Capodaglio, P.; Berselli, M.E.; Villa, V.; Corti, S.; Leonardi, M.; Raggi, A. ICF-OB: A multidisciplinary questionnaire based on the international classification of functioning, disability and health to address disability in obesity. *Eur. J. Physical. Rehab. Med.* **2018**, *54*, 119–121. [CrossRef] [PubMed]
7. Saltiel, A.R. New therapeutic approaches for the treatment of obesity. *Sci. Transl. Med.* **2016**, *8*, 323rv2. [CrossRef]
8. Chien, M.Y.; Ku, Y.H.; Chang, J.M.; Yang, C.M.; Chen, C.H. Effects of herbal mixture extracts on obesity in rats fed a high-fat diet. *J. Food. Drug. Anal.* **2016**, *24*, 594–601. [CrossRef]
9. Bessesen, D.H.; Van Gaal, L.F. Progress and challenges in anti-obesity pharmacotherapy. *Lancet Diabetes Endocrinol.* **2018**, *6*, 237–248. [CrossRef]
10. Fu, C.; Jiang, Y.; Guo, J.; Su, Z. Natural products with anti-obesity effects and different mechanisms of action. *J. Agric. Food Chem.* **2016**, *64*, 9571–9585. [CrossRef]
11. Rio, D.D.; Rodriguez-Mateos, A.; Spencer, J.P.; Tognolini, M.; Borges, G.; Crozier, A. Dietary (poly) phenolics in human health: Structures, bioavailability, and evidence of protective effects against chronic diseases. *Antioxid. Redox. Signal.* **2013**, *18*, 1818–1892. [CrossRef] [PubMed]
12. Gao, Z.; Gao, W.; Zeng, S.L.; Li, P.; Liu, W.H. Chemical structures, bioactivities and molecular mechanisms of citrus polymethoxyflavones. *J. Func. Food.* **2018**, *40*, 498–509. [CrossRef]
13. Guo, J.; Tao, H.; Cao, Y.; Ho, C.T.; Jin, S.; Huang, Q. Prevention of obesity and type 2 diabetes with aged citrus peel (Chenpi) extract. *J. Agric. Food Chem.* **2016**, *64*, 2053–2061. [CrossRef] [PubMed]
14. Zeng, S.L.; Li, S.Z.; Xio, P.T.; Cai, Y.Y.; Chu, C.; Chen, B.Z.; Li, P.; Lim, J.; Liu, E.H. Citrus polymethoxyflavones attenuate metabolic syndrome by regulating gut microbiome and amino acid metabolism. *Sci. Adv.* **2020**, *6*, eaax6208. [CrossRef]
15. Feng, K.; Zhu, X.; Xhen, T.; Peng, B.; Lu, M.; Zheng, H.; Huang, Q.; Ho, C.T.; Chenu, Y.; Cao, Y. Prevention of obesity and hyperlipidemia by heptamethoxyflavone in high-fat diet-induced rats. *J. Agric. Food Chem.* **2019**, *67*, 2476–2489. [CrossRef]
16. Lee, Y.S.; Cha, B.Y.; Saitoa, K.; Choi, S.S.; Wanga, X.X.; Choi, B.K.; Yonezawaa, T.; Teruyab, T.; Nagaia, K.; Woo, J.T. Effects of a Citrus depressa Hayata (shiikuwasa) extract on obesity in high-fat diet-induced obese mice. *Phytomedicine* **2011**, *18*, 648–654. [CrossRef]
17. He, B.; Nohara, K.; Park, N.; Park, Y.S.; Guillory, B.; Zhao, Z.; Garcia, J.M.; Koike, N.; Lee, C.C.; Takahashi, J.S.; et al. The small molecule nobiletin targets the molecular oscillator to enhance circadian rhythms and protect against metabolic syndrome. *Cell Metab.* **2016**, *23*, 610–621. [CrossRef]

18. Choi, S.Y.; Ko, H.C.; Ko, S.Y.; Hwang, J.H.; Park, J.G.; Kang, S.H.; Han, S.H.; Yun, S.H.; Kim, S.J. Correlation between flavonoid content and the NO production inhibitory activity of peel extracts from various citrus fruits. *Biol. Pharm. Bull.* **2007**, *20*, 772–778. [CrossRef]
19. Kang, S.I.; Shin, H.S.; Kim, H.M.; Hong, Y.S.; Yoon, S.A.; Kang, S.W.; Kim, J.H.; Kim, M.H.; Ko, H.C.; Kim, S.J. Immature citrus sunki peel extract exhibits antiobesity effects by β-oxidation and lipolysis in high-fat diet-induced obese mice. *Biol. Pharm. Bull.* **2012**, *35*, 223–230. [CrossRef]
20. Ko, H.C.; Jang, M.G.; Kang, C.H.; Lee, N.H.; Kang, S.I.; Lee, S.R.; Park, D.B.; Kim, S.J. Preparation of a polymethoxyflavone-rich fraction (PRF) of Citrus sunki Hort. ex Tanaka and its antiproliferactive effects. *Food Chem.* **2010**, *123*, 484–488. [CrossRef]
21. Wang, C.Y.; Liao, J.K. A mouse model of diet-induced obesity and insulin resistance. *Methods Mol. Biol.* **2012**, *821*, 421–433. [CrossRef] [PubMed]
22. Reagan-Shaw, S.; Nihal, M.; Ahmad, N. Dose translation from animal to human studies revisited. *FASEB J.* **2008**, *31*, 659–661. [CrossRef] [PubMed]
23. Jio, X.; Wang, Y.; Lin, Y.; Lang, Y.; Li, E.; Zhang, X.; Zhang, Q.; Feng, Y.; Meng, X. Blueberry polyphenols extract as a potential prebiotic with anti-obesity effects on C57BL/6J mice by modulating the gut microbiota. *J. Nutr. Biochem.* **2019**, *64*, 88–100. [CrossRef] [PubMed]
24. Matthews, D.R.; Hosker, J.P.; Rudenski, A.S.; Naylor, B.A.; Treacher, D.F.; Tuner, R.C. Homeostasis model assessment: Insulin resistance and β-cell function from fasting plasma glucose and insulin concentrations in man. *Diabetologia* **1985**, *28*, 412–419. [CrossRef] [PubMed]
25. Jang, M.G.; Oh, J.M.; Ko, H.C.; Kim, J.-W.; Baek, S.; Jin, Y.J.; Hur, S.-P.; Kim, S.-J. *Clerodendrum trichotomum* extract improves metabolic derangements in high fructose diet-fed rats. *Anim. Cells Syst.* **2021**, *25*, 396–404. [CrossRef]
26. Andrews, S. FastQC. Available online: https://www.bioinformatics.babraham.ac.uk/projects/fastqc (accessed on 7 December 2010).
27. Hannon Lab. FASTX Toolkit. Available online: http://hannonlab.cshl.edu/fastx_toolkit/ (accessed on 7 December 2021).
28. Bushnell, B. BBMap. Available online: https://sourceforge.net/projects/bbmap/ (accessed on 7 December 2021).
29. Trapnell, C.; Pachter, L.; Salzberg, S.L. TopHat: Discovering splice junctions with RNA-Seq. *Bioinformatics* **2009**, *25*, 1105–1111. [CrossRef]
30. Roberts, A.; Trapnell, C.; Donaghey, J.; Rinn, J.L.; Pachter, L. Improving RNA-Seq expression estimates by correcting for fragment bias. *Genome Biol.* **2011**, *12*, R22. [CrossRef]
31. R Development Core Team. *R: A Language and Environment for Statistical Computing*; R Foundation for Statistical Computing: Vienna, Austria, 2020.
32. Ford, E.S.; Li, C.; Zhao, G. Prevalence and correlates of metabolic syndrome based on a harmonious definition among adults in the US. *J. Diabetes* **2010**, *2*, 180–193. [CrossRef]
33. Tung, Y.C.; Chang, W.T.; Li, S.; Wu, J.C.; Badmeav, V.; Ho, C.T.; Pan, M.H. Citrus peel extracts attenuated obesity and modulated gut microbiota in mice with high-fat diet-induced obesity. *Food Funct.* **2018**, *9*, 3363–3373. [CrossRef]
34. Malkanthi, E.; Judy, W.V.; Wilson, D.; Rumberger, J.A.; Guthrie, N. Randomized, double-blind, placebo-controlled, clinical study on the effect of Diabetinol® on glycemic control of subjects with impaired fasting glucose. *Diabetes Metab. Syndr. Obes.* **2015**, *8*, 275–286. [CrossRef]
35. Lee, Y.S.; Cha, B.Y.; Choi, S.S.; Choi, B.K.; Yonezawa, T.; Teruya, T.; Nagai, K.; Woo, J.T. Nobiletin improves obesity and insulin resistance in high-fat diet-induced obese mice. *J. Nutr. Biochem.* **2013**, *24*, 156–162. [CrossRef]
36. Pan, M.H.; Yang, G.; Li, S.; Li, M.Y.; Tsai, M.L.; Wu, J.C.; Badmaev, V.; Ho, C.T.; Lai, C.S. Combination of citrus polymethoxyflavones, green tea polyphenols and Lychee extracts suppresses obesity and hepatic steatosis in high-fat diet induced obese mice. *Mol. Nutr. Food Res.* **2017**, *61*, 1601104. [CrossRef]
37. Sundarama, R.; Shanthi, P.; Sachdanandam, P. Effect of tangeretin, a polymethoxylated flavone on glucose metabolism in streptozotocin-induced diabetic rats. *Phytomedicine* **2014**, *21*, 793–799. [CrossRef] [PubMed]
38. Ghaben, A.L.; Scherer, P.E. Adipogenesis and metabolic health. *Nat. Rev. Mol. Cell Biol.* **2019**, *20*, 242–258. [CrossRef]
39. Yang, J.; Eliasson, B.; Smith, U.; Cushman, S.W.; Sherman, A. The size of large adipose cells is a predictor of insulin resistance in first-degree relatives of type 2 diabetics. *Obesity* **2012**, *20*, 932–938. [CrossRef]
40. Skurk, T.; Alberti-Huber, C.; Herder, C.; Hauner, H. Relationship between adipocyte size and adipokine expression and secretion. *J. Clin. Endocrinol. Metab.* **2007**, *92*, 1023–1033. [CrossRef] [PubMed]
41. Meyer, L.K.; Ciaraldi, T.P.; Henry, R.R.; Wittgrove, A.C.; Phillips, S.A. Adipose tissue depot and cell size dependency of adiponectin synthesis and secretion in human obesity. *Adipocyte* **2013**, *2*, 217–226. [CrossRef] [PubMed]
42. Daval, M.; Foufelle, F.; Ferré, P. Functions of AMP-activated protein kinase in adipose tissue. *J. Physiol.* **2006**, *574*, 55–62. [CrossRef]
43. Steinberg, G.R.; Kemp, B.E. AMPK in health and disease. *Physiol. Rev.* **2009**, *89*, 1025–1078. [CrossRef]
44. Greenberg, A.S.; Shen, W.J.; Muliro, K.; Patel, S.; Souza, S.C.; Roth, R.A.; Kraemer, F.B. Stimulation of lipolysis and hormone-sensitive lipase via the extracellular signal-regulated kinase pathway. *J. Biol. Chem.* **2001**, *276*, 45456–45461. [CrossRef]
45. Zechner, R.; Zimmermann, R.; Eichmann, T.O.; Kohlwein, S.D.; Haemmerle, G.; Lass, A.; Madeo, F. Fat signal-lipases and lipolysis in lipid metabolism and signaling. *Cell Metab.* **2012**, *15*, 279–291. [CrossRef] [PubMed]
46. Smith, B.K.; Marcinko, K.; Desjardins, E.M.; Lally, J.S.; Ford, R. Treatment of nonalcoholic fatty liver disease: Role of AMPK. *J. Am. J. Physiol. Endocrnol. Metabol.* **2016**, *311*, E730–E740. [CrossRef] [PubMed]

47. Fullerton, M.D.; Galic, S.; Marcinko, K.; Sikkema, S.; Pulinilkunnil, T.; Chen, Z.P.; O'Neill, H.M.; Ford, R.J.; Palanivel, R.; O'Brien, M.; et al. Single phosphorylation sites in Acc1 and Acc2 regulate lipid homeostasis and the insulin-sensitizing effects of metformin. *Nat. Med.* **2013**, *19*, 1649–1654. [CrossRef] [PubMed]
48. Steinberg, G.R.; Michell, B.J.; van Denderen, B.J.; Watt, M.J.; Carey, A.L.; Fam, B.C.; Andrikopoulos, S.; Proietto, J.; Gorgun, C.Z.; Carling, D.; et al. Tumor necrosis factor alpha-induced skeletal muscle insulin resistance involves suppression of AMP-kinase signaling. *Cell Metab.* **2006**, *4*, 465–474. [CrossRef]
49. Salmine, A.; Hyttinen, J.M.; Kaarniranat, K. AMP-activated protein kinase inhibit NF-kB signalling and inflammation: Impact on healthspan and lifespan. *J. Mol. Med.* **2011**, *89*, 667–676. [CrossRef]
50. Saponaro, C.; Gaggini, M.; Carli, F.; Gastaldelli, A. The subtle balance between lipolysis and lipogenesis: A critical point in metabolic homeostasis. *Nutrients* **2015**, *7*, 9453–9474. [CrossRef]
51. Carmen, G.Y.; Victor, S.M. Signalling mechanisms regulating lipolysis. *Cell Signal.* **2006**, *18*, 401–408. [CrossRef]
52. Benetti, E.; SA Patel, N.; Collino, M. The role of PPARβ/δ in the management of metabolic syndrome and its associated cardiovascular complications. *Endocr. Metab. Immune Disord. Drug Targets* **2011**, *11*, 273–284. [CrossRef]
53. Wilson, C.G.; Tran, J.L.; Erion, D.M.; Vera, N.B.; Febbraio, M.; Weiss, E.J. Hepatocyte-Specific Disruption of CD36 Attenuates Fatty Liver and Improves Insulin Sensitivity in HFD-Fed Mice. *Endocrinology* **2016**, *157*, 70–585. [CrossRef] [PubMed]
54. Mun, J.; Kim, S.; Yoon, H.G.; You, Y.; Kim, O.K.; Choi, K.C.; Lee, Y.H.; Lee, J.; Park, J.; Jun, W. Water Extract of *Curcuma longa* L. Ameliorates Non-Alcoholic Fatty Liver Disease. *Nutrients* **2019**, *11*, 2536. [CrossRef]
55. Luiken, J.J.F.P.; Bonen, A.; Glatz, J.F.C. Cellular fatty acid uptake is acutely regulated by membrane-associated fatty acid-binding proteins. *Prostaglandins Leukot. Essent. Fat. Acids* **2002**, *67*, 73–78. [CrossRef] [PubMed]
56. Alves-Bezerra, M.; Cohen, D.E. Triglyceride metabolism in the liver. *Compr. Physiol.* **2017**, *8*, 1–8. [CrossRef] [PubMed]
57. Furuhashi, M.; Hotamisligil, G.S. Fatty acid-binding proteins: Role in metabolic diseases and potential as drug targets. *Nat. Rev. Drug Discov.* **2008**, *7*, 489–503. [CrossRef]
58. Khristi, V.; Ratri, A.; Ghosh, S.; Pathak, D.; Borosha, S.; Dai, E.; Roy, R.; Charkravarthi, V.P.; Wolfe, M.W.; Rumi, M.K. Disruption of ESR1 alters the expression of genes regulating hepatic lipid and carbohydrate metabolism in male rats. *Mol. Cell. Endocrinol.* **2019**, *490*, 47–56. [CrossRef] [PubMed]

nutrients

MDPI

Article

Olfactory Stimulation by Fennel (*Foeniculum vulgare* Mill.) Essential Oil Improves Lipid Metabolism and Metabolic Disorders in High Fat-Induced Obese Rats

Seong Jun Hong [1], Sojeong Yoon [1], Seong Min Jo [1], Hyangyeon Jeong [1], Moon Yeon Youn [1], Young Jun Kim [2], Jae Kyeom Kim [3] and Eui-Cheol Shin [1,4,*]

1 Department of Food Science, Gyeongsang National University, Jinju 52725, Korea; 01028287383a@gmail.com (S.J.H.); dbsthwjd0126@naver.com (S.Y.); jojo9875@naver.com (S.M.J.); giddus9967@naver.com (H.J.); ringspot@naver.com (M.Y.Y.)
2 Department of Food and Biotechnology, Korea University, Sejong 30019, Korea; yk46@korea.ac.kr
3 Department of Behavioral Health and Nutrition, University of Delaware, Newark, DE 19716, USA; jkkim@udel.edu
4 Department of GreenBio Science, Gyeongsang National University, Jinju 52725, Korea
* Correspondence: eshin@gnu.ac.kr; Tel.: +82-55-772-3271; Fax: +82-55-772-3279

Citation: Hong, S.J.; Yoon, S.; Jo, S.M.; Jeong, H.; Youn, M.Y.; Kim, Y.J.; Kim, J.K.; Shin, E.-C. Olfactory Stimulation by Fennel (*Foeniculum vulgare* Mill.) Essential Oil Improves Lipid Metabolism and Metabolic Disorders in High Fat-Induced Obese Rats. *Nutrients* **2022**, *14*, 741. https://doi.org/10.3390/nu14040741

Academic Editor: George Moschonis

Received: 10 January 2022
Accepted: 7 February 2022
Published: 10 February 2022

Abstract: In this study, odor components were analyzed using gas chromatography/mass spectrometry (GC/MS) and solid-phase microextraction (SPME), and odor-active compounds (OACs) were identified using GC-olfactometry (GC-O). Among the volatile compounds identified through GC-O, *p*-anisaldehyde, limonene, estragole, anethole, and trans-anethole elicit the fennel odor. In particular, *trans*-anethole showed the highest odor intensity and content. Changes in body weight during the experimental period showed decreasing values of fennel essential oil (FEO)-inhaled groups, with both body fat and visceral fat showing decreased levels. An improvement in the body's lipid metabolism was observed, as indicated by the increased levels of cholesterol and triglycerides and decreased levels of insulin in the FEO-inhaled groups compared to group H. Furthermore, the reduction in systolic blood pressure and pulse through the inhalation of FEO was confirmed. Our results indicated that FEO inhalation affected certain lipid metabolisms and cardiovascular health, which are obesity-related dysfunction indicators. Accordingly, this study can provide basic research data for further research as to protective applications of FEO, as well as their volatile profiles.

Keywords: *Foeniculum vulgare* Mill.; fennel essential oil; metabolic health; obesity

1. Introduction

Obesity is a non-communicable disease that can induce complications, and has been recognized as a serious health problem worldwide [1]. Obesity, defined as the state of accumulation of energy and fat in the body, is caused by an increase in energy intake and a decrease in energy expenditure. This status is defined as abnormal fat accumulation [2,3]. Increased body fat can lead to leptin resistance, a status in which the level of leptin increases, while there is no variation in food intake. In addition, this status can increase the prevalence of type 1 and type 2 diabetes, and it can lead to hypertension due to an increase in blood pressure and heart rate beyond the normal status [1,3,4]. Furthermore, excessive accumulation of body fat may induce metabolic syndrome via an increased prevalence of cardiovascular diseases, which may be caused by increased levels of total cholesterol (TC), low-density lipoprotein cholesterol (LDL-cholesterol), triglyceride (TG), and decreased high-density lipoprotein cholesterol (HDL-cholesterol) [1,3]. Many researchers have reported various functional materials and substances that have been investigated for decreasing body weight and increasing energy expenditure [1,2,5,6].

A recent study reported that the development of natural functional materials has been of increasing interest to undermine the side-effects of obesity [2]. Functional foods, which are made of natural functional materials, have been reported to reduce obesity and metabolic syndrome. In addition, functional food hardly ever has side effects, unlike synthetic medicine for obesity [2,6]. Many studies on the physiological activity of functional materials and medicinal herbs have been reported, including a study on the suppression of white adipose tissue accumulation using pine bark extracts [5], anti-obesity via inhalation of citronella essential oil [7], and anti-obesity and metabolic health via inhalation of patchouli essential oil (PEO) [1].

Fennel (*Foeniculum vulgare* Mill.) originated in southern Europe and the Mediterranean, and is now a herbaceous and perennial plant mainly cultivated in China. In addition, all of its parts can be used, such as its bulbous roots, stem, and seeds. Owing to its strong scent, fennel is used in flavoring agents, medicines, cosmetics, and in the food industry [8]. Recent studies have reported that fennel and fennel essential oil (FEO) may play a role in the control of the central nervous system (CNS) and autonomic nervous system (ANS) in the body [1,8,9]. Furthermore, fennel has been demonstrated as a medicinal herb with pharmacological and physiological activities, such as suppression of food intake, body weight, white adipose tissue, blood glucose, and cardiovascular diseases [8–11].

Volatile compounds in FEO can be influenced by agro-environmental conditions (climatic conditions). Thus, FEO, which was cultivated from five different agro-environmental conditions in India, represented that the concentrations of limonene in FEO cultivated from Moradabad (latitude: 28°3′ N; longitude: 78°77′ E), Tanakpur (latitude: 29°07′ N; longitude: 80°10′ E), and Haldwani (latitude: 29°13′N; longitude: 79°31′ E) detected more than 30% of FEO; however, the concentrations of limonene in FEO cultivated from Pithoragarh (latitude: 30°08′ N; longitude: 80°36′ E) and Didihat (latitude: 29°80′ N; longitude: 80°24′ E) detected less than 10% of FEO. Additionally, the concentrations of *trans*-anethole in FEO cultivated from Moradabad and Haldwani detected less than 30% of FEO, while concentrations of *trans*-anethole in FEO cultivated from Tanakpur, Pithoragarh, and Didihat detected more than 40% of FEO [12]. Furthermore, volatile profiles of fennel in Spain have been identified where the concentrations of volatile compounds were varied according to the agor-environmental conditions [13].

Although several previous studies of fennel intake have been reported, research investigating the mechanism and observation of metabolic health in vivo after FEO inhalation is insufficient. In the present study, we investigated the changes in metabolic aspects, such as body weight, food intake, hormones, blood pressure, heart rate, and blood glucose in vivo in Sprague Dawley rats by controlling the FEO concentration and inhalation period.

2. Materials and Methods

2.1. Essential Oil

Fennel essential oil (FEO) was purchased from Aroma Care Solution (Helga Stolz GmbH Co., Grafenwoerth, Austria) and stored at 4 °C before the experiment to minimize any biological activity and chemical composition of the sample.

2.2. Volatile Compounds and Odor Description

Volatile compounds of FEO were analyzed by headspace analysis using polydimethylsiloxane (PDMS) coated with 100 µm. One gram of sample was placed in a collection bottle, sealed with an aluminum cap, and exposed to the headspace of the heated sample at 60 °C. Volatiles, which are absorbed by the SPME fiber, were analyzed using gas chromatography-mass spectrometry (GC-MS; Agilent 7890A & 5975C, Agilent Technologies, Santa Clara, CA, USA) and a HP-5MS column (30 m × 0.25 mm i.d. × 0.25 µm film thickness). The analysis was done in an oven that was first kept at a temperature of 40 °C for 5 min, and then heated to 200 °C at a speed of 5 °C. The injector temperature was 220 °C; the flow rate of the carrier gas helium was 1.0 mL/min; and the split ratio was splitless. Each compound separated from the total chromatogram (TIC) was sympathetic to the mass spectrum library (NIST 12) and literature, and each concentration of volatile compound was calculated by converting the peak area of the internal standard material (pentadecane) into a peak area of g/100 g. During volatile analysis, the odor description of each compound was measured using GC-MS, and the four levels of odor intensity and the duration of scent were measured using GC-olfactometry (GC-O) [1].

2.3. Animal Care and Experimental Design

This study was conducted with the approval of Document #: IACUC-4 of Gyeongnam National University of Science and Technology in accordance with the Animal Protection Act. In the present study, a total of 24 experimental animals, which were four-week-old male Sprague Dawley rats, were purchased from Coretec Co. (Busan, Korea). After a week of preliminary adaptation, the experimental animals were randomly separated into four groups and bred for 12 weeks. In detail, animals were randomly separated into groups: the control group was fed a normal diet (N; $n = 6$), and the control group was fed a 45% high-fat diet (H; $n = 6$). Both groups n and H were given 30 min/day inhalation of distilled water (DW) at the same time per day for 12 weeks. The low-dose inhaled FEO group (H-LFI; $n = 6$) with a high-fat diet (HFD) was given 30 min/day inhalation of 0.3% FEO for 12 weeks, and the high-dose inhaled FEO group (H-HFI; $n = 6$) with HFD was given 30 min/day inhalation of 1% FEO for 12 weeks. The body weights, food intake, and length of experimental animals were measured at a fixed time each week. After fasting for 16 h before dissection, blood was collected using a syringe containing 20 mg ethylenediaminetetraacetic acid. Subsequently, the aorta was cut, and the rats were sacrificed. The collected blood samples were stored on ice for 30 min and then centrifuged at 1008 G to separate the plasma. Next, the brain, heart, liver, kidney, white adipose tissue (WAT), brown adipose tissue (BAT), lung, adrenal glands, spleen, testis, and epididymis were extracted, and organs and plasma were stored at −80 °C [1].

2.4. Body Composition Assessment Using Dual-Energy X-ray Absorptiometry (DXA)

The body composition of rats was measured using InAlzer (Medikors Co., Seongnam, Korea) at the 12th week after fasting for 16 h. Before analyzing body composition, the rats were anesthetized with isoflurane using XGI-8 (Caliper Life Science Co., Hopkinton, MA, USA). Subsequently, body composition was measured for lean body mass, body fat mass, body fat percentage, and bone mineral density (BMD), and the analysis was conducted without dissection processing.

2.5. Plasma Biomarker Analysis

Plasma concentrations of TG, TC, HDL-cholesterol, aspartate transaminase (AST), and alanine transaminase (ALT) were analyzed using commercial kits (Asan Pharm Co., Seoul, Korea). Plasma insulin and leptin levels were analyzed using an ELISA kit (Bertin Technologies Co., Montigny le Bretonneux, France). Plasma cortisol and testosterone levels were analyzed using a hormone ELISA kit (Mybio Source Inc., San Diego, CA, USA).

2.6. Blood Glucose Assessment

The blood glucose levels in rats were analyzed after fasting for 16 h using a glucometer (Accu-Chek, Roche Diagnostic, Basel, Switzerland) and were measured at the 1st, 7th, and 12th weeks, respectively. For blood glucose analysis, blood was collected from the tail vein of rats. The experiments were performed in duplicate for each animal, and three measurements with less deviation, excluding the highest and lowest values, were calculated and recorded as the mean and standard deviation in each group [14].

2.7. Blood Pressure Assessment

Changes in the blood pressure and pulse of experimental animals were analyzed using an animal blood pressure meter (BP-2000, Visitech Systems Co., Apex, NC, USA), and the measurements were made on the tail of experimental animals at a fixed time in the 1st and 12th weeks. Blood pressure and pulse were measured 30 times per rat, including 10 preliminary tests and 20 main tests, and six measured values with small deviations, excluding the highest and lowest values, were calculated and recorded as the mean and standard deviation [1].

2.8. Statistical Analysis

In this study, the experiments were conducted in duplicate or triplicate, from which the standard deviations (SD) and mean were identified and used to present the results. Each group was non-parametrically compared using the Friedman test with the chi-square distribution, following Dunn's post-test. *P*-values less than 0.05 were considered statistically significant (SAS Institute Inc., Cary, NC, USA).

3. Results

3.1. Volatile Compounds

The results of volatile compounds in FEO using GC/MS and GC-O are shown in Table 1 and Figure 1. A total of seven odor-active compounds (OACs), including one aldehyde, five hydrocarbons, and one ketone, were detected. Among the volatiles detected by GC/MS, p-anisaldehyde, limonene, estragole, cis-anethole, and trans-anethole elicit fennel-like odors, which were detected by GC-O, whereas a-pinene and fenchone elicit herb and fragrance-like odors, which were detected by GC-O. Trans-anethole and fenchone had the highest and second-highest concentrations among FEO volatiles, respectively, and both trans-anethole and cis-anethole showed the highest odor intensity [2].

Table 1. Odor active compounds (OACs) of fennel essential oil by gas chromatography/mass spectrometry (GC/MS) and GC-olfactometry (GC-O).

Major Compound	RT [a] (min)	RI [b]	Mean ± SD (μg/100 g)	Mean ± SD (%)	Odor Intensity	Odor Description	I.D. [c]
Aldehyde (1)							
p-Anisaldehyde	22.57	1276	304.17 ± 38.79	1.18 ± 0.15	1	Fennel	MS [d]/RI
Hydrocarbons (5)							
α-Pinene	13.03	961	1635.87 ± 139.95	6.33 ± 0.54	1	Herb	MS/RI
Limonene	16.11	1056	733.77 ± 18.60	2.84 ± 0.07	1	Fennel	MS/RI
Estragole	21.16	1224	1556.21 ± 40.28	6.02 ± 0.15	1	Fennel	MS/RI
cis-Anethole	22.55	1275	382.86 ± 35.95	1.48 ± 0.16	2	Fennel	MS
trans-Anethole	23.69	1317	13,100.47 ± 3971.88	50.69 ± 15.37	2	Fennel	MS
Ketone (1)							
Fenchone	18.23	1123	4956.46 ± 117.31	19.18 ± 4.32	1	Herb, fragrance	MS

Data are given as mean ± SD values from experiments performed in duplicate. [a] RT: retention time; [b] RI: retention index; [c] I.D.: identification; [d] MS: mass spectrum.

Among fennel volatiles, trans-anethole, fenchone, and estragole, which are described as major volatiles of fennel, are known to have higher concentrations and odor activity than the other fennel volatiles, as detected by GC-O [13]. Several studies have reported trans-anethole and isomers cis-anethole and estragole as major fennel volatiles [13,15,16]. A previous study reported that volatiles in FEO were extracted from hydrodistillation, supercritical fluid extraction, and microwave extraction methods. Anethole had the most abundant content of FEO (<50%), and fenchone had the second most abundant content (approximately 20%) [17]. On the other hand, another study reported that volatiles in FEO, which were extracted from ultrasound-assisted extraction, showed the concentration of anethole more than 80% among all volatiles, and the concentration of fenchone showed to be less than 10% among all volatiles [18]. Trans-anethole, cis-anethole, and estragole elicit fennel odors and have anti-obesity, anti-cholesterol, anti-inflammatory, and anti-stress effects [15]. Fenchone has antifungal, antidepressant, local anesthesia, and wound healing effects [16]. α-Pinene, a monoterpene compound, is found in several medicinal herbs and exerts positive effects by suppressing tumors, oxidation, bacteria, inflammation, stress, convulsion, and acute pain [19]. Limonene, another major monoterpene compound, is known for its anti-tumor and anti-inflammatory effects and exerts positive effects by reducing hepatotoxicity through reduced AST and ALT levels [16]. A previous study reported that limonene is a major volatile component of grapefruit oil and lowers blood glucose levels via inhalation [20]. The above studies on the pharmacological effects of volatiles show that FEO may exert positive physiological effects in vivo via inhalation.

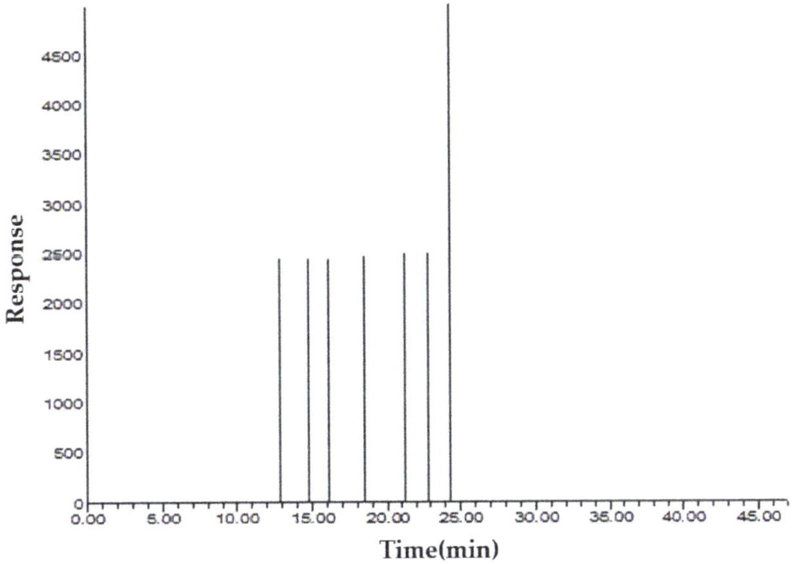

Figure 1. Representative aroma gram of odor active compounds (OACs) in fennel (*Foeniculum vulgare* Mill.) essential oil by GC-O test.

3.2. Food Intake, Body Weight, Body Composition, and Length

Changes in food intake, average daily food intake, and average daily calorie intake were measured during the experimental period (12 weeks), and the results are presented in Table 2 and Figure 2a. During FEO inhalation, the H-LFI and H-HFI groups showed no appearance of behavioral abnormalities. The food intake of group N, which was fed a normal diet, was higher than that of the other groups ($p < 0.05$). The food intake of groups H and H-LFI showed no significant differences among the high-fat diet (HFD)-fed groups. However, the H-HFI group showed decreased weeks (fourth and seventh weeks) of food intake compared to groups H and H-LFI ($p < 0.05$). In contrast, H-HFI showed no significant differences among the HFD-fed groups. Unlike food intake, the calorie intake of group N was the lowest compared to the other groups ($p < 0.05$), with no significant differences among all HFD-fed groups.

Changes in body weight were measured every week, and the results are shown in Figure 2b. Body weight and weight gain in group N showed the lowest variation compared to the other groups during the experimental period (12 weeks). In the 11th week, group H-LFI had significantly lower body weight than group H, and the H-HFI group had significantly lower body weight than group H in the 3rd, 4th, 11th, and 12th weeks ($p < 0.05$). There were no significant differences in average weight gain among the HFD-fed groups from the 1st week to the 6th week, as well as from the 7th week to the 12th week. On the other hand, the FEO-inhaled groups (H-LFI and H-HFI) had significantly lower average weight gain compared to group H from the 1st week to the 12th week ($p < 0.05$). In particular, the average weight gain of group H-HFI did not differ significantly from that of group N from the 1st week to the 12th week ($p > 0.05$). The food efficiency ratio (FER) of group N was the highest compared to the other groups ($p < 0.05$), and the FER of the FEO-inhaled groups did not differ significantly from that of group H.

Body composition was measured using the DXA system, and the results are shown in Figure 2f. The ratio of fat-free mass did not differ significantly among all groups; however, the FEO-inhaled groups showed a relatively increased ratio compared to group H. Unlike fat-free mass, the ratio of body fat increased in group H compared to group N ($p < 0.05$); the H-LFI and H-HFI groups showed a lower body fat ratio compared to group H ($p < 0.05$). Regardless of FEO inhalation and types of food intake, bone mineral density (BMD), bone area, and bone volume did not differ significantly among the groups.

Changes in the length of experimental animals were analyzed and are presented in Table 2. All animal groups showed an increase in length during the breeding period (12 weeks). In the first week, the length of the animals in group N were the highest compared to those in the other groups ($p < 0.05$); however, the FEO-inhaled groups (H-LFI and H-HFI) were significantly longer than those in groups N and H in the 12th week ($p < 0.05$). In addition, length gain increased in the FEO-inhaled groups compared to the DW-inhaled groups ($p < 0.05$). The body mass index (BMI) was higher in group H than in group N ($p < 0.05$). H-LFI and H-HFI were lower than H ($p < 0.05$). In particular, the FEO-inhaled groups did not differ significantly from group N ($p < 0.05$).

Fennel regulates food intake and causes decreased body weight by reducing food intake [21,22]. In the present study, the inhalation of high-dose FEO by group H-HFI led to the identification of the week of decreased trends of food intake; however, the daily average food intake did not decrease for 12 weeks (Table 2 and Figure 2). Bae et al. reported that the intake of fennel decreased appetite in overweight female subjects [9]. In addition, Heo et al. reported that the FEO-inhaled group had a relatively decreased food intake in rats, but with no significant difference from the control group [22]. According to previous studies, fennel inhalation may have a lower regulatory effect on food intake than on fennel intake. Weight gain is associated with changes in food intake [21]. In the present study, increased body weight in HFD-fed groups may have been associated with calorie intake (Table 2). Changes in body weight were hardly influenced by changes in food intake; however, FEO inhalation influenced the reduction in body fat mass and body weight. A previous study reported that body weight was lower in the FEO-inhaled group than in the control group ($p < 0.05$), which was similar to our results [22]. Moreover, a previous study reported that the intake of fennel affected decreases in body weight compared to the non-intake of fennel [11]. In addition, another study reported that trans-anethole, a major compound of fennel, increased the activation of AMP-activated protein kinase (AMPK) and peroxisome proliferator-activated receptor α (PPARα) in adipocytes, thereby decreasing body fat accumulation and increasing body weight [23]. In this study, reduction of body weight was associated with body fat, and the anti-obesity effect may be influenced by inhalation of FEO.

The BMI is an index of length and body weight [21]. In this study, a reduction in BMI was related to a reduction in body weight and increase in length. A previous study identified that the growth of length was influenced by the inhalation and concentration of PEO [1]; therefore, the inhalation of FEO may also be one of the various factors related to the growth of length.

Table 2. Effects of fennel essential oil inhalation on growth parameters, organ weights, and plasma biomarkers in male rats fed a normal diet and high-fat diet (HFD).

Parameters	N	H	H-LFI	H-HFI
Growth parameters				
Food intake (g/day)	18.03 ± 0.70 ^{a1}	15.09 ± 0.64 ^{b}	14.92 ± 0.37 ^{b}	15.21 ± 0.78 ^{b}
Initial length (cm)	17.3 ± 0.3 ^{a}	16.7 ± 0.1 ^{b}	16.7 ± 0.1 ^{b}	16.5 ± 0.2 ^{b}
Final length (cm)	23.4 ± 0.3 ^{b}	23.4 ± 0.1 ^{b}	24.1 ± 0.1 ^{a}	23.9 ± 0.1 ^{a}
Length gain (cm)	6.1 ± 0.2 ^{c}	6.6 ± 0.1 ^{b}	7.4 ± 0.1 ^{a}	7.4 ± 0.2 ^{a}
Energy intake (kcal/day)	50.48 ± 1.93 ^{b}	69.40 ± 2.94 ^{a}	68.64 ± 1.68 ^{a}	69.95 ± 3.58 ^{a}
BMI	7.71 ± 0.45 ^{b}	9.17 ± 0.30 ^{a}	8.13 ± 0.13 ^{b}	7.92 ± 0.15 ^{b}
FER (%)	17.65 ± 1.29 ^{b}	26.93 ± 1.83 ^{a}	25.45 ± 1.48 ^{a}	23.27 ± 1.49 ^{a}
Organ weights				
Brain (g/kg)	3.91 ± 0.17 ^{a}	3.57 ± 0.25 ^{a}	3.94 ± 0.06 ^{a}	3.98 ± 0.39 ^{a}
Liver (g/kg)	24.84 ± 0.29 ^{a}	24.81 ± 5.46 ^{a}	24.44 ± 0.64 ^{a}	25.41 ± 2.28 ^{a}
Kidney (g/kg)	5.71 ± 0.18 ^{a}	4.46 ± 0.11 ^{b}	5.46 ± 0.14 ^{ab}	5.36 ± 0.72 ^{ab}
Heart (g/kg)	3.02 ± 0.05 ^{a}	2.34 ± 0.12 ^{b}	3.01 ± 0.10 ^{a}	2.89 ± 0.47 ^{ab}
WAT (g/kg)	26.76 ± 1.71 ^{b}	41.37 ± 1.64 ^{a}	32.22 ± 4.51 ^{b}	29.30 ± 1.06 ^{b}
BAT (g/kg)	0.53 ± 0.04 ^{b}	0.62 ± 0.07 ^{b}	1.04 ± 0.06 ^{a}	0.55 ± 0.08 ^{b}
Lung (g/kg)	3.45 ± 0.11 ^{ab}	3.10 ± 0.33 ^{b}	3.87 ± 0.06 ^{a}	3.89 ± 0.32 ^{a}
Adrenal galnds(g/kg)	0.12 ± 0.01 ^{b}	0.12 ± 0.01 ^{b}	0.13 ± 0.01 ^{b}	0.15 ± 0.01 ^{b}
Spleen (g/kg)	1.67 ± 0.01 ^{a}	1.34 ± 0.03 ^{b}	1.47 ± 0.19 ^{ab}	1.75 ± 0.14 ^{a}
Testicular (g/kg)	9.81 ± 0.57 ^{a}	7.52 ± 0.45 ^{b}	9.30 ± 0.17 ^{a}	9.53 ± 0.48 ^{a}
Epididymis (g/kg)	2.87 ± 0.12 ^{a}	2.59 ± 0.20 ^{a}	2.89 ± 0.07 ^{a}	2.70 ± 0.26 ^{a}
Plasma biomarkers				
TC (mg/dL)	128.91 ± 3.63 ^{b}	131.45 ± 2.07 ^{b}	124.07 ± 4.05 ^{b}	173.44 ± 1.77 ^{a}
HDL (mg/dL)	35.50 ± 0.79 ^{d}	42.51 ± 0.44 ^{c}	49.90 ± 1.13 ^{b}	52.82 ± 0.25 ^{a}
LDL (mg/dL)	46.72 ± 1.78 ^{c}	75.62 ± 1.59 ^{b}	55.95 ± 5.41 ^{c}	109.85 ± 5.63 ^{a}
AI (mg/dL)	2.61 ± 0.05 ^{a}	2.10 ± 0.10 ^{c}	1.44 ± 0.04 ^{d}	2.27 ± 0.04 ^{b}
CRF (mg/dL)	3.61 ± 0.05 ^{a}	3.17 ± 0.10 ^{b}	2.44 ± 0.04 ^{c}	3.27 ± 0.04 ^{b}
LHR (mg/dL)	131.70 ± 7.94 ^{c}	177.92 ± 5.61 ^{b}	112.05 ± 8.29 ^{c}	207.93 ± 9.69 ^{a}
TG (mg/dL)	94.20 ± 3.03 ^{b}	102.52 ± 3.14 ^{a}	88.75 ± 2.02 ^{b}	111.91 ± 6.61 ^{a}
Cortisol (ng/dL)	4.29 ± 0.54 ^{a}	4.38 ± 0.37 ^{a}	4.50 ± 0.05 ^{a}	4.07 ± 0.07 ^{a}
Insulin (ng/mL)	0.42 ± 0.17 ^{d}	4.13 ± 0.14 ^{a}	1.79 ± 0.13 ^{c}	2.83 ± 0.04 ^{b}
HOMA-IR	2.19 ± 1.10 ^{d}	22.57 ± 2.78 ^{a}	8.54 ± 1.25 ^{c}	14.05 ± 0.39 ^{b}
Leptin (pg/mL)	2842.79 ± 173.45 ^{a}	3926.00 ± 225.35 ^{a}	3863.55 ± 790.87 ^{a}	2711.10 ± 936.96 ^{a}
Testosterone (pg/mL)	0.43 ± 0.11 ^{b}	0.29 ± 0.07 ^{b}	1.38 ± 0.19 ^{a}	0.31 ± 0.01 ^{b}
ALT (Karmen/mL)	5.16 ± 2.71 ^{a}	8.29 ± 3.73 ^{a}	10.25 ± 8.53 ^{a}	3.05 ± 0.79 ^{a}
AST (Karmen/mL)	40.00 ± 24.28 ^{a}	70.63 ± 1.36 ^{aA2}	54.08 ± 7.54 ^{aB}	56.46 ± 1.36 ^{aB}

Data are given as mean ± SD values from experiments performed in triplicate. [1] Means with different small letters (a–d) correspond to the significant differences determined among all groups using the non-parametric Friedman test followed by Dunn's test ($p < 0.05$). [2] Means with different capital letters (A and B) correspond to the significant differences determined among all HFD groups through the non-parametric Friedman test followed by Dunn's test ($p < 0.05$). N: normal diet-induced control, H: high-fat diet-induced control, H-LFI: high-fat diet-induced and 0.3% FEO-inhaled rats, H-HF: high-fat diet-induced and 1% FEO-inhaled rats. BMI (Body mass index) = Body weight/Length^2. FER (Food efficiency ratio) = Body weight gain/food intake × 100. AI (Atherogenic index) = (TC-HDL/HDL). CRF (Cardiac risk factor) = (TC/HDL). LHR = LDL/HDL × 100. HOMA-IR (Insulin resistance) = Insulin × Blood glucose/405.

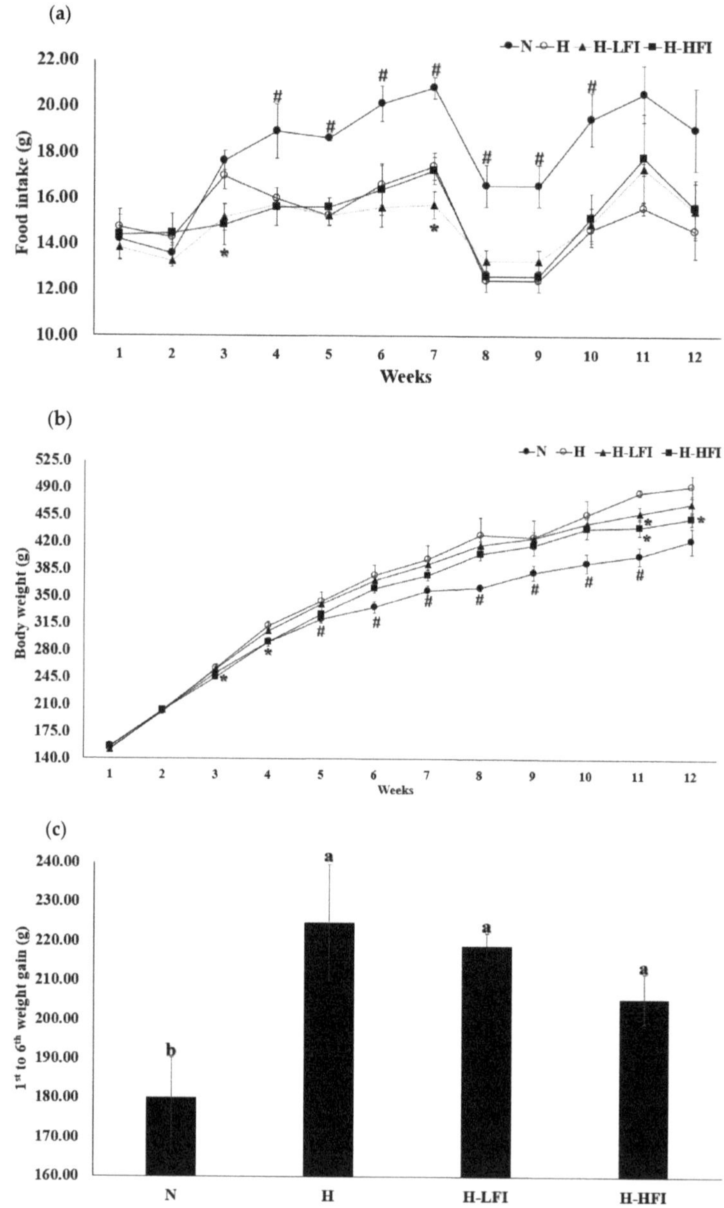

Figure 2. *Cont.*

(d)

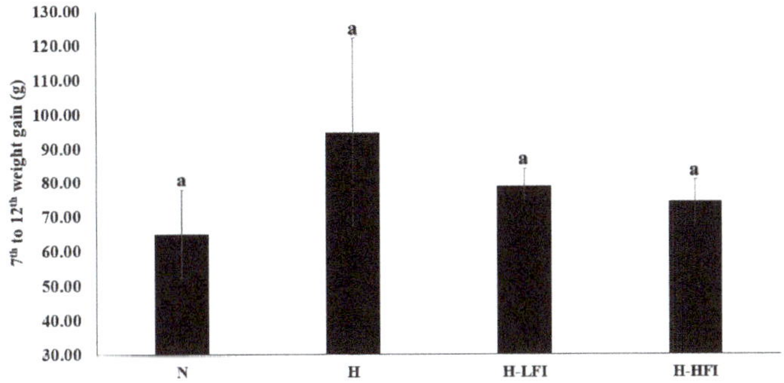

(e)

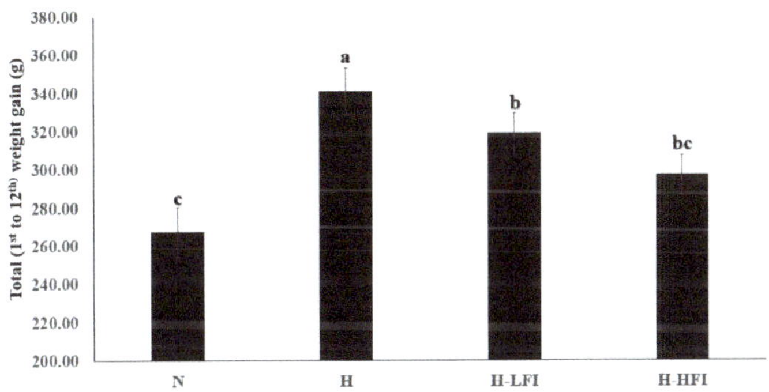

(f)

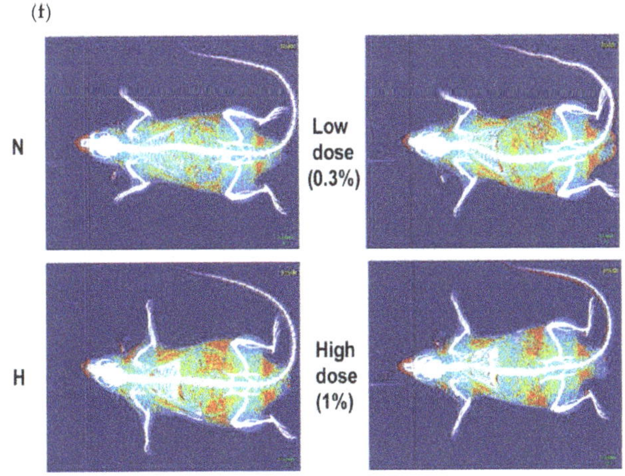

Figure 2. Measurement of blood, food intake, weight, and body composition in rats. (**a**) Change of food intake. (**b**) Change of body weight. (**c**) Comparisons of body weight gain during the period from the first to the seventh week. (**d**) Comparisons of body weight gain during the period from the 7th to the 12th week. (**e**) Comparisons of body weight gain during the period from the 1st to the 12th week. (**f**) Analysis of body composition on rats using the dual-energy X-ray absorptiometry (DXA) system (White—Bone/Blue—Lean/Red—Fat). The data are presented as the means ± SD values, # $p < 0.05$, comparison among all groups. * $p < 0.05$, compared with the high-fat diet (HFD) group. Means with different letters (a–c) corresponds to the significant differences determined through the non parametric Friedman test followed by Dunn's test ($p < 0.05$).

3.3. Organ Weight

The weights of 11 organs were measured after dissection in the experimental animals (Table 2). The brain weights ranged from 3.57 ± 0.25 to 3.98 ± 0.39 g/kg, while the liver weights ranged 24.44 ± 0.64 to 25.41 ± 2.28 g/kg. The heart weights ranged from 4.46 ± 0.11 to 5.71 ± 0.18 g/kg, while the kidney weights ranged from 2.34 ± 0.12 to 3.02 ± 0.05 g/kg. In our results, the organ weights showed no significant intergroup differences. The white adipose tissue (WAT) weight in group H was significantly higher than that in group N ($p < 0.05$), and the FEO-inhaled groups (H-LFI and H-HFI) had a significantly higher weight of WAT than group H ($p < 0.05$). Regardless of the type of food intake, the weights of brown adipose tissue (BAT) were not significantly different between groups N and H, while group LFI significantly increased compared to group H ($p < 0.05$). Lung weights were not significantly different between groups N and H, while the FEO-inhaled groups were significantly higher than those in group H ($p < 0.05$). Adrenal gland weights were not significantly different among the groups. In addition, the weights of the spleen and epididymis showed no significant differences between the normal diet-fed and HFD-fed groups. On the other hand, the testicular weights in group H dramatically decreased compared to group N ($p < 0.05$); however, the FEO-inhaled groups dramatically increased compared to group H ($p < 0.05$).

WAT is a representative index of visceral fat, and thus, its accumulation may cause obesity [3]. Accordingly, FEO-inhaled groups may have decreased body fat through the reduction of WAT (Figure 2f and Table 2). A previous study observed that trans-anethole, which accounted for the concentration of fennel, increased the activation of PPARα, PPARγ, and AMPK, and decreased pAMPK and WAT weight in animal models [23]. In addition, our previous study reported that the reduction of WAT is influenced by the inhalation of essential oil [1], and an earlier study identified that volatile compounds, which are a major compound in essential oils, caused increased activation of PPARα and energy expenditure via olfactory receptor stimulation [6]. According to the results of several studies, trans-anethole, a dominant compound in FEO, may stimulate the olfactory receptor and/or central nervous system in vivo, thereby causing a reduction in WAT [1,4,6,8].

BAT suppresses metabolic syndrome originating from obesity. In addition, it can control the suppression of visceral fat in vivo according to energy expenditure and heat generation [24]. An increase in the activation of BAT can relieve insulin resistance and/or increase body weight, and BAT activation is associated with BAT mass [23]. In this study, the BAT of the H-LFI group was higher than that of the H group ($p < 0.05$). A previous study reported that BAT mass increased via the administration of trans-anethole in vivo in mice and thus the activation of uncoupling protein-1 (UCP1), which is related to heat activation, increased in mitochondria. Accordingly, the increased BAT mass may suppress the body weight due to a reduction in visceral fat mass [23,24]. In the present study, 0.3% dose inhalation of FEO may have been one of the various factors affecting BAT activation through changes in BAT mass.

Testicles secrete sexual hormones, such as testosterone, and a change in testicular mass is related to sperm production and testosterone secretion. An earlier study reported that testosterone levels, testicular mass, and gonadosomatic index are positively correlated [25]. In addition, Nejabakhsh et al. identified that the HFD-fed group treated with fennel extracts showed a higher tendency to increase testicular mass and sperm number, with a decreased abnormal sperm ratio [26]. Our results showed that the FEO-inhaled groups had dramatically increased testicular mass compared to group H ($p < 0.05$), and showed no significant differences compared to group N. Accordingly, HFD intake may lead to decreased testicular mass, while FEO inhalation may prevent decreased testicular mass.

3.4. Plasma Analysis

Plasma biomarkers were analyzed using an ELISA kit, and the results are shown in Table 2. The total cholesterol (TC) content of the H-HFI group was the highest among various groups ($p < 0.05$), and groups H and H-LFI showed no significant differences compared to group N. The levels of HDL-cholesterol were dramatically increased in group H compared to group N ($p < 0.05$), and the FEO-inhaled groups had the highest HDL-cholesterol compared to group H ($p < 0.05$). On the other hand, the LDL-cholesterol levels were dramatically decreased in group H-LFI compared to group H ($p < 0.05$); however, group H-HFI was significantly increased compared to group H and group N ($p < 0.05$). The atherogenic index (AI) and cardiac risk factor (CRF) showed the highest levels in group N compared to the other groups ($p < 0.05$), and the H-LFI group had the lowest level among the groups ($p < 0.05$). The ratio of LDL-cholesterol to HDL-cholesterol (LHR) was significantly higher in group H than in groups N and H-LFI ($p < 0.05$), while the ratio in group H was lower than that in group H-HFI ($p < 0.05$). Triglyceride (TG) dramatically increased in group H compared to those in group N ($p < 0.05$). Group H-HFI had significantly higher TG levels than group H ($p < 0.05$). On the other hand, group H-LFI had significantly decreased TG levels than group H ($p < 0.05$), and there was no significant difference between the H-LFI and N groups. Cortisol and leptin levels showed no significant differences among all groups. The levels of insulin and insulin resistance (HOMA-IR) of group N were higher than those of the other groups ($p < 0.05$), and the FEO-inhaled groups were significantly lower than those of the H group ($p < 0.05$). The levels of testosterone were found to be relatively lower in group H than in group N, while the H-LFI group was significantly higher than groups N and H ($p < 0.05$). ALT displayed no significant intergroup differences, while AST displayed a significant increase, as observed in group H compared to group N, as well as FEO-inhaled groups ($p < 0.05$). However, AST displayed no significant differences between the N-and FEO-inhaled groups.

The concentrations of cholesterol and TG are important indicators of cardiovascular health [3]. Cholesterol is separated by TC, LDL-cholesterol, and HDL-cholesterol, and then increased LDL-cholesterol and decreased HDL-cholesterol may induce the prevalence of cardiovascular diseases, such as disorders of blood pressure and cardiovascular health in vivo [3]. In this study, the HFD-fed groups showed increased HDL-cholesterol levels. On the other hand, LDL-cholesterol levels were decreased in group H-LFI compared to group H; however, group H-HFI showed increased HDL-cholesterol levels compared to group H. Increased HDL-cholesterol levels may be influenced by an increase in HFD intake and the FEO inhalation concentration (Table 2). TG showed a decreasing effect on the inhalation of 1% FEO. A previous study reported that HDL-cholesterol decreased as the PEO inhalation concentration increased, and inhalation of PEO decreased LDL-cholesterol. Furthermore, the inhalation of 0.3% PEO decreased the accumulation of WAT; however, the inhalation of 1% PEO increased the accumulation of WAT [1]. Another previous study reported that trans-anethole, which is the main compound of fennel, decreased TG levels. According to the above-mentioned results, the inhalation of 1% PEO may show the side-effect of fat accumulation via disorders of CNS and ANS activation, such as hormone regulation and energy storage in vivo [1]. According to the present study, the inhalation of 0.3% FEO may be considered optimal for the effect of decreased LDL-cholesterol versus the inhalation of 1% FEO. Therefore, inhalation of 0.3% FEO may be considered to have positive effects on cardiovascular health, while inhalation of 1% FEO may be considered to cause negative effects, such as increasing arteriosclerosis and blood pressure [1,3].

Insulin is a hormone that regulates blood glucose and energy storage and is affected by the activation of the CNS and ANS [4]. In the present study, FEO inhalation caused a decrease in insulin as well as insulin resistance levels. The levels of insulin and insulin resistance are influenced by the accumulation of WAT, and the levels of insulin can affect the secretion of insulin-like growth factor-1 (IGF-1) [1,4]. IGF-1 is quite similar to insulin in its amino acid structure (50%); thus, the secretion and/or level of insulin in vivo may induce the synthesis or secretion of IGF-1. Furthermore, IGF-1 is a growth hormone (GH), and increased secretion of IGF-1 can increase its length gain [1]. Our results showed that the intake of HFD increased insulin levels, insulin resistance, and length. Group H showed the highest level of insulin; however, group H had a lower length gain than the FEO-inhaled groups. The FEO-inhaled groups showed lower insulin resistance than group H. A previous study showed that body weight, body fat, and insulin levels were decreased by aerobic exercise, as well as increased GH and IGF-1 levels [27]. According to a previous study and the results of this study, insulin secretion and insulin resistance may indirectly induce an increased length gain, while increasing insulin resistance, which is affected by the excessive secretion of insulin, is one of the inhibitory factors of growth [1,27].

Testosterone, which is secreted in the testicles, is an androgen-type steroid hormone that may affect the growth of bones, muscles, and skin, as well as the generation of prostaglandins and sperms [26,28]. The present study showed that HFD-fed groups had decreased levels of testosterone, and the inhalation of 1% FEO showed increased testosterone levels compared to group H ($p < 0.05$). Erdemir et al. reported that the levels of testosterone were negatively correlated with HFD intake but positively correlated with testicular weight [28]. In addition, the testicular weight increases by the intake of fennel extracts [26]. Therefore, FEO inhalation may also influence the change in testicular weight, and may prevent the decreased secretion of testosterone according to the prevalence of obesity [26,28].

The ASL and ALT enzymes are associated with hepatotoxicity, and increasing hepatotoxicity can induce a negative effect on liver health by causing diseases, such as hepatocirrhosis and liver cancer [1]. Inhalation of FEO showed no significant variation in ALT levels among all groups. On the other hand, the AST levels decreased in the FEO-inhaled groups compared to group H ($p < 0.05$), and there has been no significant negative effect of hepatotoxicity, and it may induce the suppression of hepatotoxicity according to the decreased AST levels [11,29].

3.5. Blood Glucose

Blood glucose levels were measured using a blood glucose meter and are presented in Table 3. In the initial period, the blood glucose levels showed no significant differences among all groups, while the FEO-inhaled groups (seventh week) were dramatically decreased compared to group H (seventh week). In the final period, the levels of glucose showed no significant differences among the groups; thus, the short-term effects of decreased blood glucose levels were identified via FEO inhalation.

Table 3. Effect of fennel essential oil inhalation on blood glucose.

Blood Glucose	**Weak**		
(mg/dL)	**Initial Period**	**Obesity Induced Period**	**Final Period**
N	102.3 ± 7.2^{a1}	108.7 ± 5.5^{b}	114.0 ± 12.5^{a}
H	108.0 ± 3.6^{a}	126.7 ± 6.5^{a}	122.7 ± 11.0^{a}
H-LFI	101.7 ± 8.1^{a}	107.2 ± 6.6^{b}	117.6 ± 8.0^{a}
H-HFI	112.3 ± 3.8^{a}	109.7 ± 3.1^{b}	111.7 ± 1.5^{a}

Data are expressed as mean ± SD values from experiments performed in triplicate. [1] Means with different small letters (a–b) correspond to the significant differences determined among all groups using the non-parametric Friedman test, followed by Dunn's test ($p < 0.05$). Obesity-induced period: 7 weeks.

Blood glucose levels play an important role in evaluating diabetes, and increased levels of blood glucose may cause hyperglycemia [26,29]. An earlier study observed that blood glucose levels were decreased by an intake of fennel extracts [30]; however, we identified only short-term decreasing effects via FEO inhalation.

3.6. Blood Pressure

Blood pressure, which is composed of systolic blood pressure (SBP), diastolic blood pressure (DBP), and pulse was measured, and the results are presented in Table 4. In the initial period, SBP showed no significant differences among all groups, while SBP, which was measured at the 12th week, was significantly decreased in the FEO-inhaled groups (H-LFI and H-HFI) compared to group H ($p < 0.05$). In particular, the H-LFI group had a significantly lower SBP than the N group ($p < 0.05$). DBP and pulse, which were measured in the first week, were not significantly different among the groups. In the final period, the pulse of the FEO-inhaled groups was significantly lower than that of group H ($p < 0.05$); however, DBP showed no significant intergroup differences.

Table 4. Effect of fennel essential oil inhalation on the systolic blood pressure, diastolic blood pressure, and pulse.

Systolic(mmHg)	**Weak**	
	Initial Period	**Final Period**
N	154.7 ± 3.2 [a1]	199.3 ± 3.8 [bc]
H	153.0 ± 5.3 [a]	215.0 ± 7.0 [a]
H LFI	157.3 ± 12.4 [a]	190.0 ± 2.0 [c]
H-HFI	148.7 ± 4.9 [a]	201.0 ± 1.0 [b]
Diastolic(mmHg)		
N	69.7 ± 1.5 [a]	133.3 ± 12.1 [a]
H	52.0 ± 5.3 [a]	114.0 ± 4.0 [ab]
H-LFI	60.3 ± 10.1 [a]	131.3 ± 5.5 [ab]
H-HFI	55.0 ± 19.5 [a]	113.3 ± 5.5 [b]
Pulse(beats/min)		
N	422.0 ± 13.1 [a]	377.0 ± 14.9 [b]
H	415.7 ± 18.6 [a]	415.3 ± 8.3 [aA2]
H-LFI	408.7 ± 8.1 [a]	381.3 ± 7.44 [bB]
H-HFI	424.0 ± 22.7 [a]	382.3 ± 6.4 [abB]

Data are expressed as mean ± SD values from experiments performed in triplicate. [1] Means with different small letters (a–c) correspond to the significant differences determined among all groups using the non-parametric Friedman test followed by Dunn's test ($p < 0.05$). [2] Means with different capital letters (A and B) correspond to the significant differences determined among all HFD groups through the non-parametric Friedman test followed by Dunn's test ($p < 0.05$).

Hypertension is caused by the accumulation of visceral fat and/or increased body weight, as well as increased levels of insulin, blood glucose, visceral mass, and/or decreased HDL cholesterol [1,3]. In the present study, the levels of insulin, insulin resistance, and visceral fat mass were decreased, and HDL-cholesterol was increased via FEO inhalation. In addition, SBP and pulse, which were measured at the 12th week, were simultaneously dramatically decreased in the FEO inhaled groups compared to group H ($p < 0.05$). Increased HDL cholesterol may play an important role in lowering blood pressure according to the control of the CNS and ANS (e.g., sympathetic and parasympathetic nerves), and visceral fat mass is positively correlated with insulin resistance and blood pressure [1,3]. In addition, fennel is known to suppress cardiovascular diseases (e.g., hypertension, atherosclerosis, and hyperlipidemia), and several previous studies have reported that fennel extracts prevent increasing SBP and pulse [26,29,30]. Moreover, Hong et al. reported that SBP was decreased via inhalation of volatile compounds in PEO, and various studies have identified the blood-pressure-lowering effects of inhalation of volatiles [1,20,22]. In this study, the H-LFI and H-HFI groups had decreased SBP and/or pulse via FEO inhalation; thus, FEO volatiles may also influence decreased cardiovascular diseases. In general, most in vivo experiments have conducted oral administration. However, the animal study, which investigated the physiological effects on inhalation of volatile compounds, is insufficient. Therefore, a comparative study for advantages and side effects of FEO inhalation versus oral administration has not existed. To date, hepatotoxicity of essential oil via oral administration and recommended oral doses for toxicological safety have been represented [31]. Meanwhile, there is currently no research about the hepatoxicity of essential oil inhalation.

4. Conclusions

In conclusion, this research indicates that inhalation of FEO protected lipid and metabolic dysfunction in high-fat diet-induced obese rats. Therefore, these advantages suggest that the inhalation of FEO could lead to significant metabolic health gains and improvement of diet-induced obesity. Furthermore, these findings may contribute to the initial design of future studies.

Author Contributions: Formal Analysis and Writing-Original Draft Preparation, S.J.H.; Formal Analysis and Writing-Original Draft Preparation, S.Y.; Data Curation, S.M.J. and H.J.; Writing-Review & Editing, M.Y.Y., Y.J.K. and J.K.K.; Supervision, Project Administration, and Funding Acquisition, E.-C.S. All authors have read and agreed to the published version of the manuscript.

Funding: This study was funded by the Basic Science Research Program, through the National Research Foundation of Korea (NRF) funded by the Ministry of Education (NRF-2018R1D1A1B07045431).

Institutional Review Board Statement: Not applicable.

Informed Consent Statement: Not applicable.

Data Availability Statement: The data presented in this study are available.

Conflicts of Interest: The authors declare no conflict of interest.

References

1. Hong, S.J.; Cho, J.J.; Boo, C.G.; Youn, M.Y.; Pan, J.H.; Kim, J.K.; Shin, E.C. Inhalation of Patchouli (*Pogostemon Cablin* Benth.) Essential Oil Improved Metabolic Parameters in Obesity-Induced Sprague Dawley Rats. *Nutrients* **2020**, *12*, 2077. [CrossRef]
2. Baek, H.; Kim, S.; Lee, I.; Kang, S.; Yoo, J.; Yoon, W.; Kim, Y.; Kim, H.; Kim, J. Anti-Obesity and Anti-Lipidemic Effects of Linalool in High-Fat Diet-Induced Obese Mice. *J. Biomed. Res.* **2012**, *13*, 229–235. [CrossRef]
3. Després, J.P. Is Visceral Obesity the Cause of the Metabolic Syndrome? *Ann. Med.* **2006**, *38*, 52–63. [CrossRef] [PubMed]
4. Myers, M.G.; Olson, D.P. Central Nervous System Control of Metabolism. *Nature* **2012**, *491*, 357–363. [CrossRef] [PubMed]
5. Ahn, H.; Go, G.W. Pinus densiflora Bark Extract (PineXol) Decreases Adiposity in Mice by Down-Regulation of Hepatic De Novo Lipogenesis and Adipogenesis in White Adipose Tissue. *J. Microbiol. Biotechnol.* **2017**, *27*, 660–667. [CrossRef]
6. Lee, S.J.; Depoortere, I.; Hatt, H. Therapeutic Potential of Ectopic Olfactory and Taste Receptors. *Nat. Rev. Drug Discov.* **2019**, *18*, 116–138. [CrossRef]
7. Batubara, I.; Suparto, I.H.; Sa'diah, S.; Matsuoka, R.; Mitsunaga, T. Effects of Inhaled Citronella Oil and Related Compounds on Rat Body Weight and Brown Adipose Tissue Sympathetic Nerve. *Nutrients* **2015**, *7*, 1859–1870. [CrossRef]
8. He, W.; Huang, B.K. A Review of Chemistry and Bioactivities of a Medicinal Spice: *Foeniculum vulgare*. *J. Med. Plants Res.* **2011**, *5*, 3595–3600. [CrossRef]
9. Bae, J.Y.; Kim, J.E.; Choue, R.W.; Lim, H.J. Fennel (*Foeniculum vulgare*) and Fenugreek (Trigonella Foenum-Graecum) Tea Drinking Suppresses Subjective Short-Term Appetite in Overweight Women. *Clin. Nutr. Res.* **2015**, *4*, 168–174. [CrossRef]
10. Xiao, Z.; Chen, J.; Niu, Y.; Chen, F. Characterization of the Key Odorants of Fennel Essential Oils of Different Regions Using GC–MS and GC–O Combined With Partial Least Squares Regression. *J. Chromatogr. B Analyt. Technol. Biomed. Life Sci.* **2017**, *1063*, 226–234. [CrossRef]
11. Elghazaly, N.A.; Radwan, E.H.; Zaatout, H.H.; Elghazaly, M.M.; Allam, N.E. Beneficial Effects of Fennel (*Foeniculum vulgare*) in Treating Obesity in Rats. *Obes. Manag.* **2019**, *1*, 16–33. [CrossRef]
12. Punetha, D.; Tewari, G.; Pande, C.; Bhatt, S. Effect of Climatic Conditions on the Volatile Compounds of the Aerial Parts of *Foeniculum vulgare* Mill. *J. Essent. Oil Bear. Plants* **2019**, *22*, 1093–1103. [CrossRef]
13. Diaz-Maroto, M.C.; Pérez-Coello, M.S.; Esteban, J.; Sanz, J. Comparison of the volatile composition of wild fennel samples (*Foeniculum vulgare* Mill.) from Central Spain. *J. Agric. Food Chem.* **2006**, *54*, 6814–6818. [CrossRef] [PubMed]
14. Lee, C.J.; Lee, D.S.; Kang, J.Y.; Kim, J.M.; Park, S.K.; Kang, J.E.; Kwon, B.S.; Park, S.H.; Park, S.B.; Ha, G.J.; et al. Memory Improvement Effect of *Artemisia Argyi* H. Fermented With *Monascus Purpureus* on Streptozotocin-Induced Diabetic Mice. *Korean J. Food Sci. Technol.* **2017**, *49*, 550–558. [CrossRef]
15. Akaberi, M.; Iranshahy, M.; Iranshahi, M. Review of the Traditional Uses, Phytochemistry, Pharmacology and Toxicology of Giant Fennel (Ferula communis L. subsp. communis). *Iran. J. Basic Med. Sci.* **2015**, *18*, 1050–1062. [PubMed]
16. Keskin, I.; Gunal, Y.; Ayla, S.; Kolbasi, B.; Sakul, A.; Kilic, U.; Gok, O.; Koroglu, K.; Ozbek, H. Effects of *Foeniculum vulgare* Essential Oil Compounds, Fenchone and Limonene, on Experimental Wound Healing. *Biotech. Histochem.* **2017**, *92*, 274–282. [CrossRef] [PubMed]
17. Hammouda, F.M.; Saleh, M.A.; Abdel-Azim, N.S.; Shams, K.A.; Ismail, S.I.; Shahat, A.A.; Saleh, I.A. Evaluation of the essential oil of *Foeniculum vulgare* Mill (fennel) fruits extracted by three different extraction methods by GC/MS. *Afr. J. Tradit. Complement. Altern. Med.* **2014**, *11*, 277–279. [CrossRef]

18. Mirdehghan Ashkezari, S.M.; Bahmanyar, H.; Azizpour, H.; Mohammadi, M.; Najafipour, I. Investigation of Operating Parameters on Ultrasound-Assisted Extraction of Anethole in Fennel Essential Oil. *J. Chem. Pet. Eng.* **2021**, *55*, 339–351. [CrossRef]
19. Salehi, B.; Upadhyay, S.; Erdogan Orhan, I.; Kumar Jugran, A.; Jayaweera, S.L.D.; Dias, D.A.; Sharopov, F.; Taheri, Y.; Martins, N.; Baghalpour, N.; et al. Therapeutic Potential of α- and β-Pinene: A Miracle Gift of Nature. *Biomolecules* **2019**, *9*, 738. [CrossRef]
20. Tsuji, T.; Tanaka, S.; Bakhshishayan, S.; Kogo, M.; Yamamoto, T. Olfactory Stimulation Modulates the Blood Glucose Level in Rats. *Int. J. Med. Sci.* **2018**, *15*, 269–273. [CrossRef]
21. Noratto, G.D.; Murphy, K.; Chew, B.P. Quinoa Intake Reduces Plasma and Liver Cholesterol, Lessens Obesity-Associated Inflammation, and Helps to Prevent Hepatic Steatosis in Obese Db/Db Mouse. *Food Chem.* **2019**, *287*, 107–114. [CrossRef]
22. Hur, M.H.; Kim, C.; Kim, C.H.; Ahn, H.C.; Ahn, H.Y. The Effects of Inhalation of Essential Oils on the Body Weight, Food Efficiency Rate and Serum Leptin of Growing SD Rats. *J. Korean Acad. Nurs.* **2006**, *36*, 236–243. [CrossRef]
23. Kang, N.H.; Mukherjee, S.; Min, T.; Kang, S.C.; Yun, J.W. Trans-Anethole Ameliorates Obesity via Induction of Browning in White Adipocytes and Activation of Brown Adipocytes. *Biochimie* **2018**, *151*, 1–13. [CrossRef]
24. You, M.; Fan, R.; Kim, J.; Shin, S.H.; Chung, S. Alpha-Linolenic Acid-Enriched Butter Promotes Fatty Acid Remodeling and Thermogenic Activation in the Brown Adipose Tissue. *Nutrients* **2020**, *12*, 136. [CrossRef]
25. Latif, R.; Lodhi, G.M.; Aslam, M. Effects of Amlodipine on Serum Testosterone, Testicular Weight and Gonado-Somatic Index in Adult Rats. *J. Ayub Med. Coll. Abbottabad* **2008**, *20*, 8–10.
26. Nejatbakhsh, R.; Riyahi, S.; Farrokhi, A.; Rostamkhani, S.; Mahmazi, S.; Yazdinezhad, A.; Kazemi, M.; Shokri, S. Ameliorating Effects of Fennel and Cumin Extracts on Sperm Quality and Spermatogenic Cells Apoptosis by Inducing Weight Loss and Reducing Leptin Concentration in Diet-Induced Obese Rats. *Andrologia* **2017**, *49*, e12748. [CrossRef]
27. Tsolakis, C.; Xekouki, P.; Kaloupsis, S.; Karas, D.; Messinis, D.; Vagenas, G.; Dessypris, A. The Influence of Exercise on Growth Hormone and Testosterone in Prepubertal and Early-Pubertal Boys. *Hormones* **2003**, *2*, 103–112. [CrossRef] [PubMed]
28. Erdemir, F.; Atilgan, D.; Markoc, F.; Boztepe, O.; Suha-Parlaktas, B.; Sahin, S. The Effect of Diet Induced Obesity on Testicular Tissue and Serum Oxidative Stress Parameters. *Actas Urol. Esp.* **2012**, *36*, 153–159. [CrossRef]
29. Parsaeyan, N. The Effect of *Foeniculum vulgare* (Fennel) Extract on Lipid Profile, Lipid Peroxidation and Liver Enzymes of Diabetic Rat. *Iran. J. Diabetes Obes.* **2016**, *8*, 24–29.
30. Kooti, W.; Moradi, M.; Akbari, S.A.; Ahvazi, N.S.; Samani, M.A.; Larky, D.A. Therapeutic and Pharmacological Potential of *Foeniculum vulgare* Mill.: A Review. *J. HerbMed Pharmacol.* **2015**, *4*, 1–9.
31. Tamburlin, I.S.; Roux, E.; Feuillée, M.; Labbé, J.; Aussagues, Y.; El Fadle, F.E.; Fraboul, F.; Bouvier, G. Toxicological safety assessment of essential oils used as food supplements to establish safe oral recommended doses. *Food Chem. Toxicol.* **2021**, *157*, 112603. [CrossRef] [PubMed]

nutrients

MDPI

Article

Protective Effects of Hydroxyphenyl Propionic Acids on Lipid Metabolism and Gut Microbiota in Mice Fed a High-Fat Diet

Jingling Guo [1], Pan Wang [2], Yifan Cui [1], Xiaosong Hu [1], Fang Chen [1] and Chen Ma [1,*]

[1] Key Laboratory of Fruits and Vegetable Processing, Ministry of Agriculture, Engineering Research Centre for Fruits and Vegetables Processing, National Engineering Research Center for Fruit and Vegetable Processing, College of Food Science and Nutritional Engineering, China Agricultural University, Beijing 100083, China

[2] Beijing Key Laboratory of Agricultural Products of Fruits and Vegetables Preservation and Processing, Key Laboratory of Vegetable Postharvest Processing, Ministry of Agriculture and Rural Affairs, Institute of Agri-Food Processing and Nutrition, Beijing Academy of Agriculture and Forestry Sciences, Beijing 100097, China

* Correspondence: machen21@sina.com; Tel.: +86-158-4777-3782

Abstract: Gut microbiota imbalances lead to the pathogenesis of non-alcoholic fatty liver disease (NAFLD), which is primarily accompanied by hepatic steatosis. Hydroxyphenyl propionic acids (HPP) have shown great potential in inhibiting lipid accumulation but their protective effects concerning NAFLD and intestinal microbiota have remained unclear. In this paper, we investigated the efficacies of 3-HPP and 4-HPP on hepatic steatosis and gut flora in mice fed a high-fat diet (HFD). We found that 3-HPP and 4-HPP administration decreased body weight and liver index, ameliorated dyslipidemia, and alleviated hepatic steatosis. Furthermore, 3-HPP and 4-HPP enhanced the multiformity of gut microbiota; improved the relative abundance of *GCA-900066575*, *unidentified_Lachnospiraceae*, and *Lachnospiraceae_UCG-006* at genus level; increased concentration of acetic acid, propionic acid and butanoic acid in faeces; and reduced systemic endotoxin levels in NAFLD mice. Moreover, 4-HPP upregulated the relative abundance of genera *Rikenella* and downregulated the relative abundance of *Faecalibaculum*. Furthermore, 3-HPP and 4-HPP regulated lipid metabolism and ameliorated gut dysbiosis in NAFLD mice and 4-HPP was more effective than 3-HPP.

Keywords: hydroxyphenyl propionic acid; gut microbiota; NAFLD; lipid metabolism; SCFAs

Citation: Guo, J.; Wang, P.; Cui, Y.; Hu, X.; Chen, F.; Ma, C. Protective Effects of Hydroxyphenyl Propionic Acids on Lipid Metabolism and Gut Microbiota in Mice Fed a High-Fat Diet. *Nutrients* **2023**, *15*, 1043. https://doi.org/10.3390/nu15041043

Academic Editor: Takao Kimura

Received: 12 December 2022
Revised: 12 February 2023
Accepted: 14 February 2023
Published: 20 February 2023

1. Introduction

Non-alcoholic fatty liver disease (NAFLD) is a disorder in which lipid accumulation appears in more than 5% of the liver without excessive alcohol consumption [1]. As prevalent chronic liver diseases worldwide, NAFLD harms billions of people and the disease can develop to steatohepatitis, fibrosis, cirrhosis, and hepatocellular carcinoma [2]. The hallmark of NAFLD is excessive neutral lipids such as triglyceride (TG) and cholesterol accumulation in hepatocytes [3]. The acquisition of TG and cholesterol in the liver includes biosynthesis from acetyl CoA or importation from blood; their disposal occurs through delivery to blood or conversion to other molecules [4,5]. As the liver was not served as a lipid storage organ, imbalance among these processes gradually induces hepatic steatosis and dyslipidemia [4]. However, the pathogenesis of NAFLD is intricate and several studies have suggested that gut microbiota could regulate lipid metabolism [6,7].

Gut microbiota is a community of plentiful archaea, bacteria, eumycete, and bacteriophages that coexist in the colon; dysbiosis of this community has been repeatedly observed in NAFLD [8,9]. Aron-Wisnewsky et al. summarized that the relative abundance of *Escherichia*, *Shigella*, *Ruminococcus*, and *Blautia* is increased and *Lactobacillus* is decreased at the genera level in NAFLD patients by contrast to healthy volunteers [10]. Demir et al. found that the faecal fungi composition in fibrosis patients was characterized by a higher log-ratio of Mucor sp./Saccharomyces cerevisiae [11]. In the western diet-fed mice, gradually reducing abundances of *Clostridia* and *Ruminococcaceae* followed the occurrence of NAFLD [12]. Furthermore, the connection between gut flora and NAFLD was generally explored in recent years. A high-fat diet (HFD) failed to induce dyslipidemia and lipid accumulation of liver in germ-free (GM) mice [13]. However, transplanting gut flora from HFD fed mice to GM mice reproduced phenotypes of NAFLD such as fasting hyperglycaemia, insulinaemia, and hepatic steatosis [14]. These results demonstrated that gut microbiota played a causal role in NAFLD development. Several mechanisms may interpret how gut microbiota cause the pathogenesis of NAFLD. *Desulfovibrio*, *Escherichia coli*, and other gram-negative genus in gut microbiota could release endotoxins which would enter systematic circulation through impaired gut barriers [15]. It has been demonstrated that endotoxin induced low-grade inflammation by activating Toll-like receptor 4 signal pathway and caused lipid accumulation of liver by upregulating gene expression of fatty acid synthase (*fasn*) and acetyl-coenzyme A carboxylase α (acaca) [16,17]. Besides, short chain fatty acids (SCFA), derived from gut microbiota, could inhibit lipid accumulation, and suppress inflammation [18,19]. Therefore, the maintenance of microbial homeostasis is important for protection against NAFLD.

Hydroxyphenyl propionic acids (HPPs), widespread phenolic acid, were able to alleviate inflammation and decrease lipid content of adipocytes in vitro [20,21]. They came from biotransformation of other complex polyphenol by gut microbiota [22–24]. It has been extensively confirmed that polyphenol can shift gut flora populations by stimulating beneficial bacteria and reducing pathogenic microbial species [25–27]. Nevertheless, except for parent polyphenol, the impacts of HPPs on the richness and composition of gut microbiota in NAFLD were unclear. Therefore, the aim of this study was to investigate effects of 3-HPP and 4-HPP on gut flora and related lipid metabolism in HFD induced NAFLD model. Intraperitoneal injection was selected to enhance the bioavailability of HPPs. It is demonstrated that 3-HPP and 4-HPP altered the diversity and composition of gut microbiota, decreased serum lipid contents, and ameliorated lipid aggregation of liver.

2. Materials and Methods

2.1. Materials

Both 3-HPP ($C_9H_{10}O_3$, FW 166.17, purity > 98%) and 4-HPP ($C_9H_{10}O_3$, FW 166.17, purity > 98%) were obtained from Aladdin (Shanghai, China). Poly (ethylene glycol) average Mn400 (PEG400) and physiological saline were purchased from Solarbio (Beijing, China). The 3-HPP and 4-HPP were dissolved in PEG and diluted with physiological saline to 24 mg/mL. Assay kits used for the detection of aspartate transaminase (AST), TG, high-density lipoprotein (HDL-c), free fatty acid (FFA), LDL-c, alanine transaminase (ALT), and total cholesterol (TC) levels were provided by Jiancheng (Nanjing, China). The enzyme-linked immunosorbent assay (ELISA) kit for detecting endotoxin was obtained from Genelab (Beijing, China). RNA easy fast kits (DP451) used for RNA isolation were obtained from TIANGEN (Beijing, China) and SYBR qPCR Master Mix was from Vazyme (Nanjing, China). D12450J and D12492 were purchased from Research Diets, Inc. (New Brunswick, NJ, USA).

2.2. Animal Experimental Scheme

The method of the animal experiment was adapted from a previous work with minor modifications [28]. A total of 24 C57BL/6J mice (male, a month of age) were bought from Vital River Laboratory Animal Technology Co., Ltd. (Beijing, China) and housed under the SPF environment (22 ± 2 °C, 50% ± 5 humidity) with 12 h light-dark cycle. The mice had free access to water and feed. After acclimatization for 10 days, all mice were randomly assigned to four groups (n = six per group, three mice per cage): (1) the normal diet (ND) group raised by a normal diet (D12450J), (2) the HFD group raised by an HFD (D12492), (3) the 3-HPP group raised by an HFD, (4) the 4-HPP group raised by an HFD. The dosage and frequency of 3-HPP and 4-HPP administration were 24 mg/kg body weight (BW) and twice per week, respectively. Mice received 3-HPP and 4-HPP for 13 weeks by intraperitoneal injection and PEG with saline as a control. We recorded the BW and diet intake of the mice twice per week. All experimental procedures were conducted according to the Animal Experimental Ethics Committee of China Agricultural University (reference number: AW70702202-4-1) and followed the criterion of National Research Council Guidelines.

2.3. Sample Collection

On the 87th day, all mice were softly stimulated in a sterile box to collect fecal samples (5–6 grains per mouse). The feces of each mouse were loaded into a sterile cryovial and stored in a −80 °C refrigerator. All mice fasted for 16 h with enough water. Then they were intraperitoneally injected with 1% sodium pentobarbital for anaesthesia. After sacrifice, we collected serum, liver, and colon of mice, which were immediately stored on dry ice and then transferred to a −80 °C refrigerator until further use.

2.4. Measurements of Serum Biochemicals and Liver Lipids Levels

Not only the content of TG, HDL-c, TC, FFA, LDL-c, and endotoxin, but also the activity of AST and ALT in serum were measured following instructions of test kits. TC and TG in the liver were isolated with absolute ethyl alcohol and their content was determined as described in the serum. The levels of endotoxin in the serum were measured based on the ELISA assay kit instructions.

2.5. Histopathological Examination

The histopathological procedure was performed as a previous study with slight modifications [29]. The fresh liver was fixed with 4% paraformaldehyde and sliced. After fixation, the liver sample was dehydrated with 15% (*w*/*v*) and 30% (*w*/*v*) sucrose solution successively at 4 °C. Then the liver sample was embedded by an optimal cutting temperature compound, sectioned (6–7 μm) and stained with hematoxylin-eosin (H&E) or oil red o (ORO). The slices were then photographed using a microscope (Nikon, Tokyo, Japan) at 200× magnification.

2.6. RNA Extraction and Quantitative Real-Time Polymerase Chain Reaction (qRT-PCR)

The procedure for qRT-PCR followed a previous study with few modifications [30]. Total RNA in liver samples were extracted by following assay kit instructions and then reversely transcribed to cDNA. The reaction system of qRT-PCR contained 2 μL cDNA, 0.7 μL forward primer, 0.7 μL reverse primer, 6.6 μL ddH2O, and 10 μL SYBR qPCR Master Mix. The PCR program was run on the BioRad CFX96 System, which included a first step at 95 °C for 5 min, a second step for 40 PCR cycles of 95 °C for 10 S, 60 °C for 30 S, 72 °C for 15 S, and a last step for melting curve. The primers of targeted genes are shown in Table 1 and the results of qRT-PCR were normalized to the *gapdh* expression and calculated by the $2^{-\Delta\Delta Ct}$ method.

Table 1. Primers of qRT-PCR.

Gene	Forward Primer (5′ to 3′)	Reverse Primer (5′ to 3′)
fasn	ATGAGCGCACCTTTGATGAC	GATGCCGTCAGGTTTCAGTC
acaca	ATGTCTGGCTTGCACCTAGT	ATCGCATGCATTTCACTGCT
thrsp	ACCTAGAAGCCCAGTTCCAC	CTACAGAACCTGCCCTGTCA
elovl6	GTGCAGAGGCTTGAGAAGTG	TAATCTCCGCAGGCCCTTAG
dgat1	GTGCCATCGTCTGCAAGATT	GATCAGCATCACCACACACC
dgat2	CTTCTCTGTCACCTGGCTCA	CGTGTTCCAGTCAAATGCCA
sqle	TCGCTGCCTTCTCGGATATT	CTGAGGTAGCTGCTCCTGTT
acly	GCCAAGACCATCCTCTCACT	GAAGTTTGCAATGCTGCCT
ppara	TACTGCCGTTTTCACAAGTGC	AGGTCGTGTTCACAGGTAAGA
gapdh	AACGGATTTGGCCGTATTGG	CATTCTCGGCCTTGACTGTG

2.7. Gut Microbiota Analysis by 16S rRNA Sequencing

Procedures of 16S rRNA sequencing and analysis were developed from a published study with some modifications [31]. The DNA in mice feces was extracted by TIANamp Stool DNA Kit (TIANGEN, Beijing, China). After isolation, the quality and concentration of DNA were detected using Nanodrop 2000 (Thermo Fisher, Darmstadt, Germany) and 0.75% agarose gel electrophoresis (AGE). The V3-V4 regions of bacterial 16S rDNA were amplified by barcode primers (341F: CCTAYGGGRBGCASCAG; 806R: GGACTACNNGGGTATCTAAT) and DNA polymerase (NEB, Massachusetts, USA). Then, the 2% AGE was used to examine the quality of 16S rDNA amplification and recycled using a DNA Gel Extraction Kit with Magnetic Beads (Beyotime, Nanjing, China). The library was constructed according to the instructions of the TruSeq® DNA PCR-Free Sample Preparation Kit (Illumina, San Diego, CA, USA) and amplicon sequencing was performed on the NovaSeq 6000 platform.

2.8. Measurement of Short Chain Fatty Acids (SCFAs) Content in Feces

Analysis concentrations of SCFA was performed with reference to the method of Hsu et al. with slight modifications [32]. About 20 mg of mice feces were mixed with 800 μL 0.5% phosphoric acid solution which contained 2-ethyl butyric acid (internal standard) at the concentration of 10 μg/mL. After ultrasound and centrifugation, the supernatant was extracted with n-butanol. The qualitative and quantitative analysis of SCFA was accomplished using a GC/MSD system with an Agilent HP-FFAP capillary column (30 m × 0.25 mm × 0.25 μm) (Agilent Technologies Inc., Palo Alto, CA, USA). A mix of propionic acid, acetic acid, butyric acid, isobutyric acid, valeric acid, isovaleric acid, hexanoic acid, and isohexanoic acid (25 μg/mL each, Sigma, Saint Louis, MO, USA) was used for calibration. We calculated the concentration of these SCFA based on standard solutions corrected by internal standard and the integral area of sample curves.

2.9. Statistical Analysis

All data were expressed as means ± standard deviations (SDs). The difference between the four groups was analysed by one-way analysis of variance using SPSS 20.0. Duncan's test was used for the comparison between groups. $p < 0.05$ was deemed significant for all tests. Non-metric multidimensional scaling (NMDS) was used to analyse discrepancies among samples. The key genera between the HFD and HPP (or ND) groups were analysed by the Wilcoxon rank sum test at $p < 0.05$. Phylogenetic Investigation of Communities by Reconstruction of Unobserved States 2 (PICRUSt2) and Kyoto Encyclopedia of Genes and Genomes (KEGG) databases were selected to predict the metabolic function of gut microbiota.

3. Results

3.1. HPPs Reduced BW and Liver Index in HFD Mice

To explore the influence of 3-HPP and 4-HPP on NAFLD, C57BL/6J mice were intraperitoneally administrated with 3-HPP and 4-HPP and continuously fed an HFD for 13 weeks (Figure 1A). The BW of mice in four groups was without a marked difference in the initial experimental period. After being raised by HFD for 13 weeks, the BW, BW gain, and liver index of HFD mice were significantly increased compared to ND mice (Figure 1B–D). The 3-HPP and 4-HPP treatments attenuated these characteristics without affecting energy intake, suggesting that the efficacy of 3-HPP and 4-HPP did not depend on decreased food intake (Figure 1E).

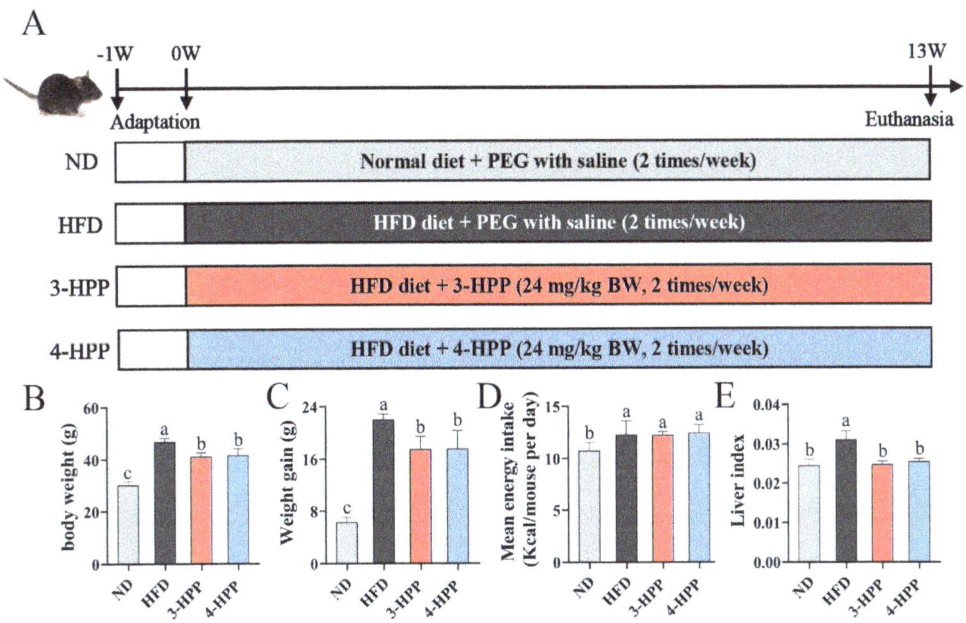

Figure 1. The influence of 3-hydroxyphenyl propionic acid (3−HPP) and 4−HPP on physiological indexes. (**A**) Schematic diagram of animal experiment period. (**B**) Final body weight (BW). (**C**) BW gain. (**D**) Energy intake. (**E**) Liver index. Data are expressed as mean ± SD. n = 6 if not specified. According to Duncan's test, mean values with different letters mean significant differences at $p < 0.05$.

3.2. HPPs Improved Hepatic Steatosis and Ameliorated Liver Injury

As the major organ involved in NAFLD, the histologic feature of the liver was examined by H&E and ORO staining. H&E staining revealed that an HFD-induced ballooning degeneration and large areas of lipid droplet vacuole were observed in HFD mice livers. Similarly, the stained ORO sections of HFD mice livers showed plenty of red lipid droplets, which illustrated lipid aggregation. However, areas of lipid droplet vacuole and red lipid droplet were reduced, and ballooning degeneration disappeared in histologic results following 3-HPP and 4-HPP interventions (Figure 2A). We then used test kits to detect liver lipid levels and found that hepatic TG and TC contents in HFD mice were obviously more than those in ND mice; this was decreased by 3-HPP and 4-HPP (Figure 2B,C). These results demonstrated that HPPs protected HFD mice from hepatic steatosis and hepatic lipid accumulation.

Generally, ALT and AST contents in serum are deemed to be indicators of liver damage [33]. The serum contents of AST and ALT were sharply increased after HFD feeding for 13 weeks and ameliorated by 3-HPP and 4-HPP administration (Figure 2D,E). This indicates that HPPs can protect mice from hepatic injury.

Then we examined hepatic gene expression of de novo lipogenesis (DNL) (Figure 3A–D), including *fasn*, *acaca*, thyroid hormone responsive (*thrsp*), and elovl fatty acid elongase 6 (*elovl6*). Except for DNL, we also explored hepatic gene expression of TG and TC synthesis, including diacylglycerol O-acyltransferase 1 (*dgat1*), diacylglycerol O-acyltransferase 2 (*dgat2*), squalene epoxidase (*sqle*), and ATP citrate lyase (*acly*) (Figure 3E–H). Gene expression of *thrsp* and *dgat2* was improved in HFD mice while 3-HPP and 4-HPP failed to reverse their upward tendency. However, 3-HPP and 4-HPP enhanced gene expression of peroxisome proliferator activated receptor alpha (*ppara*), which acts upstream of lipid metabolism (Figure 3I).

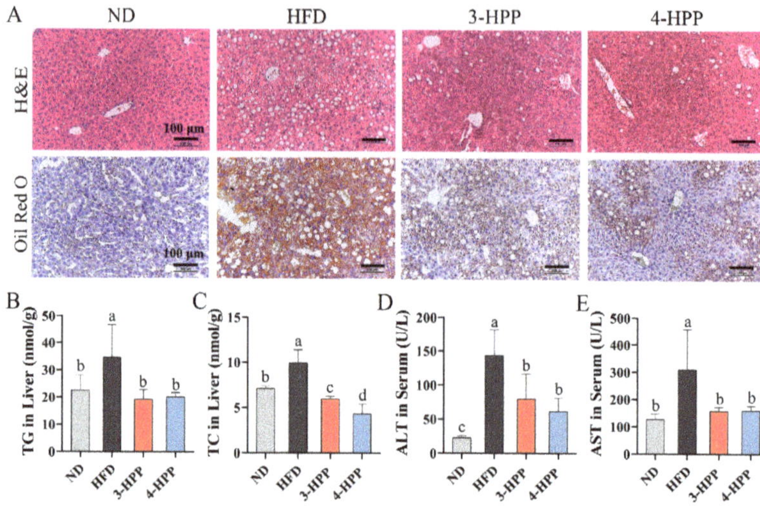

Figure 2. The impacts of 3-HPP and 4-HPP on hepatic steatosis and liver injury. (**A**) Graphs of H&E staining and ORO staining. (**B**) Hepatic TG levels. (**C**) Hepatic TC levels. (**D**) Serum contents of ALT. (**E**) Serum contents of AST. Data are shown as the mean ± SD. n = 6 if not specified. According to Duncan's test, mean values with different letters mean significant differences at $p < 0.05$.

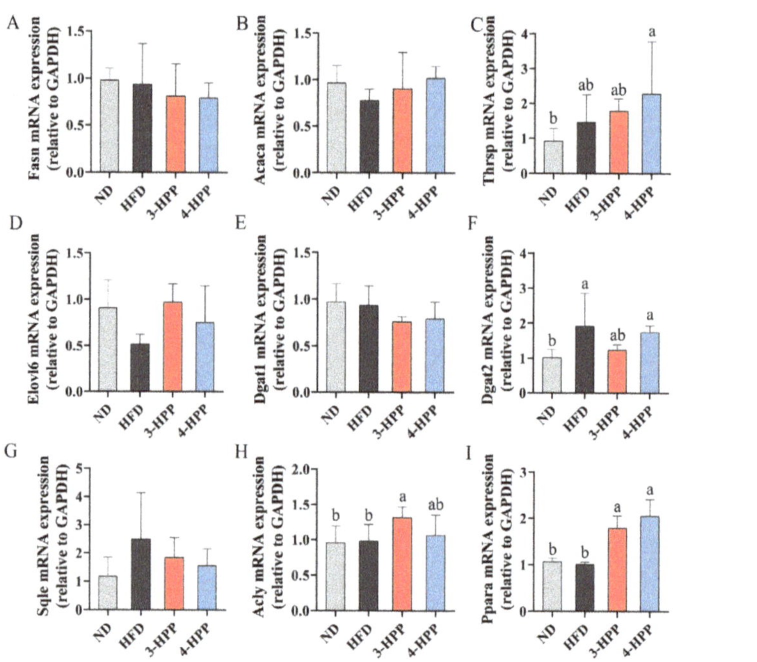

Figure 3. The impacts of 3−HPP and 4−HPP on gene expression of hepatic lipid metabolism. mRNA expression of (**A**) Fasn, (**B**) Acaca, (**C**) Thrsp, (**D**) Elovl6, (**E**) Dgat1, (**F**) Dgat2, (**G**) Sqle, (**H**) Acly, and (**I**) Ppara in liver. Data are shown as the mean ± SD. n = 6 if not specified. According to Duncan's test, mean values with different letters mean significant differences at $p < 0.05$.

3.3. HPPs Affected the Maker Levels in Serum

Clinical studies have found that the majority of NAFLD patients suffer from dyslipidemia [34]. Similarly, levels of lipids such as FFA, TG, and TC and lipoproteins (LDL-c and HDL-c) in HFD mice serum were significantly increased compared to ND mice (Table 2). Nevertheless, serum levels of TG, FFA, and TC were reduced following 3-HPP and 4-HPP treatment. The 4-HPP treatment also decreased serum levels of LDL-c. These results confirm the protective efficacy of 3-HPP and 4-HPP for dyslipidemia.

Table 2. The influence of 3−HPP and 4−HPP on serum biochemical levels.

Parameter	ND	HFD	3-HPP	4-HPP
TC (mg/dL)	2.07 ± 0.32 ^{c}	6.26 ± 0.13 ^{a}	5.31 ± 0.12 ^{b}	4.99 ± 0.18 ^{b}
TG (mg/dL)	0.81 ± 0.19 ^{c}	1.37 ± 0.07 ^{a}	1.02 ± 0.05 ^{b}	0.99 ± 0.03 ^{b}
HDL-c (mmol/L)	1.25 ± 0.39 ^{b}	4.06 ± 0.08 ^{a}	3.59 ± 0.06 ^{a}	3.59 ± 0.05 ^{a}
LDL-c (mmol/L)	0.42 ± 0.06 ^{c}	1.18 ± 0.24 ^{a}	1.04 ± 0.36 ^{a,b}	0.92 ± 0.07 ^{b}
Endotoxin (ng/L)	179.28 ± 5.28 ^{c}	359.61 ± 8.29 ^{a}	240.05 ± 13.66 ^{b}	236.83 ± 14.51 ^{b}
FFA (mmol/L)	1.27 ± 0.20 ^{a,b}	1.44 ± 0.07 ^{a}	1.17 ± 0.14 ^{b}	1.12 ± 0.18 ^{b}

Data are expressed as mean ± SD. n = 6 if not specified. According to Duncan's test, mean values with different letters mean significant differences at $p < 0.05$.

Increased endotoxin levels in serum derived from intestinal microbiota have been considered a risk factor for NAFLD [35]. Long-term HFD feeding induced a marked increment of serum endotoxin in HFD mice, which was decreased by 3-HPP and 4-HPP treatment (Table 2).

3.4. HPPs Influenced the α and β Diversity of Gut Microbiota

Recent evidence indicates that gut flora is related to the pathogenesis of NAFLD [36]. Therefore, the α and β diversity of gut flora was investigated in this study. HFD feeding significantly decreased ACE and Chao indexes in comparison with ND feeding, which inflected the richness and abundance of gut microbiota. The Simpson's index of diversity was reduced in HFD mice, illustrating that an HFD decreased gut microbiota diversity. HPPs significantly increased the Shannon and Simpson's indexes of diversity and had an upload tendency for the ACE and Chao indexes (Figure 4A–D). Additionally, the results of NMDS analysis illustrated that confidence circles among the ND, HFD, and HPP groups were not overlapping, illustrating differences among the gut microbiota composition of these groups (Figure 4E). This result suggests that 3-HPP and 4-HPP administration affected the diversity and composition of gut flora.

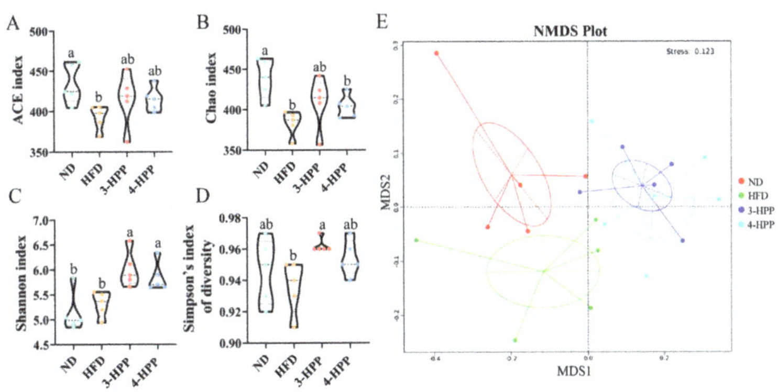

Figure 4. The impacts of 3−HPP and 4−HPP on α and β diversity of gut flora. ACE (**A**), Chao (**B**), and Shannon (**C**) indexes. (**D**) Simpson's index of diversity ACE. (**E**) NMDS analysis based on Weighted Unifrac distance. Data are expressed as mean ± SD. n = 5 if not specified. According to Duncan's test, mean values with different letters denote significant differences at $p < 0.05$.

3.5. HPPs Modulated the Composition of Gut Flora

The results of taxonomic analysis at phylum and genus levels further displayed gut microbiota composition changes (Supplementary Figure S1). The relative abundance of five main phyla is shown in Figure 5. Among the four groups of mice, phylum *Firmicutes* (50.29%) and *Bacteroidetes* (32.25%) were most widespread, followed by *unidentified_Bacteria* (6.16%), *Verrucomicrobia* (2.17%) and *Deferribacteres* (1.5%). HFD feeding not only increased the relative abundance of *Firmicutes* and the ratio of *Firmicutes* to *Bacteroidetes* (F/B) but also decreased the relative abundance of *Bacteroidetes* and *unidentified_Bacteria*, which was partially reversed by 3-HPP and 4-HPP administration (Figure 5A–D). These treatments did not affect the relative abundance of *Deferribacteres* or *Verrucomicrobia* (Figure 5E,F).

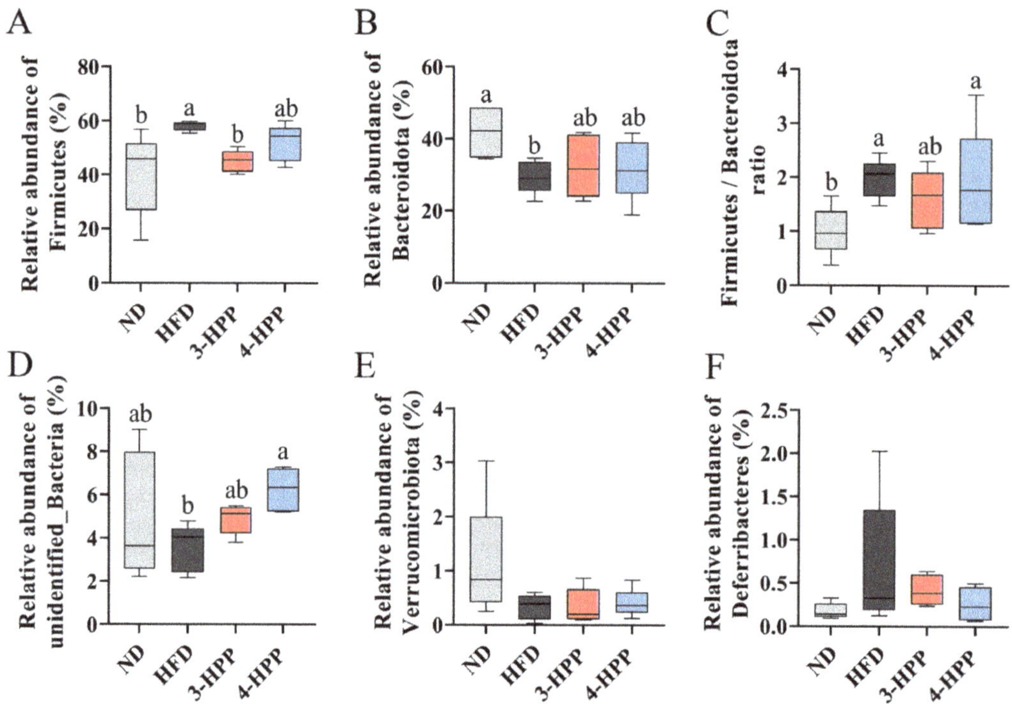

Figure 5. The effects of 3−HPP and 4−HPP on the relative abundance of gut microbiota at the phylum level. Relative abundance of *Firmicutes* (**A**), *Bacteroidetes* (**B**), *unidentified_Bacteria* (**D**), *Verrucomicrobita* (**E**), and *Deferribacteres* (**F**). (**C**) The ratio of *Firmicutes* to *Bacteroidetes*. Data are expressed as means ± SDs. n = 5 if not specified. According to Duncan's test, mean values with different letters denote significant differences at $p < 0.05$.

The relative abundance of the top 10 genera accounted for 43% of the total; the top 10 genera were *Blautia*, *Rikenellaceae_RC9_gut_group*, *Dubosiella*, *Bacteroides*, *Akkermansia*, *Ruminococcus*, *Lactobacillus*, *Ligilactobacillus* and *Alistipes* (Supplementary Figure S1B). The key genera between the HFD and HPP groups were selected according to a Wilcoxon rank sum test. HFD feeding significantly increased the relative abundance of *Desulfovibrio*, which is related to inflammation (Figure 6A). With 3-HPP and 4-HPP treatment, the relative abundance of *GCA-900066575*, *unidentified_Lachnospiraceae*, and *Lachnospiraceae_UCG-006* increased, which belong to *Lachnospiraceae* family (Figure 6B–D,G–I).

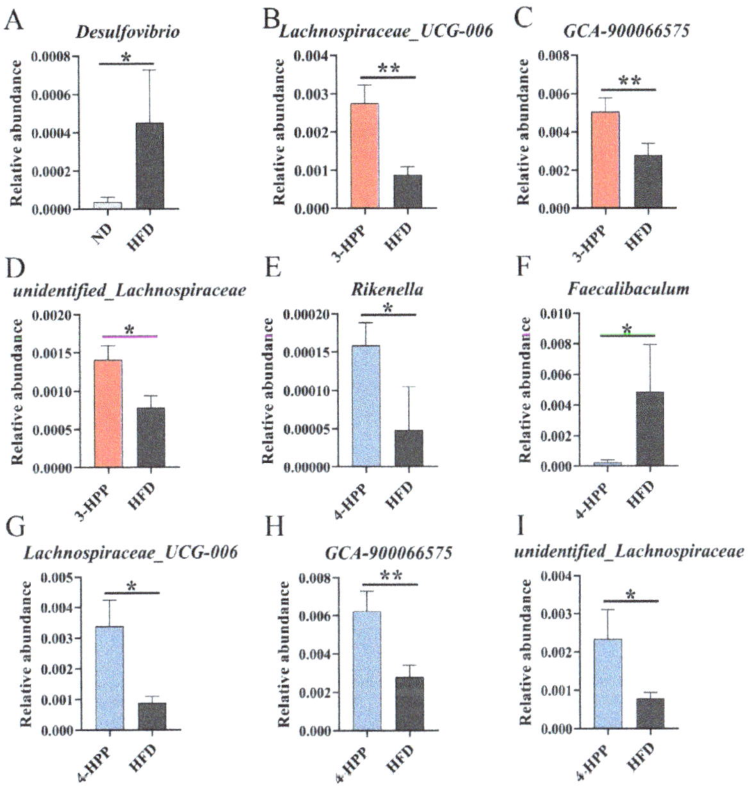

Figure 6. The effects of 3−HPP and 4−HPP on the relative abundance of key genera in different groups. Relative abundance of *Desulfovibrio* (**A**) between the ND and HFD groups. Relative abundance of *Lachnospiraceae_UCG-006* (**B**), *GCA-900066575* (**C**), and *unidentified_Lachnospiraceae* (**D**) between the 3−HPP and HFD groups. Relative abundance of *Rikenella* (**E**), *Faecalibaculum* (**F**), *Lachnospiraceae_UCG-006* (**G**), *GCA-900066575* (**H**), and *unidentified_Lachnospiraceae* (**I**) between the 4−HPP and HFD groups. Data are expressed as means ± SDs. n = 5 if not specified. The results were analysed by a Wilcoxon rank sum test. * $p < 0.05$ and ** $p < 0.01$ compared to the HFD group.

3.6. HPPs Affected the Predicted Functions and SCFAs Production of Gut Flora

OTUs from 16S rRNA sequencing were selected to predict the distinction in metabolic function between HFD and HPP groups by combining the PICRUSt2 and KEGG databases. HFD feeding upregulated nitrogen metabolism and enriched amino acid metabolism in comparison with ND feeding (Supplementary Figure S2). By contrast to HFD group, the 3-HPP group markedly increased valine, leucine and isoleucine biosynthesis (Figure 7A). The administration of 4-HPP inhibited lipid biosynthesis and glycosyltransferases and increased lipid metabolism, which is useful for explaining the beneficial efficacy of 4-HPP for hepatic and serum fat contents (Figure 7B).

In addition to the disruption of metabolic function, the content of acetic acid, propanoic acid, hexanoic acid, and isohexanoic acid in HFD mice was distinctly reduced when compared to ND mice. Nevertheless, 3-HPP and 4-HPP obviously enriched concentrations of most SCFAs except for isohexanoic acid (Figure 8A–H). These data suggested that HPPs could effectively regulate the content of SCFAs.

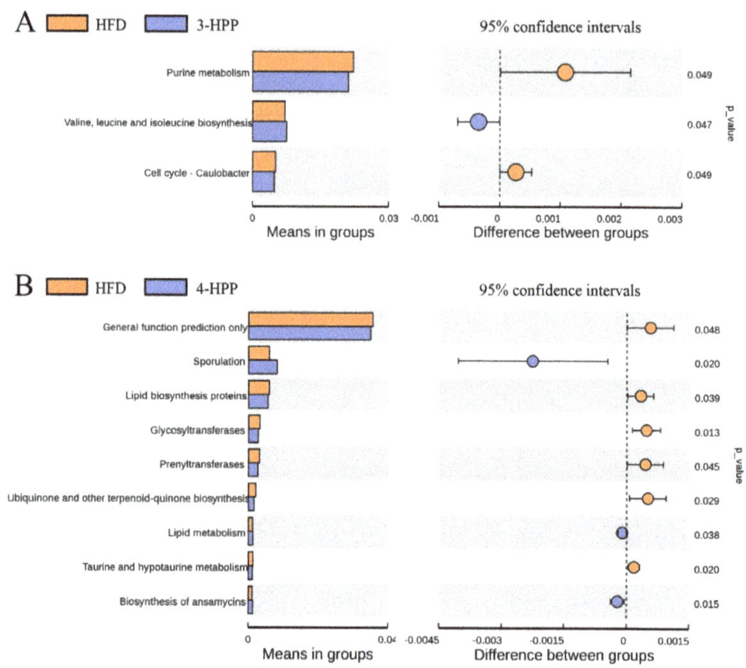

Figure 7. The effects of 3−HPP and 4−HPP on the predicated metabolic profile of gut microbiota. (**A**) Comparison between HFD and 3−HPP. (**B**) Comparison between HFD and 4−HPP. The key metabolic pathways predicted by PICRUSt2 and KEGG were selected based on a t-test at $p < 0.05$. n = 5 if not specified. The colour of the circle is the same as the group with a higher function. The line segment indicates the 95% confidence interval.

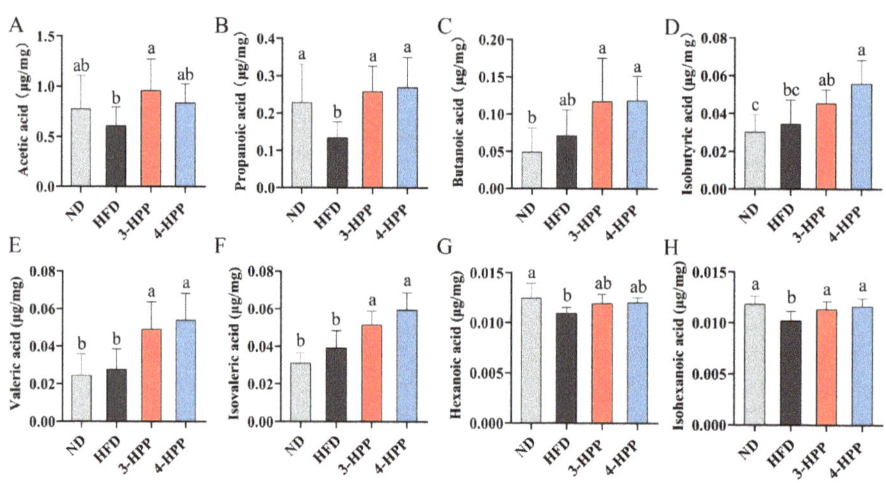

Figure 8. The effects of 3−HPP and 4−HPP on SCFAs content of feces. Fecal concentrations of acetic acid (**A**), propanoic acid (**B**), butanoic acid (**C**), isobutyric acid (**D**), valeric acid (**E**), isovaleric acid (**F**), hexanoic acid (**G**), and isohexanoic acid (**H**). n = 6 if not specified. According to Duncan's test, mean values with different letters mean significant differences at $p < 0.05$.

4. Discussion

NAFLD is becoming an increasingly serious public health problem, with high prevalence among both children and adults [37]. Long-term HFD-induced mice models have been widely used to investigate NAFLD development. In this study, HFD feeding for 13 weeks increased BW and liver index and induced dyslipidemia, which is parallel to NAFLD development in humans and other models [38,39]. The use of 3-HPP and 4-HPP supplementation not only significantly reduced liver index and BW but also lessened the contents of TG, LDL-c, and TC in serum. We found 4-HPP was more effective at reducing lipid levels in the liver and serum. These efficacies of HPPs were similar to those of other phenolic acids such as caffeic and chlorogenic acid [40,41]. Although HPPs had few impacts on gene expression of lipogenesis, they improved mRNA expression of PPARα. PPARα is a nuclear receptor with pivotal regulation functions in fatty acid oxidation, transport, and ketogenesis [42]. Chi et al. found that zinc complex of ulvan oligosaccharide enhanced fatty acid oxidation in the liver via the metal regulatory transcription factor 1/PPARα pathway [43]. As fatty acid is the main material for TG synthesis, HPPs might alleviate hypertriglyceridemia by activating hepatic PPARα gene expression to reduce fatty acid levels.

The liver is the major organ for lipid metabolism and excessive hepatic lipid accumulation can induce NAFLD. Liver biopsies of NAFLD patients show pathological changes such as lipid vacuole, lobular inflammation and hepatocellular ballooning [44]. Increments in ALT and AST levels in serum are associated with NAFLD and liver damage [33]. Interestingly, we found that 3-HPP and 4-HPP prevented pathological changes in the livers of HFD mice and increased serum ALT and AST levels. These results suggest that HPPs treatment ameliorates hepatic steatosis and liver injury. Accordingly, Cui et al. found that 3-HPP from dietary acteoside could ameliorate oxidative stress, decreasing lipid peroxidation, and reducing inflammation in the acute liver injury mice model, which demonstrated the hepatoprotective effect of 3-HPP [21]. Based on beneficial impacts of HPPs on NAFLD, their mechanism of improving NAFLD was further investigated by 16S rRNA sequencing.

Gut microbiota played critical roles in host metabolism and health. Recent evidence indicates that gut microbiota is related to NAFLD and could be a targeted therapy [9,45]. Shen found that the diversity and abundance of gut flora are lower in NAFLD patients compared to healthy subjects [46]. In our study, an HFD reduced the diversity and richness of gut microbiota, and this was reversed by 3-HPP and 4-HPP treatment. The results of NMDS showed that HFD feeding and HPPs supplement influenced the composition and structure of intestinal microbiota, indicating that HPPs can modulate gut microbiota. In this study, HFD feeding upregulated the relative abundance of Firmicutes and downregulated the relative abundance of Bacteroidetes, which caused the F/B ratio to increase. These changes in phylum have also been observed in other NAFLD models and obese humans [36,47]. HPPs administration tended to reduce the ratio of F/B. The relative abundance of *Dubosiella* genera was lower and the relative abundance of *Desulfovibrio* genera was higher in HFD mice compared to ND mice. *Dubosiella*, part of the *Erysipelotrichaceae* family, has a negative correlation with the gene expression of tumor necrosis factor-α and interleukin-1β [48]. *Desulfovibrio*, a kind of sulphate-reducing bacteria, is enriched in gastric cancer patients and its metabolite (hydrogen sulphide) induces the production of NO and IL-1β to promote inflammation [49]. The change in the relative abundance of *Dubosiella* and *Desulfovibrio* in HFD mice reflected gut dysbiosis during NAFLD. 3-HPP and 4-HPP administration improved the relative abundance of genera such as *GCA-900066575*, *unidentified_Lachnospiraceae*, and *Lachnospiraceae_UCG-006*, part of the *Lachnospiraceae* family. With 4-HPP treatment, the relative abundance of *Rikenella* genera and *Faecalibaculum* genera increased and decreased, respectively. The *Lachnospiraceae* family and *Rikenella* genera are the producers of SCFA [50,51]. *Faecalibaculum*, a proinflammatory genus, can induce depression-like phenotypes [52,53]. The regulation of 3-HPP and 4-HPP concerning the structure and composition of gut microbiota partly explained their amelioration of gut dysbiosis, and 4-HPP modulated gut microbiota composition to a stronger extent than

3-HPP. Cueva et al. discovered that 4-HPP was more effective than 3-HPP in inhibiting the growth of *Lactobacilli*, *E. coli*, and *Staphylococcus aureus* strains, which could explain the better impacts of 4-HPP on gut microbiota [54].

The regulation of gut microbiota on host metabolism majorly depended on its metabolic functions and metabolites, except for relying on its composition and abundance. SCFAs, such as propionate, butyrate, and acetate, are microbial products from carbohydrate fermentation. After utilization of enterocytes, SCFAs could enter the liver by portal vein transport to regulate hepatic lipid metabolism [55]. den Besten et al. discovered that sodium of propionate, acetate, and butyrate shifted hepatic lipogenesis to lipid oxidation by activating AMP-activated protein kinase/PPARγ signal pathway [56]. In our results, HPPs enhanced colonic production of acetic acid, propanoic acid, and other SCFAs, which partly explained the lipid-lowering effect of HPPs.

Endotoxins come from gram-negative bacteria and are a causative agent of NAFLD [35]. Wu et al. found that an HFD increased endotoxin levels in the serum and livers of NAFLD mice [57]. In our study, long-term HFD feeding also increased endotoxin levels in serum and this was reversed by 3-HPP and 4-HPP administration. In addition, HPPs enriched fecal content of butanoic acid. It has been suggested that butyrate could ameliorate gut barriers impairment by upregulating the expression of zonula occluden-1 and decrease inflammation by promoting generation of interleukin-22 through regulating histone deacetylase and G-protein receptor 41 [58,59]. Therefore, the systematic inflammation induced by endotoxin might be ameliorated by enrichment of butanoic acid through HPPs. The mechanism by which 3-HPP and 4-HPP reduce endotoxin content remains unclear and requires further investigation.

5. Conclusions

In conclusion, 3-HPP and 4-HPP significantly ameliorated the degree of NAFLD in mice models by reducing liver and serum lipid contents, thereby alleviating hepatic steatosis, inhibiting liver injury, and reducing endotoxin levels in serum. The results of high-throughput 16S rRNA gene sequencing showed that HPPs administration enhanced the diversity of gut microbiota and had a tendency to reduce the ratio of F/B. We found that 4-HPP treatment enriched *Rikenella* genera and reduced the relative abundance of *Faecalibaculum*. Furthermore, 4-HPP was more active than 3-HPP in decreasing lipid content and modulating gut microbiota. The causality between gut microbiota and the protective effects of HPPs for NAFLD is an interesting topic for future research. Our findings provide a valuable foundation for NAFLD prevention by HPPs.

Supplementary Materials: The following supporting information can be downloaded at https://www.mdpi.com/article/10.3390/nu15041043/s1. Figure S1: The relative abundance of microbiota in ND, HFD, 3-HPP and 4-HPP groups; Figure S2: A comparison between the ND and HFD groups of the predicated metabolic profiles of gut microbiota.

Author Contributions: J.G.: Methodology, Conceptualization, Validation, Writing—original draft, Writing—review and editing. P.W.: Writing—review and editing, Data curation, Formal analysis, Supervision. Y.C.: Visualization, Software, Writing—review and editing. X.H.: Writing—supervision, Project administration, Validation. F.C.: Writing—review and editing, Resources, C.M.: Investigation, Funding acquisition, Supervision, Writing—review and editing. All authors have read and agreed to the published version of the manuscript.

Funding: This research was funded by Beijing Innovation Consortium of Agriculture Research System (BAIC01-2022); Beijing Municipal Science and Technology Project (D161105005400000); and National Key Research and Development Program of China (2017YFD0400705).

Institutional Review Board Statement: The protocols of this study were approved by the Animal Care and Ethics Committee of China Agricultural University (Ethics reference number: AW70702202-4-1) and followed the guidelines of the National Research Council Guidelines.

Informed Consent Statement: Not applicable.

Data Availability Statement: Data are available from the corresponding author on reasonable request.

Conflicts of Interest: The authors declare no conflict of interest.

References

1. Powell, E.E.; Wong, V.W.S.; Rinella, M. Non-alcoholic fatty liver disease. *Lancet* **2021**, *397*, 2212–2224. [CrossRef] [PubMed]
2. Friedman, S.L.; Neuschwander-Tetri, B.A.; Rinella, M.; Sanyal, A.J. Mechanisms of NAFLD development and therapeutic strategies. *Nat. Med.* **2018**, *24*, 908–922. [CrossRef] [PubMed]
3. Loomba, R.; Friedman, S.L.; Shulman, G.I. Mechanisms and disease consequences of nonalcoholic fatty liver disease. *Cell* **2021**, *184*, 2537–2564. [CrossRef] [PubMed]
4. Ipsen, D.H.; Lykkesfeldt, J.; Tveden-Nyborg, P. Molecular mechanisms of hepatic lipid accumulation in non-alcoholic fatty liver disease. *Cell. Mol. Life Sci.* **2018**, *75*, 3313–3327. [CrossRef]
5. Li, H.; Yu, X.H.; Ou, X.; Ouyang, X.P.; Tang, C.K. Hepatic cholesterol transport and its role in non-alcoholic fatty liver disease and atherosclerosis. *Prog. Lipid Res.* **2021**, *83*, 101109. [CrossRef]
6. Sonnenburg, J.L.; Bäckhed, F. Diet–microbiota interactions as moderators of human metabolism. *Nature* **2016**, *535*, 56–64. [CrossRef]
7. Schoeler, M.; Caesar, R. Dietary lipids, gut microbiota and lipid metabolism. *Rev. Endocr. Metab. Disord.* **2019**, *20*, 461–472. [CrossRef]
8. Sanyal, A.J. Past, present and future perspectives in nonalcoholic fatty liver disease. *Nat. Rev. Gastroenterol. Hepatol.* **2019**, *16*, 377–386. [CrossRef]
9. Kolodziejczyk, A.A.; Zheng, D.P.; Shibolet, O.; Elinav, E. The role of the microbiome in NAFLD and NASH. *Embo Mol. Med.* **2019**, *11*, e9302. [CrossRef]
10. Aron-Wisnewsky, J.; Vigliotti, C.; Witjes, J.; Le, P.; Holleboom, A.G.; Verheij, J.; Nieuwdorp, M.; Clement, K. Gut microbiota and human NAFLD: Disentangling microbial signatures from metabolic disorders. *Nat. Rev. Gastroenterol. Hepatol.* **2020**, *17*, 279–297. [CrossRef]
11. Demir, M.; Lang, S.; Hartmann, P.; Duan, Y.; Martin, A.; Miyamoto, Y.; Bondareva, M.; Zhang, X.; Wang, Y.; Kasper, P.; et al. The fecal mycobiome in non-alcoholic fatty liver disease. *J. Hepatol.* **2022**, *76*, 788–799. [CrossRef]
12. Zhuge, A.; Li, S.; Lou, P.; Wu, W.; Wang, K.; Yuan, Y.; Xia, J.; Li, B.; Li, L. Longitudinal 16S rRNA sequencing reveals relationships among alterations of gut microbiota and nonalcoholic fatty liver disease progression in mice. *Microbiol. Spectr.* **2022**, *10*, 47–52. [CrossRef]
13. Rabot, S.; Membrez, M.; Bruneau, A.; Gerard, P.; Harach, T.; Moser, M.; Raymond, F.; Mansourian, R.; Chou, C.J. Germ-free C57BL/6J mice are resistant to high-fat-diet-induced insulin resistance and have altered cholesterol metabolism. *Faseb J.* **2010**, *24*, 4948–4959.
14. Safari, Z.; Gerard, P. The links between the gut microbiome and non-alcoholic fatty liver disease (NAFLD). *Cell. Mol. Life Sci.* **2019**, *76*, 1541–1558. [CrossRef]
15. Whitfield, C.; Trent, M.S. Biosynthesis and export of bacterial lipopolysaccharides. *Annu. Rev. Biochem.* **2014**, *83*, 99–128. [CrossRef]
16. Kessoku, T.; Kobayashi, T.; Tanaka, K.; Yamamoto, A.; Takahashi, K.; Iwaki, M.; Ozaki, A.; Kasai, Y.; Nogami, A.; Honda, Y.; et al. The role of leaky gut in nonalcoholic fatty liver disease: A novel therapeutic target. *Int. J. Mol. Sci.* **2021**, *22*, 8161. [CrossRef]
17. Chen, X.; Zhang, C.; Zhao, M.; Shi, C.-E.; Zhu, R.-M.; Wang, H.; Zhao, H.; Wei, W.; Li, J.-B.; Xu, D.-X. Melatonin alleviates lipopolysaccharide-induced hepatic SREBP-1c activation and lipid accumulation in mice. *J. Pineal Res.* **2011**, *51*, 416–425. [CrossRef]
18. Ohtani, N.; Hara, E. Gut-liver axis-mediated mechanism of liver cancer: A special focus on the role of gut microbiota. *Cancer Sci.* **2021**, *112*, 4433–4443. [CrossRef]
19. Kimura, I.; Ozawa, K.; Inoue, D.; Imamura, T.; Kimura, K.; Maeda, T.; Terasawa, K.; Kashihara, D.; Hirano, K.; Tani, T.; et al. The gut microbiota suppresses insulin-mediated fat accumulation via the short-chain fatty acid receptor GPR43. *Nat. Commun.* **2013**, *4*, 1829. [CrossRef]
20. Wang, P.; Gao, J.; Ke, W.; Wang, J.; Li, D.; Liu, R.; Jia, Y.; Wang, X.; Chen, X.; Chen, F.; et al. Resveratrol reduces obesity in high-fat diet-fed mice via modulating the composition and metabolic function of the gut microbiota. *Free Radic. Biol. Med.* **2020**, *156*, 83–98. [CrossRef]
21. Cui, Q.L.; Pan, Y.N.; Zhang, W.; Zhang, Y.A.; Ren, S.M.; Wang, D.M.; Wang, Z.Z.; Liu, X.Q.; Xiao, W. Metabolites of dietary acteoside: Profiles, isolation, identification, and hepatoprotective capacities. *J. Agric. Food Chem.* **2018**, *66*, 2660–2668. [CrossRef] [PubMed]
22. Selma, M.V.; Espín, J.C.; Tomás-Barberán, F.A. Interaction between phenolics and gut microbiota: Role in human health. *J. Agric. Food Chem.* **2009**, *57*, 6485–6501. [CrossRef] [PubMed]
23. Monagas, M.; Urpi-Sarda, M.; Sánchez-Patán, F.; Llorach, R.; Garrido, I.; Gómez-Cordovés, C.; Andres-Lacueva, C.; Bartolomé, B. Insights into the metabolism and microbial biotransformation of dietary flavan-3-ols and the bioactivity of their metabolites. *Food Funct.* **2010**, *1*, 233–253. [CrossRef] [PubMed]

24. Ward, N.C.; Croft, K.D.; Puddey, I.B.; Hodgson, J.M. Supplementation with grape seed polyphenols results in increased urinary excretion of 3-hydroxyphenylpropionic Acid, an important metabolite of proanthocyanidins in humans. *J. Agric. Food Chem.* **2004**, *52*, 5545–5549. [CrossRef]
25. Khairudin, M.A.S.; Mhd Jalil, A.M.; Hussin, N. Effects of polyphenols in tea (*Camellia sinensis* sp.) on the modulation of gut microbiota in human trials and animal studies. *Gastroenterol. Insights* **2021**, *12*, 202–216. [CrossRef]
26. Shen, L.; Ji, H.-F. Reciprocal interactions between resveratrol and gut microbiota deepen our understanding of molecular mechanisms underlying its health benefits. *Trends Food Sci. Technol.* **2018**, *81*, 232–236. [CrossRef]
27. Makarewicz, M.; Drozdz, I.; Tarko, T.; Duda-Chodak, A. The interactions between polyphenols and microorganisms, especially gut microbiota. *Antioxidants* **2021**, *10*, 188. [CrossRef]
28. Guo, J.L.; Wang, P.; Cui, Y.F.; Hu, X.S.; Chen, F.; Ma, C. Alleviation effects of microbial metabolites from resveratrol on non-alcoholic fatty liver disease. *Foods* **2023**, *12*, 94. [CrossRef]
29. Wang, J.; Wang, P.; Li, D.; Hu, X.; Chen, F. Beneficial effects of ginger on prevention of obesity through modulation of gut microbiota in mice. *Eur. J. Nutr.* **2020**, *59*, 699–718. [CrossRef]
30. Ke, W.X.; Wang, P.; Wang, X.H.; Zhou, X.L.; Hu, X.S.; Chen, F. Dietary platycodon grandiflorus attenuates hepatic insulin resistance and oxidative stress in high-fat-diet induced non-alcoholic fatty liver disease. *Nutrients* **2020**, *12*, 480. [CrossRef]
31. Li, D.; Feng, Y.; Tian, M.; Ji, J.; Hu, X.; Chen, F. Gut microbiota-derived inosine from dietary barley leaf supplementation attenuates colitis through PPAR gamma signaling activation. *Microbiome* **2021**, *9*, 83. [CrossRef]
32. Hsu, Y.-L.; Chen, C.-C.; Lin, Y.-T.; Wu, W.-K.; Chang, L.-C.; Lai, C.-H.; Wu, M.-S.; Kuo, C.-H. Evaluation and optimization of sample handling methods for quantification of short-chain fatty acids in human fecal samples by GC-MS. *J. Proteome Res.* **2019**, *18*, 1948–1957. [CrossRef]
33. He, M.M.; Fang, Z.; Hang, D.; Wang, F.; Polychronidis, G.; Wang, L.; Lo, C.H.; Wang, K.; Zhong, R.; Knudsen, M.D.; et al. Circulating liver function markers and colorectal cancer risk: A prospective cohort study in the UK Biobank. *Int. J. Cancer* **2021**, *148*, 1867–1878. [CrossRef]
34. Martin, A.; Lang, S.; Goeser, T.; Demir, M.; Steffen, H.M.; Kasper, P. Management of dyslipidemia in patients with non-alcoholic fatty liver disease. *Curr. Atheroscler. Rep.* **2022**, *24*, 533–546. [CrossRef]
35. Fei, N.; Bruneau, A.; Zhang, X.J.; Wang, R.R.; Wang, J.X.; Rabot, S.; Gerard, P.; Zhao, L.P. Endotoxin producers overgrowing in human gut microbiota as the causative agents for nonalcoholic fatty liver disease. *Mbio* **2020**, *11*, e03263-19. [CrossRef]
36. Backhed, F.; Ding, H.; Wang, T.; Hooper, L.V.; Koh, G.Y.; Nagy, A.; Semenkovich, C.F.; Gordon, J.I. The gut microbiota as an environmental factor that regulates fat storage. *Proc. Natl. Acad. Sci. USA* **2004**, *101*, 15718–15723. [CrossRef]
37. Younossi, Z.; Tacke, F.; Arrese, M.; Sharma, B.C.; Mostafa, I.; Bugianesi, E.; Wai-Sun Wong, V.; Sun, W.; Yilmaz, Y.; George, J.; et al. Global perspectives on nonalcoholic fatty liver disease and nonalcoholic steatohepatitis. *Hepatology* **2019**, *69*, 2672–2682. [CrossRef]
38. Ryan, M.C.; Wilson, A.M.; Slavin, J.; Best, J.D.; Jenkins, A.J.; Desmond, P.V. Associations between liver histology and severity of the metabolic syndrome in subjects with nonalcoholic fatty liver disease. *Diabetes Care* **2005**, *28*, 1222–1224. [CrossRef]
39. Zhu, M.Z.; Hao, S.J.; Liu, T.; Yang, L.L.; Zheng, P.Y.; Zhang, L.; Ji, G. Lingguizhugan decoction improves non-alcoholic fatty liver disease by altering insulin resistance and lipid metabolism related genes: A whole trancriptome study by RNA-Seq. *Oncotarget* **2017**, *8*, 82621–82631. [CrossRef]
40. Mu, H.N.; Zhou, Q.; Yang, R.Y.; Tang, W.Q.; Li, H.X.; Wang, S.M.; Li, J.; Chen, W.X.; Dong, J. Caffeic acid prevents non-alcoholic fatty liver disease induced by a high-fat diet through gut microbiota modulation in mice. *Food Res. Int.* **2021**, *143*, 110240. [CrossRef]
41. Shi, A.M.; Li, T.; Zheng, Y.; Song, Y.H.; Wang, H.T.; Wang, N.; Dong, L.; Shi, H.T. Chlorogenic Acid Improves NAFLD by Regulating gut Microbiota and GLP-1. *Front. Pharmacol.* **2021**, *12*, 693048. [CrossRef] [PubMed]
42. Montaigne, D.; Butruille, L.; Staels, B. PPAR control of metabolism and cardiovascular functions. *Nat. Rev. Cardiol.* **2021**, *18*, 809–823. [CrossRef] [PubMed]
43. Chi, Y.; Wu, Z.; Du, C.; Zhang, M.; Wang, X.; Xie, A.; Wang, P.; Li, R. Regulatory effects mediated by ulvan oligosaccharide and its zinc complex on lipid metabolism in high-fat diet-fed mice. *Carbohydr. Polym.* **2023**, *300*, 120249. [CrossRef] [PubMed]
44. Nalbantoglu, I.; Brunt, E.M. Role of liver biopsy in nonalcoholic fatty liver disease. *World J. Gastroenterol.* **2014**, *20*, 9026–9037.
45. Leung, C.; Rivera, L.; Furness, J.B.; Angus, P.W. The role of the gut microbiota in NAFLD. *Nat. Rev. Gastroenterol. Hepatol.* **2016**, *13*, 412–425. [CrossRef]
46. Shen, F.; Zheng, R.D.; Sun, X.Q.; Ding, W.J.; Wang, X.Y.; Fan, J.G. Gut microbiota dysbiosis in patients with non-alcoholic fatty liver disease. *Hepatobiliary Pancreat. Dis. Int.* **2017**, *16*, 375–381. [CrossRef]
47. Hu, W.B.; Gao, W.Y.; Liu, Z.M.; Fang, Z.F.; Wang, H.C.; Zhao, J.X.; Zhang, H.; Lu, W.W.; Chen, W. Specific strains of Faecalibacterium prausnitzii ameliorate nonalcoholic fatty liver disease in mice in association with gut microbiota regulation. *Nutrients* **2022**, *14*, 2945. [CrossRef]
48. Wan, F.; Han, H.; Zhong, R.Q.; Wang, M.Y.; Tang, S.L.; Zhang, S.F.; Hou, F.J.; Yi, B.; Zhang, H.F. Dihydroquercetin supplement alleviates colonic inflammation potentially through improved gut microbiota community in mice. *Food Funct.* **2021**, *12*, 11420–11434. [CrossRef]
49. Liu, S.L.; Dai, J.J.; Lan, X.; Fan, B.B.; Dong, T.Y.; Zhang, Y.; Han, M.Y. Intestinal bacteria are potential biomarkers and therapeutic targets for gastric cancer. *Microb. Pathog.* **2021**, *151*, 104747. [CrossRef]

50. Vacca, M.; Celano, G.; Calabrese, F.M.; Portincasa, P.; Gobbetti, M.; De Angelis, M. The controversial role of human gut Lachnospiraceae. *Microorganisms* **2020**, *8*, 573. [CrossRef]
51. Shi, H.J.; Chang, Y.G.; Gao, Y.; Wang, X.; Chen, X.; Wang, Y.M.; Xue, C.H.; Tang, Q.J. Dietary fucoidan of Acaudina molpadioides alters gut microbiota and mitigates intestinal mucosal injury induced by cyclophosphamide. *Food Funct.* **2017**, *8*, 3383–3393. [CrossRef]
52. Cai, W.; Xu, J.X.; Li, G.; Liu, T.; Guo, X.L.; Wang, H.J.; Luo, L.P. Ethanol extract of propolis prevents high-fat diet-induced insulin resistance and obesity in association with modulation of gut microbiota in mice. *Food Res. Int.* **2020**, *130*, 108939. [CrossRef]
53. Wang, S.M.; Ishima, T.; Qu, Y.; Shan, J.J.; Chang, L.J.; Wei, Y.; Zhang, J.C.; Pu, Y.Y.; Fujita, Y.; Tan, Y.F.; et al. Ingestion of Faecalibaculum rodentium causes depression-like phenotypes in resilient Ephx2 knock-out mice: A role of brain-gut-microbiota axis via the subdiaphragmatic vagus nerve. *J. Affect. Disord.* **2021**, *292*, 565–573. [CrossRef]
54. Cueva, C.; Victoria Moreno-Arribas, M.; Martin-Alvarez, P.J.; Bills, G.; Francisca Vicente, M.; Basilio, A.; Lopez Rivas, C.; Requena, T.; Rodriguez, J.M.; Bartolome, B. Antimicrobial activity of phenolic acids against commensal, probiotic and pathogenic bacteria. *Res. Microbiol.* **2010**, *161*, 372–382. [CrossRef]
55. Schoenfeld, P.; Wojtczak, L. Short- and medium-chain fatty acids in energy metabolism: The cellular perspective. *J. Lipid Res.* **2016**, *57*, 943–954. [CrossRef]
56. den Besten, G.; Bleeker, A.; Gerding, A.; van Eunen, K.; Havinga, R.; van Dijk, T.H.; Oosterveer, M.H.; Jonker, J.W.; Groen, A.K.; Reijngoud, D.-J.; et al. Short-chain fatty acids protect against high-fat diet-induced obesity via a PPAR-dependent switch from lipogenesis to fat oxidation. *Diabetes* **2015**, *64*, 2398–2408. [CrossRef]
57. Wu, S.S.; Hu, R.Z.; Nakano, H.; Chen, K.Y.; Liu, M.; He, X.; Zhang, H.F.; He, J.H.; Hou, D.X. Modulation of gut microbiota by *Lonicera caerulea* L. Berry polyphenols in a mouse model of fatty liver induced by high fat diet. *Molecules* **2018**, *23*, 3213. [CrossRef]
58. Zhou, D.; Pan, Q.; Xin, F.-Z.; Zhang, R.-N.; He, C.-X.; Chen, G.-Y.; Liu, C.; Chen, Y.-W.; Fan, J.-G. Sodium butyrate attenuates high-fat diet-induced steatohepatitis in mice by improving gut microbiota and gastrointestinal barrier. *World J. Gastroenterol.* **2017**, *23*, 60–75. [CrossRef]
59. Yang, W.; Yu, T.; Huang, X.; Bilotta, A.J.; Xu, L.; Lu, Y.; Sun, J.; Pan, F.; Zhou, J.; Zhang, W.; et al. Intestinal microbiota-derived short-chain fatty acids regulation of immune cell IL-22 production and gut immunity. *Nat. Commun.* **2020**, *11*, 4457. [CrossRef]

MDPI
St. Alban-Anlage 66
4052 Basel
Switzerland
Tel. +41 61 683 77 34
Fax +41 61 302 89 18
www.mdpi.com

Nutrients Editorial Office
E-mail: nutrients@mdpi.com
www.mdpi.com/journal/nutrients

www.ingramcontent.com/pod-product-compliance
Lightning Source LLC
Chambersburg PA
CBHW061622210926
51850CB00009B/2402
* 9 7 8 3 0 3 6 5 6 9 7 0 3 *